Endangerment, Biodiversity and Culture

The notion of Endangerment stands at the heart of a network of concepts, values and practices dealing with objects and beings considered threatened by extinction, and with the procedures aimed at preserving them. Usually animated by a sense of urgency and citizenship, identifying endangered entities involves evaluating an impending threat and opens the way for preservation strategies.

Endangerment, Biodiversity and Culture looks at some of the fundamental ways in which this process involves science, but also more than science: not only data and knowledge and institutions, but also affects and values. Focusing on an "endangerment sensibility," it brings to light tensions between the normative and the utilitarian, the natural and the cultural. The chapters situate that specifically modern sensibility in historical perspective, and examine central aspects of its recent and present forms.

This timely volume offers cutting-edge insights into the Environmental Humanities for researchers working in Environmental Studies, History, Anthropology, Sociology and Science and Technology Studies.

Fernando Vidal is ICREA Research Professor (Catalan Institution for Research and Advanced Studies) at the Center for the History of Science of the Autonomous University of Barcelona, Spain.

Nélia Dias is Associate Professor at the Department of Anthropology, University Institute of Lisbon (ISCTE-IUL), Portugal.

There are thousands of endangered species and hundreds of human cultures facing extinction along with the languages they have spoken. This fascinating book takes the reader along to delve into the reasons we are losing diversity and the many kinds of knowledge it could give us. How has politics made endangerment worse, or tried to prevent it? The wise authors of these chapters find examples from around the world and look at ways to preserve and revive what we might otherwise lose. This book raises interesting questions and is a dependable key to understanding.

J. Donald Hughes, University of Denver, USA

In this era of rapidly accelerating climate change, species extinction, and cultural vulnerability, endangerment has come to shape the science, politics, and emotions mobilized to archive and defend the fatally condemned. *Endangerment, Biodiversity, and Culture* is a timely volume that makes visible the undercurrent of loss animating work across the human and life sciences.

Gregg Mitman, University of Wisconsin-Madison, USA

Routledge Environmental Humanities

Series editors: Iain McCalman and Libby Robin

The *Routledge Environmental Humanities* series is an original and inspiring venture recognizing that today's world agricultural and water crises, ocean pollution and resource depletion, global warming from greenhouse gases, urban sprawl, over-population, food insecurity and environmental justice are all *crises of culture*.

The reality of understanding and finding adaptive solutions to our present and future environmental challenges has shifted the epicenter of environmental studies away from an exclusively scientific and technological framework to one that depends on the human-focused disciplines and ideas of the humanities and allied social sciences.

We thus welcome book proposals from all humanities and social sciences disciplines for an inclusive and interdisciplinary series. We favor manuscripts aimed at an international readership and written in a lively and accessible style. The readership comprises scholars and students from the humanities and social sciences and thoughtful readers concerned about the human dimensions of environmental change.

Endangerment, Biodiversity and Culture

Edited by Fernando Vidal and
Nélia Dias

LONDON AND NEW YORK

First published 2016
by Routledge

2 Park Square, Milton Park, Abingdon, Oxon OX14 4RN
711 Third Avenue, New York, NY 10017, USA

Routledge is an imprint of the Taylor & Francis Group, an informa business

First issued in paperback 2017

British Library Cataloguing-in-Publication Data
A catalogue record for this book is available from the British Library

Library of Congress Cataloging-in-Publication Data
Endangerment, biodiversity and culture / edited by Fernando Vidal and
    Nélia Dias.
        pages cm
    Includes bibliographical references and index.
    1. Human ecology—Philosophy.   2. Endangered ecosystems—
Philosophy.   3. Biodiversity—Philosophy.   I. Vidal, Fernando.
II. Dias, Nélia.
    GF21.E55 2015
    304.2'7 —dc23
    2015001805

ISBN: 978-1-138-84741-5 (hbk)
ISBN: 978-1-138-74356-4 (pbk)

Typeset in Goudy
by Apex CoVantage, LLC

Contents

Acknowledgements

Endangerment, Biodiversity and Culture results from the Working Group "Endangerment and Its Consequences," part of a larger project on the "sciences of the archive" organized in Department II of the Max Planck Institute for the History of Sciences, Berlin. We thank the Department's director, Lorraine Daston, for her generous support and the many opportunities offered at the Institute. The composition of the Working Group changed over time; we owe a debt of intellectual and personal gratitude to its members, both those who are present in this volume and those who are not. Finally, we would like to thank Josephine Fenger for editorial assistance, and Vasco Araújo for generously ceding his artwork for the book cover.

Contributors

Stefan Bargheer is assistant professor, Department of Sociology, University of California, Los Angeles, USA. He studied the emergence and transformation of bird conservation in Great Britain and Germany (1790–2010), and currently investigates the changing valuation of complexity in nature and culture in the League of Nations and the United Nations.
http://www.sociology.ucla.edu/faculty/stefan-bargheer

Etienne Benson is assistant professor, Department of History and Sociology of Science, University of Pennsylvania, USA. He has published on surveillance technologies in wildlife management, global infrastructures of environmental monitoring, the legal regulation of endangered species research, and the history of human-animal relations in urban areas.
www.etiennebenson.com, hss.sas.upenn.edu/people/benson

Nélia Dias is associate professor at the Department of Anthropology (ISCTE-IUL Lisbon), Portugal. She has published on the history of anthropology, museums and collections, and on French colonialism.
www.degois.pt/visualizador/curriculum.jsp?key=0394108532005053#Produca ocientifica

Stefanie Gänger is assistant professor at the Institute for Iberian and Latin American History, University of Cologne, Germany. Her work focuses on the Andes in the late-colonial and early-Republican period, with a particular interest in the history of botany, indigeneity, medicine and antiquarianism.
www.ihila.phil-fak.uni-koeln.de/839.html

Rodney Harrison is Reader in Archaeology, Heritage and Museum Studies at the UCL Institute of Archaeology, UK. He has published widely on the history and politics of heritage, museums and collections, drawing on the perspectives of colonial and post-colonial governmentality and assemblage theory.
www.ucl.ac.uk/archaeology/people/staff/harrison

Rebecca Lemov is associate professor, Department of the History of Science, Harvard University, USA. She is currently writing a book about the history of

coercive interrogation techniques. Her work examines experiments in collecting and storing "subjective materials" as data in once-futuristic data-storage devices and traces the emergence of new forms of self-fashioning in the information age.
www.fas.harvard.edu/~hsdept/bios/lemov.html

Shaylih Muehlmann holds the Canada Research Chair in Language, Culture and the Environment, University of British Columbia. She studies environmental politics, linguistic anthropology, drug trafficking, indigeneity, water scarcity, the anthropology of the awkward, multiculturalism and US-Mexico borderlands, focusing on the intersections between environmental conflict, language and identity.
anth.ubc.ca/faculty-and-staff/shaylih-muehlmann/

José Augusto Pádua is professor of Environmental History, Institute of History, Federal University of Rio de Janeiro, Brazil. After extensive research into the Brazilian political debate on environmental destruction during the colonial and post-colonial periods, he has turned to contemporary issues related to tropical forests and territorial politics in Brazil.
http://lattes.cnpq.br/1619560388482953

Joanna Radin is assistant professor in the History of Medicine, of Anthropology and of History, Yale University, USA. Her research examines the conditions of possibility for contemporary systems of biomedicine and biotechnology, with a focus on the history of biomedical technology, scientific collections, anthropology, public health, humanism and research ethics.
medicine.yale.edu/intranet/facultybydept/joanna_radin.profile

David Sepkoski is Research Scholar at the Max Planck Institute for the History of Science, Berlin, Germany. He has published widely on topics in the history of evolutionary biology, paleontology and natural history, and is currently writing an intellectual and cultural history of extinction.
www.mpiwg-berlin.mpg.de/en/staff/members/dsepkoski

Julia Adeney Thomas is associate professor, Department of History, University of Notre Dame, USA. Her research concerns concepts of nature in Japanese political ideology, the impact of the climate crisis on historiography, and photography as a political practice. Currently, she is writing a book on the environmental conditions of the historian's craft.
history.nd.edu/faculty/directory/julia-adeney-thomas/

Fernando Vidal is ICREA Research Professor (Catalan Institution for Research and Advanced Studies) and professor at the Center for the History of Science, Autonomous University of Barcelona, Spain. A former Guggenheim Fellow, he has published broadly in the history of the mind/body sciences from the 16th to the 21st century.
icrea.academia.edu/FVidal

Introduction

The endangerment sensibility

Fernando Vidal and Nélia Dias

Anything we leave untouched we have already touched.

Bernard Williams (1995)

Endangerment: the notion stands at the heart of a network of concepts, values and practices dealing with entities considered threatened by extinction and destruction, and with the techniques aimed at preserving them.[1] An entity's "endangered" status crystallizes by way of its incorporation into various documentary devices – archives, catalogues, databases, inventories and atlases. These devices materialize values that inspire an urge to perpetuate, but they do so through the concrete objects and information they choose to archive, and the techniques they use to do so. Usually animated by a sense of urgency and citizenship, both among scientists and the general public, cataloguing an endangered entity involves evaluating the intensity of the impending threat and opens the way for preservation strategies. The resulting instruments function as mechanisms to produce knowledge, pursue ideals and enact policies. Looking into the contexts in which they are rooted and the processes whereby they come about involves examining the construction of data deemed significant, the kind of knowledge such data constitutes, the structures into which it is organized, the affects that permeate it, and the moral and epistemic values it incarnates across the vast range of entities, both "natural" and "cultural," which have come to be perceived as essentially vulnerable. "Endangerment," then, not only refers to states of the world that the sciences may identify and describe, but also names an individual and collective resource for apprehending the world at the level of symbols and action.

Architectural patrimony fixed in photographs, extinguished species in museum displays, or dead dialects in recordings may also nurture nostalgia for a more varied and beautiful world, and give rise to resuscitation fantasies. Together with dramatized depictions of surviving but imperiled places and organisms, they proclaim that those endangered things matter, and that we should care about them. They thus acquire significant political and moral valence inside and outside science. In the endangerment regime, turned as it is toward preservation, irreversible loss and definitive forgetting are ultimate forms of negativity, anti-values par excellence.

An indispensable step for coping with endangerment consists of inventorying and ranking. Although the list may be seen as a result of diagnosing entities as endangered, listing is itself both a diagnostic tool and a way of anticipating the failure of the cure: if the endangered entity came to disappear, it will persist as a record. In both capacities (aimed at creating records of what should be saved and at keeping traces of what disappeared), the cognitive and material practices of endangerment rest on sciences that "depend on data and specimens preserved by past observers and project the needs of future scientists in the creation of present collections" (Daston 2012, 156). What distinguishes these "sciences of the archive" is not the phenomena they study nor the methods they employ, but the fact that the data they gather, at vast temporal and spatial scales, are deliberately stored as material for later investigators.

These scientific practices belong in a complex of knowledge, values, affects and interests characterized by a particularly acute perception that some organisms and things are "under threat," and by a purposeful responsiveness to such a predicament. We characterize this complex as an *endangerment sensibility*. We do not use *sensibility* in a psychologistic sense focused on individual "feelings," nor exclude initiatives that predate the rise of *endangerment* as a contemporary keyword. Neither would we claim that such sensibility is universally pervasive and animates in the same way every single project related to the protection of biodiversity and culture. But we are interested in the emergence and features of the perception that vast portions of the human and non-human world are in danger of extinction or destruction: how, and in which contexts and circumstances, did the adjective "endangered" come to be so widely applied – to biological and cultural entities, to species and peoples, to languages and buildings? *That* perception *is* widespread and recurrent, and the historical overtones of "sensibility" capture some of its central features: a certain sentimental impulse, empathy extending to animals, landscapes and marginalized humans, as well as attention to one's own affective experience (Todd 1986; compare historian of anthropology George W. Stocking's 1989 use of "anthropological sensibility").

Basic in the endangerment sensibility are diverse forms of an emotional, moral, political, cognitive and institutional imperative to preserve – to protect the threatened so that it will not disappear, to register the fatally condemned, even to revive (a species "extinct in the wild," a language about to die out). How did that imperative evolve and become constitutive of perceptions of endangerment? Both conservation and memorializing assume that the objects to be protected or remembered are valuable; this sense of value is often accompanied by aesthetic appreciation, and associated with a supposedly natural state or with a condition of primeval authenticity. Which values, emotions and interests are entailed in the endangerment sensibility? How do they compare across fields and break down the barriers between "nature" and "culture"? What are their genealogies? And what are the particular "regimes" – in the Foucauldian sense of sets of power mechanisms and knowledge structures – that are associated with them?

The ancient feeling for the shortness of life and the transience of all things, as captured by Koheleth's memorable incipit *vanitas vanitatum omnia vanitas* and its numerous variations, was accompanied by a sense for the futility of human ambition, the inexorability of time and fate, and the pointlessness of wishing to correct God's works (Ecclesiastes 1:2, 1:10, 3:1, 7:13). Loss and forgetting were part of life, and of course they remain so. Yet, in contrast to the ancient sensibility, the late-modern feeling for the fragility of things is permeated by the sense of future-oriented responsibility that comes from humanity's living in the Anthropocene – in other words, from an awareness that humans are responsible for the decimation of biological and cultural diversity, for the devastating consequences and extreme inequities of global warming, and ultimately for jeopardizing the survival of their own species by bringing about the sixth mass extinction (Kolbert 2014). In contrast to earlier forms of understanding extinction, current thinking is fraught with guilty feelings and a sense of responsibility. Both are made more tragic by the many ways in which present technologies make the natural world visible and compel us to recognize how much is at risk: as Gillian Beer (2009, 326) put it, "Instead of taxidermy we have film."

A major goal for humans in the 21st century will consist of preventing the fateful *de* of the long-documented phenomenon of *deforestation* and the more recently christened *defaunation* (Dirzo et al. 2014) from becoming, funereally, a privative *a*. Most societies seem not to have known anything analogous until recent times (see for example the overviews in McNeill and Mauldin 2012). It is nevertheless a basic principle of environmental history that humans have always transformed the "natural communities" to which they belong and that inhabited spaces are to a large measure shaped by human action (see for example Hughes 2009). It should therefore come as no surprise that environmental reflexivity and awareness of anthropogenic impact did not arise suddenly around the mid-20th century. Such reflexivity, however, was not the same as a sensibility dominated by a concern for endangerment.

To speak only of the Western tradition, awareness of human impacts predates the emergence of the modern concern for the environment, which can be traced to the 18th and 19th century (Lowenthal 1990). Ancient authors noted how quarrying destroyed mountains, complained about urban air pollution, and described how water, air and earth were contaminated with the waste products of domestic and industrial activities (Hughes 2014). They remarked on environmental depletion and deterioration, and correlated changes in local weather with human-induced processes they also depicted, such as drainage, deforestation, or soil erosion and exhaustion. Greek and Roman farmers would sometimes combat such processes efficiently at the local level, but neither their practices nor philosophers' comments on how humans destroyed the environment modified attitudes, which remained essentially predatory and eventually contributed to the downfall of ancient societies. Ancient civilizations, Greco-Roman and others, lacked environmental awareness or a deliberate sense for "sustainability" – how to use natural resources so that they are not devastated or completely used up. As for the conservation and restoration of monuments and objects, they happened and could express a desire to understand the past and sometimes even a vision of the future (Schnapp 2013). Nevertheless, they had fundamentally political, celebratory and authenticating functions, and in any case did not derive from any sensibility to "heritage," considered either locally or in a sense akin to that of UNESCO's 1972 Convention Concerning the Protection of the World Cultural and Natural Heritage, which defines it as sites or monuments "of outstanding universal value" (whc.unesco.org/en/conventiontext).

Forest rarefaction was apparent in many regions of Medieval Europe, and lords and communities drew up rules to reconcile conservation and exploitation. However, then as today, attempts at regulating the management of resources to prevent their exhaustion gave rise to conflicts of interest (see for example Leroy 2007). The large areas of woodland that had become hunting reserves for the ruling elites were treated in the same exploitative spirit that guided land management (Jones 2013). In short, throughout the Old Regime, the handling of resources and the conflicts over them do not reveal anything like an "environmental awareness," including, as one component element, the idea that nature must be protected for its own sake.

Purposeful preservation of nature was envisioned as newly colonized lands were exploited (Grove 1992). The consequences of the plantation economy in the European colonies were clear by the second half of the 18th century. In Europe itself, early in the 19th century, the French government asked departmental prefects to conduct surveys on climate modification, which it detected in such phenomena as atmospheric cooling or abrupt seasonal changes, and imputed in part to deforestation. What the utopian socialist Charles Fourier characterized as the "material deterioration of the planet" was widely attributed to human actions such as land clearing. In the 18th and 19th century, climate became an eminently political topic, providing "a matrix for environmental reflexivity and was used to reflect upon people, objects, and processes . . . within a perspective that was attentive to their common future" (Locher and Fressoz 2012, 598).

At different points during the 19th century, in lands as distant from each other as Brazil (Pádua 2002 and this volume), Spain (Casado de Otaola 2010) and the United States (whose national and wildlife refuges sparked an international movement; Gissibl, Höhler and Kupper 2012), the protection of forests and mountain landscapes brought together aesthetic sensibilities, poetic and sometimes mystical feelings, ideals for the consolidation of national identities, hopes for political and moral regeneration of the elites and the people, glimpses of solace from insalubrious cities, scientific inquiry in the field and the lab, and analyses of economic consequences and resource management. In parallel, both in Europe and in the Americas, a variety of professional and amateur scholars recorded vanished or vanishing cultures, both proximate and distant. Early "salvage anthropology" collected materials from human groups whose disappearance it anticipated, with a focus on assembling them for museums while seeing their makers as members of inevitably vanishing races (Gänger this volume; Gruber 1970; Lemov this volume). Not all of these elements were present everywhere and at the same time; they sometimes contradicted each other, and could be pursued by individuals and groups of incompatible political orientations. Nevertheless, even in its most melancholy forms (as in late 19th-century lamentations about Spain's decline), anguish about human environments and natural land-scapes always came together with attention to resource management in local and national contexts.

What has been, historically, the connection between concern for imperiled natural environments and concern for threatened human cultures (Sepkoski this volume)? By the time Charles Darwin wrote in *The Descent of Man* (1871, Ch. 7) that "[w]hen civilised nations come into contact with barbarians the struggle is short, except where a deadly climate gives its aid to the native race," a consider-able literature had been devoted to the decay of indigenous populations as a con-sequence of "civilised" violence and of their own "savagery" (Brantlinger 2003). In the 1830s, the painter George Catlin (1841, Letter n° 31, 262), known for his portraits of American Indians and their culture, called for a park to preserve the Indian and the buffalo, "man and beast," together, "in all the wild and freshness of their nature's beauty." Later, for example in India (Rashkow 2008) and the Bel-gian Congo (Adams 2004, 5), it was sometimes foreseen to preserve indigenous groups inside natural parks.

Paul Sarasin, the Swiss pioneer of world nature conservation in the years pre-ceding World War I, called for threatened primitive populations to "be preserved as an 'anthropological natural monument' in the same way as endangered non-human forms of nature. They should be isolated in reserves and thus safeguarded 'in the greatest untouched purity for science, for ourselves, and for posterity'" (Kupper 2014, 58). Nevertheless, *we*, *science* and *posterity* are different from *they*, *their forms of knowledge* and *their descendants* – none of which were contemplated. Most realized national parks fenced off their original inhabitants, thus becoming exclusively "natural." Moreover, early interest in the fate of indigenous groups did not imply the will to perpetuate them as living cultures, nor manifested the sensibility captured by the late 20th-century term *biocultural diversity*. And even in this new framework, critics have contended that treating peoples (with their languages and cultural practices) as "endangered" maintains an analogy to non-human organisms and a top-down conservationist approach that may have racist and colonial overtones, occludes the causes of endangerment and deprives of agency the populations concerned (Gqola 2005).

* * *

The discursive, scientific and political practices related to endangerment are about changing the present for the sake of the future. This future, however, is imagined less as a continuation of the present than as a time in which the pres-ent (in the form of currently existing DNA, species, languages or cultures) will no longer be. The endangerment outlook is both proleptic and regretful. It seems to involve some of the longing Svetlana Boym (2001, 41) calls "restorative" nostalgia. Contrary to the "reflective" nostalgia that emphasizes loss, ruins and "the patina of time and history," the restorative seeks "to rebuild the lost home and patch up the memory gaps." The people committed to salvaging biocultural diversity know that the lost home was not in all respects better and that, anyhow, it cannot be regained; they work for the future, largely by combating destructive forces in the present. Yet insofar as it is turned toward a future shaped by a past that the present destroys, and by a present intrinsically under threat, the endangerment sensibility reveals a deep nostalgic undercurrent.

The contributions to this book concur to explicate that disposition and its apparently paradoxical proactive nature by putting the endangerment sensibility in historical perspective, and examining by way of concrete cases central aspects of its emergence and present forms. Focusing on issues of endangerment and extinction in relation to languages, species, cultures and scientific data, a first group of chapters documents the interweaving of affects, values and science in bringing about those very issues. A second group illustrates the political dimensions of endangerment-related discourses and practices with studies on threatened indigenous groups in the late 19th century, 20th-century biosphere reserves and changing views of tropical forests in Brazilian history. A third group explores technologies of preservation in connection with the California condor, cultural heritage and biodiversity collections. The book closes with a reflection on the "we" endangered by climate change, and gives the humanistic disciplines the task of ultimately defining the objects of endangerment. While each chapter discusses distinct empirical materials, they all touch, to various extents, on affects, values, politics and preservation technologies – these are all fundamental to their topics.

The present introduction sketches some of the main issues associated with the endangerment sensibility in three areas that are crucial across all domains: *diversity* as a constitutive value, *listing* as a fundamental epistemic practice, and *emotions* as integral to both the values and the sciences of endangerment. It should be clear that exploring endangerment as an idea set deeply in a sensibility does not mean it lacks concrete referents in the world. Species go extinct due to human action, buildings are demolished, languages disappear and landscapes change beyond recognition. Yet they do not qualify as "endangered" until someone perceives them as such; and although that perception involves scientific knowledge, it emerges only in connection with certain values, feelings, interests and views about the present and the future of communities and humanity at large. Individuals and institutions involved in the defense of natural and cultural heritages use the term mainly in adjectival form, and convey the message that if an entity is designated as "endangered," then it must, ipso facto, be protected and preserved. Focusing on the noun introduces critical distance, and directs our attention to the processes that bring about the complex cultural phenomenon we call "endangerment sensibility."

Diversity and endangerment

We just gave, in passing, an idea of the range of entities that are considered endangered. Although that range is potentially as large as there are things in the world, the particular objects of solicitude ultimately incarnate one of its aspects: diversity. In contrast to its synonym *variety*, *diversity* has ceased to be a descriptive term. It has become the name of a supreme late-modern value; and at the same time that diversity acquired the status of inherently endangered attribute, the concept's extension grew to encompass both "nature" and "culture," and came to be widely used to derive the value of both (see Maier 2012 for a critique of such uses). Many endangerment-related undertakings illustrate these characteristics, but they are particularly well displayed in large-scale endangerment-related projects.

The Svalbard Global Seed Vault, for example, opened in 2008 on the island of Spitsbergen, in the Norwegian Svalbard archipelago, about 1,300 kilometers from the North Pole. It consists of a tunnel protected with a sleeve against erosion and climate changes, which takes us first through an office and handling area, and then, at the end, beyond airlock doors, to the seed vaults themselves. There stand rows of metal shelves with boxes stacked on them; the boxes contain sealed envelopes, and the envelopes, seeds of all major food crops from around the world. These seeds are duplicates of samples held in other seed banks, but because of its location, Svalbard is expected to be the safest of them all. The local permafrost conditions – that is to say, an environment that is permanently at or below the freezing point of water – should aid preservation of the seeds for hundreds of years. They will be thus protected from natural and non-natural disasters, including, according to a *National Geographic* article about the Vault, "global warming, asteroid strikes, plant diseases, nuclear warfare, and even earthquakes."[2]

It is no wonder that the facility is also known as Doomsday Vault, and that Norway's prime minister described it as "our insurance policy." The Vault, he proclaimed, is "the Noah's Ark for securing biological diversity for future generations." Due to the massive loss of food varieties, we have come to depend on a small number of plants and animals, and if disease, climate change or human actions decimate them, "we might desperately need one of those varieties we've let go extinct" (Siebert 2011). The justification for the enterprise is in the first place utilitarian: humanity needs crop diversity in order to feed itself.

Its most distant horizon, however, is the threat of total extinction – a threat that has been incorporated into a worldview permeated with the perception of omnipresent risk and the need to manage it. Thus, Cary Fowler, director of the Global Crop Diversity Trust, which launched the Seed Vault project, explains that the facility was not created with doomsday in mind, but that "in the case of a regional or global catastrophe," it "would prove to be very, very useful."[3] When in February 2012, 25,000 seed samples from America, Colombia, Costa Rica, Tajikistan, Armenia and Syria arrived in the Vault, Fowler felt especially happy about the Syrian chickpeas and fava beans:

> the Philippines' national seed bank was destroyed by fire in January, six years after it was damaged by flooding. Those of Afghanistan and Iraq were destroyed in recent wars. Should the conflict in Syria reach that country's richest store, in Aleppo, the damage would now be less. Some 110,000 Syrian seed samples are now in the Svalbard vault, out of around 750,000 samples in all. "When I see this," says Mr Fowler, looking lovingly at his latest consignment, "I just think, 'thank goodness, they're safe.'"
>
> ("Banking Against Doomsday")

Yet it is not enough to keep seeds alone. Since locally adapted cultivars are in most cases superior to imported breeds, storing seeds in banks "is only a halfway measure."

> Equally worthy of saving is the hard-earned wisdom of the world's farmers, generations of whom crafted the seeds and breeds we now so covet. Perhaps the most precious and endangered resource is the knowledge stored in farmers' minds.
>
> (Siebert 2011)

As the Norwegian minister of agriculture put it on the Seed Vault's first anniversary, "field-level knowledge of our farmers continues to be the most important element of crop conservation."[4]

Hoarding nature and hoarding culture go here hand in hand, and indigenous knowledge emerges as a form of experiential "wisdom" that provides a better basis for interventions than those derived from imported science. Against such perception of closeness to nature, critics point out that autochthonous knowledge systems have long interacted with Western ones, and that their "native" character partly results from the very structures that marginalize them (Dove 2006). The imagined proximity of the natural and the indigenous tends to obliterate the history and form of the connections between local and imported knowledge, while at the same time proclaiming that culture and nature – peoples and forests (Brosius 1997) – are endangered together, and that it is only as a unity that they can be saved.

By its aims, functioning and structure, as well as by the discourses that support them and the media coverage it receives, the Svalbard Vault highlights the extent to which biological and cultural diversity has come to be characterized as a preeminently endangered property of living systems, and its endangerment, as both consequence and potential cause of major cataclysms. The Vault thus incarnates the built-in relationship between a celebratory passion for diversity and an anxious fixation on endangerment. Although the two are ontologically distinct, since the 1970s, they have become sociological and cognitive correlatives through processes that include the professionalization of environmental consciousness, the development of various sciences of the environment (conservation biology, for example) and other interdisciplinary areas (such as environmental anthropology, ecological or environmental humanities, political and cultural ecology, environmental ethics or climate ethics), the internationalization of environmental politics and human rights regimes, the multiplication of specialized NGOs, and the rise of a global biocultural perspective.

In the framework of the United Nations Environment Program, established in 1972, governments signed agreements to deal with such specific issues as the protection of wetlands and the regulation of international trade in endangered species. In 1987, the World Commission on Environment and Development (known as the Brundtland Commission) called for making development "sustainable," that is to say, capable of meeting "the needs of the present without compromising the ability of future generations to meet their own needs" (www.un-documents.net/ocf-02.htm). A few years later, in response to the growing awareness that sustainability was far from advancing and that a dramatically accelerated anthropogenic reduction of biodiversity had reached alarming levels, the Convention on Biological Diversity signed at the 1992 Rio de Janeiro Earth Summit emphasized that protecting biodiversity is in humanity's own interest:

> Biological resources are the pillars upon which we build civilizations. . . . The loss of biodiversity threatens our food supplies, opportunities for recreation and tourism, and sources of wood, medicines and energy. It also interferes with essential ecological functions.
>
> (www.cbd.int/convention)

Biodiversity loss, the Convention explained, not only destabilizes and reduces the productivity of ecosystems, "thereby shrinking nature's basket of goods and services," but also endangers cultures, since, it claimed, "[o]ur cultural identity is deeply rooted in our biological environment." Hence the commitment not merely to pursue biodiversity conservation, sustainability and the "fair and equitable sharing" of benefits, but also to "[p]rotect and encourage customary use of biological resources in accordance with traditional cultural practices that are compatible with conservation or sustainable use requirements" (ibid., articles 1 and 10c).[5]

These quotations are taken from some of the main international documents in the domain of biodiversity protection. Thousands of initiatives throughout the world echo their spirit, and corroborate the fundamentally pragmatic and anthropocentric justification of endangerment/conservation initiatives across the biological and cultural domains. For Luisa Maffi, the prolific co-founder and director of Terralingua, an NGO that works "to sustain the biocultural diversity of life" (www.terralingua.org), such initiatives aim at "the continued viability of humans on Earth." In this regard, she writes, "issues of linguistic and cultural diversity conservation may be formulated in the same terms as for biodiversity conservation": both help keep open a multiplicity of options, resist the impoverishing monocultures of land and mind, and thus enhance or maintain the human species' adaptive strength (Maffi 2001a, 38). These claims are matters of debate since, as philosopher David Heyd (2010, 170) pointed out, it may well be that "the chances of survival of cultural communities are better advanced by policies of adjustment and adaptation to modern conditions than by protective policies which strive to perpetuate them in their traditional form."

Not all elements of a culture lend themselves to assessment in terms of utility. Yet in the framework of biocultural diversity, the *cultural* has become valuable on account of its reputed contribution to protecting biodiversity as condition for humankind's survival. While views about biocultural diversity have long been biased toward more "exotic," remote and isolated communities, there have been calls to integrate into conservation initiatives all types of local groups (for example, urban ones whose traditions have been heavily rearticulated; Cocks 2006).

One of the central claims of the Declaration of Belem, issued in 1988 during the First International Congress of Ethnobiology, is that "there is an inextricable link between cultural and biological diversity."[6] The Declaration was an important moment in the history of the rise of the notion of *biocultural diversity* as comprising "the diversity of life in all of its manifestations – biological, cultural, and linguistic – which are interrelated (and likely co-evolved) within a complex socio-ecological adaptive system" (Maffi 2010, 5). By the early 2000s, the *biocultural axiom* – according to which effectively preserving biodiversity requires protecting cultural diversity, and vice versa – seemed confirmed by two types of evidence: on the one hand, geographical overlaps between biological richness and linguistic diversity, and between indigenous territories and "biologically high-value regions"; on the other hand, observations about the significance of indigenous peoples "as main managers and inhabitants of well-preserved habitats," and claims that their ecologically sustainable behavior derives "from their pre-modern belief/knowledge/practices complex" (Toledo et al. 2002, 561). All such research naturally assumes the joint "high social value of biological and cultural diversity" (Sodikoff 2011, 13).

Yet the bio-cultural axiom and the evidence presented for it have generated considerable debate, some of it updating the venerable controversy around the "ecologically noble savage" (Hames 2007; Krech 1999; Nadasdy 2005; Redford 1991). Such "savages" may have arisen chiefly as a result of the bad conscience of the ecologically ignoble civilized (Hames 2007; Kuper 2003). But historically marginalized communities have come "to recognize the political potency of strategically deployed essentialisms" (Brosius 1999, 281). Nineteenth-century Euro-Americans believed that Indians practiced a cult of the Mother Earth, and this fantasy ended up shaping the American Indian experience (Gill 1987). By the late 20th century, indigenous groups had adopted the ecological stance as part of their resistance strategies. As anthropologist Philippe Descola (2008) notes, they have assimilated

> the reasoning behind the pre-eminence of universal interest over local interests and how they can make the best of it. Accordingly they have begun to present themselves as the keepers of nature – an abstract notion which does not appear in their languages or cultures – to whom the international community should entrust the mission to keep watch at their level on environments which it is becoming clearer everyday have been shaped by their practices.

Of course, the point is not that the peoples called "indigenous" lack the sense of a larger or transcendent entity to which they belong, but, rather, that such an entity, if contemplated, does not necessarily correspond to "nature," with its dichotomous associate "culture." Whether or not they ancestrally had notions and practices translatable as *biocultural diversity* or *environmental awareness* (often called TEK or "traditional ecological knowledge"), they have actively entered into those "makeshift links across distance and difference" that, in Anna L.

Tsing's (2005, 2) depiction, shape "globalization." (On TEK, see the conceptual discussion by Whyte 2013, and Berkes 2012, a landmark book first published in 1999 and now in its third edition.) Just as we write these lines, a study carried out by a collaborative network of Amazonian indigenous organizations, NGOs, scientists and policy experts computes for the first time the carbon storage capacity of Amazonian indigenous territories and protected natural areas, highlights the contribution of indigenous territorial management to the preservation of ecosystems, and sees the recognition of indigenous cultural identities – and "land and resource rights" with them – as crucial for ecosystem integrity and climate stability (Walker 2014). At the same time, and in the same spirit, the Palangka Raya Declaration on Deforestation and the Rights of Forest Peoples criticizes top-down international anti-deforestation initiatives and calls for the full participation of forest peoples in environmental decision-making (Forest Peoples Programme 2014a, 2014b).

In short, debates notwithstanding, negotiation and decision-making with regard to biocultural endangerment have become integral to global governance. The bio-cultural axiom has turned into official doctrine, with linguistics as a major site for its defense and illustration. UNESCO's position postulates that precious environmental knowledge is embedded in indigenous languages, and can be lost if speakers shift to another language.[7] The assumption seems reinforced by the fact that areas with high biological diversity are home to about 70% of the world's languages, and that endangered languages and endangered species coincide geographically. From this, increasing numbers of scholars and activists conclude that biodiversity conservation necessitates language preservation (Gorenflo et al. 2012). As the endangered language enterprise grew, it moved from documentation as rescue work to documentation as a participatory process (Turin 2011), and saw its premises, frameworks and procedures extended to music (Grant 2014). At the same time, some authors have criticized the analogy of languages and organisms, and pointed out that the "ecologization" of language endangerment depoliticizes cultural issues by turning culture into an extension of nature, subsuming political interests into a global celebration of diversity, and merely asserting that languages must be saved without asking why and for whom (Cameron 2007; Dobrin and Berson 2011; Heller and Duchêne 2007; Muehlmann this volume).

By the second decade of the 21st century, the bio-cultural axiom has come to underpin most conservation theory and advocacy, "biocultural diversity" has developed into a transdisciplinary research area (e.g. Maffi 2001b, 2005), attempts have been made to quantify global biocultural diversity by means of an index (Loh and Harmon 2005), and the notion has been assimilated into international phraseology in such a way that earlier conventions on heritage and biological diversity are now perceived as part of an integrated mode of action (e.g. Bargheer this volume; Harrison this volume; UNESCO 2008). But this phraseology is the crystallization of a process in which the notion that biodiversity and cultures are interrelated and mutually supportive grew piecemeal through projects anchored in a multiplicity of disciplines, from conservation biology to linguistics (Maffi

2010), and concerning myriad organisms and cultures whose aggregation and reciprocal interactions actualize diversity.

These developments are rooted in the rise of endangerment as a process constitutive of contemporary perceptions of the world. The Cold War played here a decisive role. NATO scientists and military officers anticipated "environmental warfare" and worked on various projects, dealing for example with radiological contamination, biological weapons and weather control, "that united scientific knowledge of the natural environment with the strategic goal of killing large numbers of people" (Hamblin 2013, 4). As "the environment" became a global issue after World War II, the prospect of human-made apocalypse nourished the style and worldview of the emerging environmentalist movement and its attention to large-scale anthropogenic changes. The languages of global military crisis and that of environmental crisis were similar; military and environmental questions were "often identical and pursued by the same people," and the worldviews of the military and the environment experts "went hand in hand, both attempting projections of catastrophic environmental consequences on a massive scale" (ibid., 8–10). The environmental monitoring that gave rise to the sense of global environmental vulnerability "began as an explicitly Cold War activity" (ibid., 86). In short, those who planned how most efficiently to provoke a worldwide catastrophe fostered the sensibility that makes us see our planet as a constitutionally endangered place.

Although more recent apocalyptic scenarios derive from empirically based projections and real scenes of hurricanes and floods, they are also heirs to that Cold War mind-set. The "silence" of Rachel Carson's 1962 *Silent Spring*, caused by the chemical poisoning of all birds, was that of established literary and filmic nuclear imagery (Buell 2010). At the same time, contrary to the nuclear holocaust, the environmental apocalypse is not pictured as a sudden event, but as a slow (though possibly accelerating) process, as a sort of asymptotic approach to a "tipping point." It therefore demands urgent action to stop humanity's march toward an irreversible doom marked not by a definitive end, but by ever-increasing social misery and environmental deterioration. Even Naomi Klein, who in her 2014 *This Changes Everything: Capitalism vs. the Climate* imagined that the planetary emergency of global warming might bring about worldwide "regeneration" by spurring alternatives to the economic system that has caused it, told *Vogue* that she had "seen the future, and it looks like New Orleans after Hurricane Katrina" (Powers 2014).

The catastrophic imagination thus remains energizing – and all the more so because, as noted, humans themselves are the main source of endangerment for their own species, and that makes perceptions of extinction more unsettling than ever. But the apocalyptic viewpoint does more than merely equate the most likely future to the worst-case scenario. The apocalyptic mode of expression embodies a "dialectic of peril and recovery" that may be regenerative and allow for some recuperation of the loss (Turner 2007). Similarly, realizing that the biocultural persuasion is the product of a "crisis narrative" does not imply denying the many legitimate reasons for worrying about the erosion of diversity on a global scale;

rather, it offers a starting point for analyses of endangerment that are less defined with regard to crisis, and more oriented toward the possibilities that emerge from assuming human agency and from the "processes of hybridity" that characterize globalization (Brosius and Hitchner 2010, 142, 143).

The important point here is that diversity is seen as valuable through the lens of its being at risk, and that, in the final analysis, endangerment takes center stage because it affects diversity and has become implicitly incorporated into its very definition. To the extent that diversity is the value, its defense is the moral imperative, endangerment the most ominous risk, and the extinction of life or the destruction of cultural heritage the paradigmatic anti-values. This process of reciprocal shaping underlies the moral and emotional tone of preservation initiatives, and motivates a particular way of doing science. But what kind of value is diversity, and what kind of duty is its protection?

The duty to preserve

We noted above that the ultimate justification for protecting biocultural diversity appears to be generally pragmatic and anthropocentric, aimed at ensuring humanity's viability. The feeling of crisis, as put by Terralingua co-founder David Harmon, is

> driven by the conviction that we soon will reach a momentous threshold, a point of no return beyond which a critical amount of biological and cultural diversity will have been lost, never to be regenerated on any time scale significant *to the development of humanity.*
>
> (Harmon 2001, 61; our emphasis)

What seems to be at stake, therefore, is less diversity as an intrinsically valuable feature of the world, than diversity as a practical condition of human survival. The widespread vocabulary of "natural capital," "ecosystem services," "sustainable management" and "biodiversity-based economies" attests to a range of pursuits aimed at balancing the exploitation and preservation of resources. And yet, when not the words, then the tone, style and moral passion of those engaged in protecting biocultural diversity convey the inherent value they attribute to it, and that they find materialized in the particular objects of their concern.

At the bottom of this apparent tension is the familiar distinction, widely discussed in environmental ethics but also applicable to issues of heritage, between the instrumental value given to things as means to other ends, and the intrinsic value conferred upon them as ends in themselves (typically, only human persons in the Western tradition; see Brennan and Lo 2011 for a concise overview of environmental ethics). What are the foundations of the duty to preserve biodiversity when its value is not conceived as following from the mere fact of its existence? If biodiversity is primarily considered as useful for human survival, then utility tells us to protect it. If it is regarded as part of the good life (for example because it provides aesthetic pleasure), then preserving it is a virtue but not a duty. Finally, if it is understood as affecting other peoples and future generations, then defending it is a matter of justice and therefore a moral obligation. These conceptual configurations combine variously with views about whether the problems to be faced are human-induced or not. Moreover, from biodiversity, they have expanded to biocultural diversity, and moved beyond the somewhat one-sided assumption that cultural diversity is one of the tools for preserving natural diversity. Thus, the Chilean ecologist Ricardo Rozzi (2012, 27) has coined the term "biocultural ethics" to name the project of "recovering the vital links between biological and cultural diversity, between the habits and the habitats of the inhabitants" (which, he believes, are acknowledged by early Western philosophy, Amerindian traditional ecological knowledge, and contemporary ecological and evolutionary sciences), and thereby of combating "biocultural homogenization" on the basis of a detailed field knowledge of specific causes, agents and practices.

Such a project intends to go deeper than UNESCO's well-meaning cosmopolitanism, which considers cultural diversity as beneficial for security, peace and development (Labadi 2010; UNESCO 2001). The ultimate goal, again, is to safeguard the interconnected organic and cultural systems that will ensure humanity's survival. The process requires the identification of the appropriate cultural entities. While this implies the risk of actually excluding local communities, exclusion is supposedly compensated by the fact that diversity itself emerges as the value to be cultivated and protected. In practice, only "selected objects of localized descent heritage" get transformed "into a translocal consent heritage," which is in turn defined as the heritage of humanity (Kirshenblatt-Gimblett 2006, 183). By producing cultural assets that are universalized as world heritage, difference is downplayed to the advantage of diversity, and diversity is given intrinsic value.

Intrinsic value, in turn, can be of different sorts. *Subjective* intrinsic value is conditional and recognized as resulting from evaluative judgments. Ecosystems, as well as "tangible" and "intangible" heritages, may be valued for their beauty, variety or spiritual and cultural significance.[8] For example, at the beginning of the United States Endangered Species Act (section 2), the organisms to be protected are declared to be "of esthetic, ecological, educational, historical, recreational, and scientific value to the Nation and its people." Yet these valuations and the objects to which they apply evolve, and the range of what is to be protected may expand or contract. In contrast, *objective* intrinsic value is not conditional, and is perceived as being identified, rather than created by humans. The most radical view in this respect assumes an analogy between human persons and non-human natural entities, such that these have value by virtue of an immanent good or interest that must be respected.

It may seem obvious that, both in nature and in culture, things can have value only through us (Thomas this volume) – and even that they *are* only by virtue of human action in the sense meant by the English moral philosopher Bernard Williams (1995, 240) when he insisted that "[a] natural park is not nature, but a park," and that "[a]nything we leave untouched we have already touched." He thus formulated "the inescapable truth that our refusal of the anthropocentric must itself be a human refusal" (ibid.). The existence of views to the contrary leads in principle to considerable quandaries, for, as philosopher Elliot Sober (1986, 351–352) put it, "to the degree that 'natural' means anything biologically, it means very little ethically. And, conversely, to the degree that 'natural' is understood as a normative concept, it has very little to do with biology." Thus, even in settings dominated by bleak perceptions of an ongoing "crisis of life," the prevalent attitude has become increasingly pragmatic.[9] While the philosophical discussion keeps going on and anthropocentrism remains contested, while the language of intrinsic worth is retained as useful in various contexts, and while a sense of inherent value lies at the bottom of many a conservationist passion and communication strategy, humankind's survival remains the ultimate purpose of preserving biocultural diversity.[10]

Such pragmatism also infuses a particular way of doing science, or at least calls for doing science in a new manner that no longer aims at separating it from morality and emotion, but embraces them as part of knowledge-production and policy-making. In 1996, on the basis of interviews with major actors in the research and defense of biodiversity, environmental law specialist David Takacs (1996, 99) observed that the term

> makes concrete – and promotes action on behalf of – a way of being, a way of thinking, a way of feeling, and a way of perceiving the world. It encompasses the multiplicity of scientists' factual, political, and emotional arguments in defense of nature, while simultaneously appearing as a purely scientific, objective entity. In the term *biodiversity*, subjective preferences are packaged with hard facts; eco-feelings are joined to economic commodities; deep ecology is sold as dollars and sense to more pragmatic, or more myopic, policy makers and members of the public.

What Takacs noted of *biodiversity* in the mid-1990s can be said of *biocultural diversity* in the first decade of the 21st century, and points to a central feature of the endangerment regime.

Activism by scientists is not a new phenomenon. In 1945, for example, Manhattan Project scientists urged President Truman not to authorize the use of atomic bombs, warning of a future atomic arms race and of the dangers it would represent for humanity.[11] In this and other instances, the actors involved assumed that truth was sought for its own sake, and was in principle separate from the subsequent uses of scientific discoveries. In 1972, however, the manifesto "Toward a Science for the People" declared "truth for truth's sake" defunct, emphasized that science does not exist in a political vacuum, encouraged "scientific workers" to put their skills "at the service of the people and against the oppressors," and advocated removing decision-making from experts' hands.[12] This still represented science *for* the people, but pointed toward participatory and citizen science, and even "extreme citizen science" – a bottom-up practice involving "broad networks of people to design and build new devices and knowledge creation processes that can transform the world" (www.ucl.ac.uk/excites).

Insofar as the ultimate goal of such an approach to science is to transform the world, knowledge production is inspired and guided by values that are primarily moral and socio-political. That science is not "value-free" has long been known (see Kincaid, Dupré and Wylie 2007 for recent studies). However, science as understood and practiced in the framework of the endangerment sensibility tends not only to break down the boundaries between research and application, but also to assert itself as being neither value-neutral nor producing value-neutral facts.

On the one hand, preserving biocultural diversity requires not only factual evidence and expert knowledge, but above all "value orientation" (Eser 2009). Such orientation, it is sometimes claimed, can be achieved by means of participatory processes that incorporate local and lay knowledge, develop "regimes of environmental governmentality" that are no longer dictated mainly by "major organizations located in the global North," and broaden the foundations of credibility beyond the scientific community (Brosius and Hitchner 2010, 155, 157). On the other hand, the rise of the "participatory paradigm" in science policy and the idea of a "civic science" born in the context of global environmental politics (Bäckstrand 2003) have come together with a "general shift in conservation organizations from advocacy to management" (Brosius and Hitchner 2010, 153). What such a shift has meant is not that there is no more advocacy, but, on the contrary, that it has become integrated into research. By the 21st century, as Maffi (2001a, 28) put it, "theoretical and applied issues are two sides of the same coin, as are scientific and ethical issues" – and "there must be a genuine commitment to dealing with the two sides together." In truth, the metaphor of the coin does not convey the intended message, which is that activism consists of doing a certain type of science, and that scientifically studying biocultural diversity is a militant move.

Such reconfiguration of the theory and practice of science is perhaps one of the most important correlates of the endangerment sensibility in the early 21st century – one that permeates cognitive forms down to the level of the most apparently basic of them all: the list.

Listing endangerment

The very act of defining an entity as endangered entails the duty to find instruments and techniques to protect it. Conversely, practices of preservation and documentation presuppose the previous identification of an entity as "endangered." Tools associated with those practices – lists and repertoires to begin with – redescribe certain objects as endangered entities, and are by the same token expected to prompt awareness and trigger interventions. Such tools are not neutral catalogues, but result from processes of interpretation and selection, and function as calls for action and forms of argumentation (Schuster 2004). Officially labeling an entity as *endangered* is both an outcome and a beginning (Benson this volume). As has been noted in connection with the United States Endangered Species Act, simply listing a species as endangered virtually precludes any cost-benefit analysis and makes the species eligible for funds to support actions aimed at removing it from the list (Metrick and Weitzman 1996). Similarly, in the field of historic preservation, counting and indexing transform the listed items into potential candidates for the allocation of resources (Harrison 2012). The biggest challenge is that of making the list. Ideally, in addition to fulfilling the quantitative or qualitative criteria for being "endangered," the entities to be listed should have indisputable boundaries (in the case of living species) or unequivocally possess certain features (in the case of tangible or intangible heritage). They rarely do.

In 2013, for example, lemurs were declared the most endangered mammal group in the world. Such an assessment was based on documentation and taxonomic standards that are constantly reviewed by biologists working with the International Union for Conservation of Nature.[13] It required mapping all existing lemur species in order to define their exact distribution according to degrees (*critically endangered, endangered, vulnerable*).[14] Understanding the taxonomy, diversity and distribution of lemurs is thus essential for their conservation. Now, field and genetic research have contributed to the creation of new species, as well as to the splitting of some species "into smaller and sometimes more threatened taxonomic groups" (Yoder and Welch 2012; see also Davies and Schwitzer 2013). The classificatory operation of dividing an established unit into taxa of the same or lower rank therefore multiplied the number of endangered groups. By the same token, disagreements about the validity of the named forms and the attributed taxonomic rankings imply disagreements about the groups that are to be considered endangered (Mittermeier et al. 2008).

Although "endanger**ment**" emphasizes the process of something being threatened, the cognitive device that captures it is a list of states or conditions. Precisely because endangerment takes place in time, the definitions and contents of its categories are (with the possible exception of *extinct*) not merely revisable, but essentially a matter of problematic consensus. The Red List of Threatened Species (www.iucnredlist.org), elaborated since 1963 by the International Union for Conservation of Nature, places taxa in several categories ranging from *extinct* to *least concern* (for organisms under no immediate threat) through *extinct in the wild, critically endangered, endangered, vulnerable* and *near threatened*. Below *extinct* and *extinct in the wild*, which designates a taxon "known only to survive in cultivation, in captivity or as a naturalized population (or populations) well outside the past range," the categories reflect varying degrees of threat (probability of extinction in the immediate future, in the near future, or in the mid-term); *near threatened* applies to a taxon that "is close to qualifying for or is likely to qualify for a threatened category in the near future." Moving a taxon from one category to another depends on factors such as criteria revision, taxonomic change, or increase in the rate of decline. The vocabulary thus embeds temporality and the process of endangerment at the very heart of the list.

The United States 1973 Endangered Species Act (ESA) establishes only four categories: *extinct, endangered, threatened*, and *delisted* – a rank that emphasizes the list's ontological power (Benson this volume). Indeed, as Geoffrey C. Bowker

(2000, 675) noted, the biodiversity database "will ultimately shape the world in its image. If we are only saving what we are counting, and if our counts are skewed in many different ways, then we are creating a new world in which those counts become more and more normalized." The tension between the continuity of the endangerment process and the normalizing fixation operated by the list is palpable here. Between *extinct* and *delisted*, which are the opposite ends of a continuum, ESA retains only *endangered* and *threatened*, which are seamlessly connected by a continuum of probability. While *endangered species* labels a group that is "in danger of extinction throughout all or a significant portion of its range," the term *threatened species* designates one that is "likely to become an endangered species within the foreseeable future throughout all or a significant portion of its range" (ESA section 3; Barrow 2009, Ch. 10).

In the case of languages too, endangerment is characterized by a gradient and by combinations of criteria. Thus, the six degrees of UNESCO's Language Vitality and Endangerment Framework, from *safe* to *extinct* ("no speakers left"), are based on nine criteria, of which intergenerational transmission is the most important.[15] Distinguishing levels of linguistic obsolescence from *potentially endangered* to *moribund*, or establishing "taxonomies of fatality" (Heller-Roazen 2005, 57), implies seeing language transformation as a threat to an original yet essentially elusive tongue, which must be protected from its own evolution. The usual mechanisms whereby languages metamorphose and are sometimes replaced (by contact or divergence, by hybridization, by absorption or by more or less gradual and more or less forced or voluntary abandonment) tend to be seen less as part of a "natural cycle of change" than as threats to precious repositories of thoughts and worldviews that humanity cannot afford to lose (Austin and Sallabank 2011, 6). Some linguists recognize that the extravagant valorization of endangered languages turns these "into objects which seem better suited for museum showcases than for everyday usage by everyday people" (ibid., 18). This suggests that the fundamental activities in which they are themselves engaged – documenting, archiving, counting, labeling, listing, classifying and eventually revitalizing – may be more useful for the linguistic profession than for the communities of speakers it studies.

Commenting on the endangerment/diversity logic that correlates cultural richness with linguistic multiplicity, the Spanish philosopher Fernando Savater (2009) mordantly observed that its terminus is the idiolect, a language unique to a single person; and he questioned the analogy between species and language extinction: while no dinosaur wants to become extinct, some speakers do want to switch language when their own offers only disadvantages. Such intentionally contentious remarks share in University of Chicago linguist Salikoko Mufwene's (2005, 41) irritation vis-à-vis language rights advocacy for its tendency to wish "traditional culture" to survive in a supposedly "pristine form," as if European cultures themselves "were unaffected by the cultures of the populations that adopted them," and as if there were an ingrained connection between "endangered languages" and "endangered peoples." The question is not only whether there can be such a thing as "culture in its pristine form," but also, "What's the point of maintaining diversity if it appears to be adverse to one's adaptation to the new socio-economic ecology?" (ibid., 39).

Mufwene's position is controversial, but it raises fundamental questions about the status of the knowledge devices used to label cultural entities as endangered. As Daniel Heller-Roazen (2005, 65) has noted, since linguistic decay mostly involves the transformation of tongues, "all documents of decease bear witness to the same obstinate will to set aside the one possibility the experts in the health and the sickness of tongues would rather not ponder: that in language there may be no dead ends, and that the time of the persistent passing of speech may not be that of living beings." Lists and gradients, with *extinct* as one extreme ontological status, epitomize the denial of such possibility.

The most common criterion of extinction remains the death of the last individual. Such a choice seems to identify a group clearly, and turns a process into an irreversible condition. Yet, in spite of appearances, extinction is not a precisely datable event. For example, in August 2014, the International Union for Conservation of Nature declared the giant earwig *Labidura herculeana* officially extinct. The insect, however, had been last seen in Saint Helena Island in 1967 and had been assessed as "critically endangered" in 1996 (www.iucnredlist.org/details/11073/0). Similarly, although Lonesome George, the last known specimen of a giant Galapagos tortoise, died in June 2012, its subspecies is still considered as "extinct in the wild" (www.iucnredlist.org/details/9017/0). The reason is that endangerment is a process, and therefore so is extinction and the procedures that make it official. Moreover, extinction is not one thing alone. The statement that the thylacine or Tasmanian tiger "became extinct on 7 September 1936 when the last known specimen died in captivity in Beaumaris Zoo, Hobart" (Paddle 2000, 1) echoes heroic views about "the last of" (the Mohicans, the Tasmanian Aborigines, the Wild Indian, and so on), which typically conflate the individual with the culture (Clifford 2013; Gänger this volume). Yet the figure of "the last" does not represent the only valid way of characterizing extinction (Delord 2007; Sepkoski this volume). A population may be considered functionally extinct when there are no more relations between its members. Or an extinct species may be considered as virtually or potentially non-extinct if it were possible to recreate it on the basis of genetic information (Radin this volume; see Sandler 2013 on the ethics of de-extinction). Or, finally, individuals may persist, while the species disappears as a cognitive category.

The domain of heritage offers an analogous situation. In the United Kingdom, the Heritage at Risk Register, began in 1998, includes not only buildings, but also assets of other sorts (monuments and archeological sites, parks, battlefields, wreck sites and conservation areas), provided they are included in the Statutory List of Buildings of Special Architectural or Historic Interest (risk.english-heritage.org.uk/register.aspx). The List classifies hundreds of items in "grades" according to "interest," and listed sites are assessed for inclusion in the Heritage at Risk Register on the basis of condition and, if applicable, use or occupancy. Condition ranges from *very bad* to *poor*, *fair* and (occasionally) *good*, but the register – in a characteristic attempt to fixate evolving situations – also includes buildings "vulnerable to becoming at risk." Once a building is thus identified and included in the Register, priority for action is assessed on a scale. As Harrison shows in his chapter in this volume, such gradients "not only establish and articulate with other systems of valuation, but are strategic in the sense in which they imply different actions (or regimes of management) that should be taken if sites are threatened."

A snowball effect is built into the logic of those procedures. It is not only that the very concept of tangible and intangible world heritage originates in the West and implies attitudes toward culture and history that are also of European origin, but also that beneath the universalist rhetoric of the Authorized Heritage Discourse (Smith 2006), the practices at work are national and driven by national interests. As heritage becomes in itself a market economy of goods and services, countries compete to get on the list to gain prestige, publicity and economic advantage; the criteria for heritage (never the object of a workable operational definition) weaken under the pressure of public and private interests, and the list, incarnation of a growing "heritage glut" (Lowenthal 1998), acquires a life of its own as a space for trading economic and symbolic values (Frey and Steiner 2011; Harrison and Hitchcock 2005; Rizzo and Mignosa 2013).[16]

Defining the relevant gradients and their contents involves constant negotiations at the scientific, political and normative levels. On the one hand, categorizations manifest conflicting interests whose resolution, as substantiated in conventions, regulations and policies, is intrinsically fragile. Since the category "endangered species" functions as a legal concept, placing or maintaining particular species on a list may spark lawsuits – for example, when conservation groups seek to block oil drilling in order to protect whales. Such legal proceedings highlight the fact – consistent with the already-noted primacy of pragmatism – that the conservation of biocultural diversity is in practice inseparable from resource management. For example, the International Convention for the Regulation of Whaling, signed in 1946, explicitly seeks to "provide for the proper conservation of whale stocks and thus *make possible the orderly development of the whaling industry*" (iwc.int/convention).

On the other hand, the very existence of management and conservation negotiations throws light on more specifically scientific and conceptual challenges. Two major ones relate to uncertainty and categorization. First, as Michael Heazle (2006) has shown, the factious politics of whaling (to continue the example) is riddled by scientific uncertainty – over population estimates, the effects of environmental change, maximum sustainable yield levels and the monitoring of catches. The perceptions and disputes about uncertainty itself play an essential role in shaping policy.

Second, as mentioned, endangerment is a process, but it is dealt with by means of categories that denote states of being. The use of adverbs in labels such as "*critically* imperiled" or "*apparently* secure" reflects the negotiable and temporary nature of decisions, and points to the difficulties involved. The Red List grade *extinct in the wild* is a good example. As we saw, the category designates species of which only captive individuals survive. Sometimes these captive-bred animals are reintroduced into ecosystems from which they had disappeared, and this is said to play a vital role in conservation (www.iucnsscrsg.org). The process, however, begins when the surviving individuals of a critically endangered species are moved from an area characterized as "the wild" or as a "natural ecosystem" into a non-natural human-made facility (Benson this volume). This redescribes the species as "extinct in the wild." Re-introduction specialists do not equate breeding individuals in captivity with domesticating them in the strict sense. Nevertheless, the Convention on Biological Diversity defines domesticated species as those "in which the evolutionary process has been influenced by humans to meet their needs."[17] This definition suggests that reintroduction is in fact a form of domestication. Indeed, human needs are not limited to subsistence and organic survival. They are also moral, ideological, political or aesthetic; and breeding captive animals to reintroduce them into nature may obey to what is felt as a moral necessity. Moreover, the act of releasing a saved species "into the wild" may imply particular beliefs about wilderness – those expressed, for example, in the 1964 law that created the USA National Wilderness Preservation System when it defined wilderness as an area that retains its "primeval character" and "where man himself is a visitor who does not remain." However, as William Cronon put it (1996a, 69), wilderness "hides its unnaturalness behind a mask that is all the more beguiling because it seems so natural." The ideas of "primeval character" and "natural condition" obscure the fact that wilderness was invented in the 19th century in opposition to the urban-industrial civilization that was then seen as contaminating it.

Through wilderness, nature retains its traditional moral authority (Daston and Vidal 2004) and remains a vessel into which people pour "all their most personal and culturally specific values: the essence of who they think they are, how and where they should live, what they believe to be good and beautiful, why people should live in certain ways" (Cronon 1996b, 51). These beliefs, values and norms are imbued with intense emotions, which are to be considered neither foundational nor tributary, but constitutive of the endangerment sensibility.

The emotions of endangerment

It's time to kill off the extinction message. Such is the commandment found in a brochure entitled "Branding Biodiversity."[18] According to its author, the British "sustainability communications agency" Futerra, biodiversity messages appeal to Loss, Love, Need or Action. The first ones, based on the threat of extinction, are virtually intrinsic to the notion of biodiversity. Yet they are the least efficient. In most people, awareness of endangerment may induce no more than guilt and head-shaking, but no action; except for the "biocentric" few, "extinction comes across more as an empty threat than a lifestyle threat." The most powerful of the other messages is the one appealing to Love, which trades on empathy, and builds upon feelings of "awe, fascination and wonder for the natural world." Need messages, which convey the economic value of biodiversity, work well in policy-making and business, but partner badly with Love. Both have to be combined with Action messages aimed at everyone beyond those who already demonstrate or donate. The goal, to make sustainable development "so desirable it becomes normal," is to be achieved by turning biodiversity into a brand where emotions take center stage and concepts play a minor role.

Some may object to the means advocated by Futerra, but the fact is that the branding agency applies principles academic psychology seems to confirm: that rationality and emotions work together, that behaviors are not guided by principles of "perfect rationality," that humans rarely make choices on the basis of conscious calculations, and that emotional dispositions are not bothersome factors that "bias" decisions, but integral and useful components of practical reason. "Emotional rationality" and "rational emotions" no longer are oxymora, but philosophical, psychological and neuroscientific commonplaces.[19] Inevitably, conflicting interpretations subsist. Embedded in the "turn to affect" in the human sciences is the opposition between those who separate affect from meaning and emphasize the pre-personal character of emotional responses, and those for whom emotions are states directed toward objects and depend on beliefs, meanings and desires (Leys 2011).

It should come as no surprise that, in arenas of dispute, emotionality plays its traditional villain role. Michael Crichton, the late author of *Jurassic Park* and other fiction best sellers, became notorious for his skepticism about global warming and his attacks on environmentalism. He described the latter as "the religion of choice for urban atheists," and sustainability as "salvation in the church of the environment." In Crichton's (2003) understanding, to say that environmentalism is a religion implied that it is "generated by our emotional state," by individual hopes and fears – in short, that it is driven by faith, apocalypticism and guilt rather than by clear-headed judgment based on scientific knowledge. Similarly, the conservative philosopher Roger Scruton (2007) accused the environmental movement of having "crystallized into a faith" and of being driven by an *odium theologicum* typical (in his view) of leftist movements. By reducing religion (and its allegedly deleterious impact) to pure affect, such comments once again demonstrate emotions' bad reputation. Naturally, their very style attests to the role of emotions in driving argument and sustaining values. In a more positive vein, physicist Freeman Dyson (2008) described environmentalism as a "worldwide secular religion . . . of hope and respect for nature" with which most people may agree. He nonetheless complained that some environmentalists adopted "as an article of faith the belief that global warming is the greatest threat to the ecology of our planet," and therein found "one reason why the arguments about global warming have become bitter and passionate." In fact, a more essential reason is to be found in the strategies conservative and industrial groups use to present anthropogenic climate change as a matter of dispute (see Kitcher 2010 for a concise review of the debates).

The condemnation of emotions can be understood as part of the "scientization of politics" – ultimately a process of depoliticization in which "matters of concern" are transformed into "matters of fact" (Bowman 2010, 182). Nevertheless, there can be no such thing as depoliticization, and those who push for that transformation are themselves animated by political passions. Whether seen as positive motivators or as obstacles to clear thinking, emotions are recognized as crucial for conveying meanings, shaping perceptions, sustaining values and driving action. Takacs' (1996) already-mentioned interviews with researchers and advocates of biodiversity eloquently illustrate the integration of scientific, moral and emotional dimensions well beyond utilitarian considerations. Emotions figure prominently in conservation psychology, a field that emerged in the 1990s to bring psychological research and theory to bear on discussions of environmental issues, as well as to promote "healthy and sustainable relationships with nature" (Clayton and Saunders 2012, 3). The very development of a discipline that studies why humans hurt or help the environment and wants to make their behavior more ecological is part of the history and sociology of the endangerment sensibility. Conservation psychology itself exhibits the sensibility whose structure, mechanisms, limits and conditions of possibility it investigates. Revealingly, one of the features usually highlighted as a novel contribution of the field is precisely that it gives prominence to the emotions.

Research carried out since the 1990s has suggested that both positive and negative emotions serve as predictors of conservation behavior; insofar as emotional affinity with nature is a strong individual motivation for ecological attitudes, negative emotions arise when treasured natural resources and environments are perceived as threatened; though less studied, self-conscious emotions such as pride, shame and guilt have also been shown to play a key role as conservation incentives; negative self-evaluative emotions can be used to encourage pro-environmental behaviors; and emotions have been studied as structural behavioral factors, as well as for their role in communication and persuasion (Vining and Ebreo 2002).

Especially important in this respect are emotions rooted in feelings of connection with nature. The endangerment landscape is dominated by "charismatic" species like the panda, the tiger and the koala, whose protection is more likely to attract funding than the protection of other species. A sign of how deeply emotional identification is implicated in the endangerment sensibility is that most "extinction stories" deal with animals, on whom humans can more easily project themselves, than with plants (Heise 2010). There has been debate on the animal preference and the choice of charismatic organisms. At the same time, it is recognized that "flagship species" help generate support for less glamorous biodiversity conservation projects. In the form of such species, nonhuman charisma may help people who hold "different understandings of nature to work together for a common conservation cause" (Lorimer 2006, 2007). It can also separate those who share views of nature, as illustrated by the 2009 flurry after a British TV naturalist suggested that pandas should be left to die out (Benedictus 2009; the debate goes on).

Emotion, then, is explicitly given a decisive role as vehicle for values. Many examples demonstrate the extent to which members of local communities may be animated by interests and feelings that conflict with those of political authorities and environmental organizations (see Boitani 2000 for the case of European wolves and Dowie 2009 for extra-European cases), or how different types of organizations oppose each other fiercely in spite of apparently sharing the final goal of environmental justice, as for example in the clash of cultural rights and animal rights advocates over Makah whale hunting (Sullivan 2002; on the social effects of protected areas, see West, Igoe and Brockington 2006). The same takes place in the universe of cultural heritage, where the "UNESCOisation" of sites or the protection of native forms of art may conflict with the wishes and values of their makers and users (see for example Berliner 2010 on the old center of the Laotian city of Luang Prabang after it became a World Heritage Site; or Mowaljarlai et al. 1988 and O'Connor, Barham and Woolagoodja 2008 on the aboriginal repainting of ancient rock art in northwest Australia).

There might be differences in the quality of the affect involved, depending on whether the object to which it attaches is nearby and connected to one's life history (plant or animal species in one's region, a wind farm in one's favorite landscape, an abandoned medieval castle in one's birthplace) or distant and prompted by the media or NGO communication (Madagascar lemurs, deforestation in the Amazon, the Djinguereber Mosque in Timbuktu). But in all cases, spanning the local and the global, the immediate present and the distant future, people's response is – in addition to anything else – fundamentally emotional. Organizations have understood this, and apply it in their social marketing strategies, which are aimed at evoking positive emotional responses rather than at communicating the complex concepts (such as *ecosystem* or *biodiversity*) that underlie the justification and development of conservation programs (Wilcove 2010). (This is not to say that the public is uninformed. Grassroots organizations and national institutions support information exchange systems, and many legislations ensure free access to official environmental information. Finally, as mentioned, various initiatives promote the "participatory" paradigm in environmental science and policy-making.)

In environmental education too, *management* has become the key notion with regard to emotion. Optimistic "emotion talk" and "emotion discourse" (Reis and Roth 2009) are basic resources for generating the bodily and psychological states that, it is hoped, will lead to awareness of endangerment, and from there to a sense of care and a will to act. In both their discursive and experiential aspects, emotions are simultaneously tools and outcomes, means and ends. We have seen that some critics accused environmentalism of being "religious," by which they meant: irrational, subjective, emotion-driven. Environmentalists, in contrast, vindicate affect; some pursue their cause as a religious quest (see for example Dunlap 2004 and Kearns 2004) and use religious emotions as a pedagogical tool.

For example, in *A Greener Faith*, Roger S. Gottlieb (2006, 160), a professor of philosophy at Worcester Polytechnic Institute in Massachusetts, argues that "environmentalism can function as a religion because it begins with religious emotions and connects them to an articulated set of beliefs about our place in the universe."[20] In the introduction to his anthology *This Sacred Earth: Religion, Nature, Environment*, Gottlieb (2004, 2–3) recounts that on the first day of his environmental philosophy class, he tells students of his own "fear, grief, and anger about the ecological crisis."

I then ask them to speak in turn about what they feel. They respond hesitantly, emboldened by my example but still unsure that a university classroom is the proper place for emotions. As the hour progresses, however, their statements become more revealing.

"I'm terribly angry," one will say. . . .
"I'm scared," a young woman admits. . . .

> It helps to begin not with a long list of environmental problems, but with the acknowledgment that our anguish over the fate of the earth is a real element in our everyday emotional lives. Bury these emotions as we may, they surface whenever we hear of another oil spill. . . . Before we can take in or effectively act in response to the environmental crisis we must admit just how deeply we feel for the earth. . . .
> There is nothing shameful or "weak" in the pain we feel about the environment. Grief and fear are rational responses to our losses and perils. And sorrow over what we have done is a hopeful sign that despite everything we can still love and mourn.

In this scene, emotions appear as consubstantial with values, are placed at the very root of decision-making and action, and come before information. Its style and contents express a romantic pathos of human sympathy with the natural world, the sense of nature as the dwelling of the divine and the sacred, and the stance of the "worshipper of Nature" who (as William Wordsworth wrote in his 1798 "Lines Composed a Few Miles Above Tintern Abbey") recognizes in it

> The anchor of my purest thoughts, the nurse,
> The guide, the guardian of my heart, and soul
> Of all my moral being.

While certainly such feelings are not universally shared, they highlight a strain of religious sentimentality that has not been without effectiveness in fostering environmentalism. The *Canticle of the Creatures* is about praising God, not celebrating nature for itself; and the stories about Francis of Assisi and the animals are meant to express his love of God through the Creation, not of animals for themselves. Yet when in 1979, John Paul II named Francis patron saint of ecology, he was not merely rejuvenating the traditional role of saints as models of life, but specifically encouraging Catholics to behave better toward the environment (francis35.org/pdf/papal_declaration.en.pdf).

Beyond the individual level, emotions play a major role in conflicts over environmental issues and in the psychological and social dynamics of activism.[21] Research highlights the emotions' significance as sources of diverging views and attitudes, as motivating forces, and as central factors in the processing of information and the escalation of protests (see for example Buijs and Lawrence 2013 on conflicts in forestry). It comes as no surprise that high arousal emotions such as anger manifest themselves in activism and sustain protest and resistance, nor – as we saw with Crichton and Scruton – that some contenders seek to delegitimize them. A study of conflicts over wind farm developments shows that industry and policy actors systematically appeal to rationality and dismiss positions and decisions that are contrary to theirs as being emotionally influenced: "Emotion is rejected strategically, neutralized substantively and is seen as in need of manage-ment" (Cass and Walker 2009, 68).

Yet affect in general plays a crucial functional and structural role in all areas involving perceptions of endangerment and their political and scientific con-sequences (see for example Carrus, Passafaro and Bonnes 2008 on choices in recycling and use of public transportation). Saving the environment may require emotional and cognitive empathy across space and time (Krznaric 2010). As for language endangerment, it "is generally presented in emotive and moral-istic terms" and "[a]s with most phenomena which become the focus of moral panic, there is no attempt to present a 'balanced' argument about whether or not endangered languages should be preserved" (Cameron 2007, 269). In the French-speaking world, where the notions of *patrimoine* and "national monument" were institutionalized in the early 1790s in the wake of Revolutionary violence (Pou-lot 1997), the anthropologist Daniel Fabre launched around 2000 the concept of "patrimonial emotions" to designate a key element in the dynamics of heritage (Fabre 2013).

Heritage-bound emotions obviously differ from nature-bound emotions in the objects to which they attach. Yet they are likely to fulfill equivalent motiva-tional, expressive and appraising functions, and convey different modes of the same values, such as diversity and authenticity, playing them in different keys. Commonalities and specificities remain to be empirically investigated. It seems nonetheless legitimate to extend to "non-natural" contexts the definition of "environmentally relevant emotions" as emotions referring directly to the "nat-ural" environment or to entities linked to the environment (Kals and Müller 2012), and to imagine that they present substantive parallels. In the environ-mental context, "burdens and worries" are caused by (generally local) threats; affective connections to nature come in myriad forms, from loving one's pet to a sense of unity with the cosmos; and moral emotions, such as anger and indigna-tion, typically arise when short-term individual interests collide with long-term societal ones and give rise to behaviors or decisions that are judged contrary to certain moral standards. Although the emotional and behavioral configura-tions that ensue from such disputes are diverse in their details, internal emotional states and dispositions always carry meanings and contribute to motivate and drive both cooperation and conflict at the interpersonal and collective level. In fact, any emotion can be "environmentally relevant" and any environment, emotionally significant.

Endangerment-related decisions ultimately derive from values, and it is these values themselves that are ultimately at stake in conflicts and negotiations. This is not to deny that emotions can signal or motivate the adoption of particular values; but since parties who sit on opposite sides of a negotiating table may share those emotions (e.g. anger), it is not feelings that are at stake, but values and their practical consequences. While the place of emotions in morality has been the subject of controversy since the beginnings of Western philosophy, and while the issue of emotions and values is philosophically and psychologically complex, it is safe to say that, in the contexts we deal with here, emotions are understood as expressing values and involving cognitive appraisals of consequences.

As French sociologist Nathalie Heinich (2009, 2010–11, 2012) has documented, the expert assessment of tangible heritage appeals to features regarded as values, such as the "typicality" or "representativeness" of an object or site with respect to the category to which it belongs. Although these values are actualized via inventorial and descriptive techniques supposed to ensure emotional distance, feelings of harmony and pleasure associated with beauty play a role in expert judgments even when beauty is proscribed as an official assessment criterion. As for the public, it reacts openly with high arousal negative and positive emotions: defensive ones, such as outrage when a monument is seen as disfigured or in danger of being unduly modified; approving ones, such as admiration at the state of preservation of a site.

Heritage-related emotions reveal three basic values: authenticity, the very core of heritage; "presence," a sense of entering in contact with people and realities formerly attached to the valued object; and beauty. These values can be positively or negatively modulated by two others: age and rarity. Rarity, for example, might work in favor of a site if eccentricity is appreciated, but against it if the objects are judged by the extent to which they represent traditions or common collective habits. In short, the sense of heritage subsumes a multiplicity of emotions, a limited set of values, and their interplay in particular contexts. That is why, to avoid substantializing "heritage," Heinich prefers to speak of a "patrimonial function" whose goal is to conserve objects considered to be a common good and to have everlasting value. Like beauty, "being patrimony" is a relational property rather than an inbred quality; thus, "it is not the object which makes heritage, but it is the patrimonial function which makes a patrimonial good out of an object" by way of emotional experiences that embody values (Heinich 2010–11, 127).

* * *

As we have seen, the values involved in the endangerment sensibility range from the intrinsic to the instrumental. To the extent that even the attribution of intrinsic value to biocultural diversity depends on human valuation, it might lead to expressions of cultural relativism that contradict the bio-cultural axiom as well as the preservation imperative. In both its affective and cognitive dimensions, the endangerment sensibility thus partakes in one of the thorniest endeavors since the end of World War II, namely the reconciliation of human rights and cultural rights. In the context of what Charles Taylor (1994) calls the "politics of recognition," such an undertaking has involved trying to work the respect for the particular cultural identities of citizens, often based on race, gender, ethnicity, religion or disadvantage, into the universalism of the law and the neutrality of public institutions. This is the challenge of multiculturalism, which, as Amy Gutmann (1994, 3) has noted, "is endemic to liberal democracies because they are committed in principle to equal representation of all." The tensions inherent in the democratic or democratizing outlook – between toleration and respect, difference and equal dignity, particularism and universalism, ethnocentric standards and the "homogenizing demand for recognition of equal worth" (Taylor 1994, 72) – are built into UNESCO's doctrine of diversity. In its biocultural form, this doctrine has come to imply that all natural and cultural entities deserve in principle to be preserved, and functions as a cosmopolitan and secular soteriology (Stoczkowski 2009), as a vision of redemption and salvation that receives meaning and legitimacy from crisis narratives and a "doomsday" perspective.

The equation between the Svalbard Vault and Noah's Ark is not merely metaphorical, for both raise the question, "Salvation of what, by whom, and for whom?" Philippe Descola (2008) gives the question a shorter secular form – "Who Owns Nature?" – and calls for admitting "that there are no absolute, scientifically founded criteria on which to justify universally recognized values concerning the preservation of natural and cultural assets." If this is so, then negotiating and managing togetherness is the only way to face the challenges of cultural homogenization, environmental degradation and resource depletion. These challenges derive from forces that predate the endangerment sensibility, such as industrialization, commerce and imperialism; the sensibility itself entails a grasp of consciousness and eventually a will to act. As highlighted by the contributions to this volume, it involves perceptions of the world, conceptual choices, ethical and aesthetic attitudes, and emotional and scientific practices that have evolved in history and taken different forms in different contexts. It thus promises to keep offering a rich entanglement of possibilities around the crucial issue of human responsibility.

The idea of the human stewardship of nature has distant roots in the book of Genesis (2:15), according to which "the LORD God took the man, and put him into the Garden of Eden to dress it and to keep it." In the Christian Middle Ages, it was connected to the figure of Nature as *vicaria Dei* (Economou 2002). This allegorical and symbolic personification not only represented the benignity of a providential order to be preserved and obeyed, but, in some narratives, also involved the possibility that nature itself could disturb the divine order – that the *contra naturam* actually resulted from natural processes (Cadden 2004, 2013). Starting in the 1990s, the notion of stewardship was revived in the environmentalist framework (Worrell and Appleby 2000), combining the themes of moral accountability for the present and the future with an emphasis on the sustainable management of natural resources. By the 2010s, stewardship had become "biocultural" (Beckford et al. 2010; Caston 2013). In its religious versions, God is the ultimate authority to whom humans are answerable; in the secular ones, stewardship is exerted on behalf of present society and future generations. In both cases, however, ulterior reasons coexist with the attribution of intrinsic value to nature as such – for God, as Genesis (1:31) proclaims, "saw every thing that he had made, and, behold, it was very good." The essential element here is not the belief in a transcendent God (who may or may not stay in the picture as first cause), but the assertion of an immanent good.

Yet the kind of goodness the medieval *Natura* signified and the kinds of threats it faced differ profoundly from those that motivate modern endangerment discourses and practices. Acting *against nature* consisted of behaving in ways that infringed rules based on Biblical precepts and empirical regularities – hence homosexuality as the *vitium contra naturam* par excellence. It entailed a violation of norms applicable to humans, not damage to other species and their habitats. In contrast, the forms of reflexivity that developed since the 18th century were prompted by the perception of humans' destructive action on natural and cultural environments.

As the "environmental crisis" took on worldwide proportions, so did this perception. Thus, since the mid-20th century, environmentalism – with all its diversity, contradictions, competing agendas, and manifold interactions with social, political and economic factors and interests – emerged as a global phenomenon (Oosthoek and Gills 2008; Radkau 2014). *Global* it is, not only due to its presence across the globe, but also by virtue of its ultimate concern for the fate of the planet as a whole in which events at the local level have worldwide significance. While some mid-19th-century conservation initiatives were inspired by a recognition of ecological interdependence (based, for example, on the observation of links between deforestation and erosion, or fish depletion and logging), that recognition now applies globally, and has, as we repeatedly saw, broadened to the domain of culture. Such expansion has given prominence to knowledge practices where documentation and preservation are collective and mobilize a vast number of scientists and non-scientists at the international scale.

Those features befit their ultimate object, diversity, understood as an object of global accountability. Such accountability implies the duty to preserve not only the biological and cultural heritages identified as endangered, but also the information that enables their protection. Thus, knowledge production, preservation and transmission are undertaken for the benefit of present and future humanity, but also designed for the scientists and conservationists of posterity. In anthropology, for example, the materials that record traces of the existence of unique lost peoples are themselves "unique and unrecoverable"; yet many of them may be destroyed, scattered or deteriorated, making it critical to take measures for preserving them "for future generations" (Lemov this volume; Silverman 1995). Such urge to produce traces of traces and to keep records of records is characteristic of an endangerment regime that, beyond concrete persons, looks with the sciences of the archive toward "an imagined community that transcends time" (Daston 2012, 184).

While earlier conservation sensibilities responded to perceived perils, they were sustained by visions of material and moral progress. The global preoccupation with biocultural diversity, in contrast, embodies a sensibility that is saturated with *endangerment*. When it turns to action, it certainly aims at improving humanity's present and future lot, but it does so in a race against time, focusing on preventing further damage. The tasks it inspires, creative and admirable as they often are, seem mainly remedial, recuperative, therapeutic, even palliative. Given how fast, in its own description, the threats it combats keep advancing, and how powerful the forces that drive them remain, this can probably not be otherwise.

Notes

1 Special thanks to Etienne Benson and Rodney Harrison, who generously acted as consultants on many intellectual and practical aspects of the project embodied in *Endangerment, Biodiversity and Culture*; and to Lorraine Daston for her sharp reading of the present text. Any signs of ineptitude are of course our own.
2 news.nationalgeographic.com/news/2008/02/photogalleries/seedvault-pictures.
3 Transcript of conversation with Cary Fowler, 26 February 2008, www.washingtonpost.com/wp-dyn/content/discussion/2008/02/26/DI2008022601020.html. Crop Diversity Trust website: www.croptrust.org.
4 www.regjeringen.no/en/dep/lmd/whats-new/Speeches-and-articles/speeches-and-articles-by-the-minister/speeches-and-articles-/one-year-anniversary-seminar-of-the-sval/one-year-anniversary-seminar-of-the-sval.html?id=547254#.
5 Over two decades after the Rio Summit, the 2014 Synthesis Report of the Intergovernmental Panel on Climate Change still notes that local and indigenous knowledge is not consistently taken into account in planning and implementing policy (www.ipcc.ch/report/ar5/syr) – a goal promoted by the Indigenous Peoples' Biocultural Climate Change Assessment Initiative (ipcca.info).
6 ethnobiology.net/what-we-do/core-programs/global-coalition/declaration-of-belem.
7 www.unesco.org/new/en/culture/themes/endangered-languages/biodiversity-and-linguistic-diversity. Those beliefs and values are wonderfully displayed in the 2008 documentary *The Linguists*, which is completely independent from UNESCO (thelinguists.com). For interactive documentation on endangered languages, see www.ethnologue.com (considered the most comprehensive source of its kind) and the UNESCO Atlas of the World's Languages in Danger (www.unesco.org/culture/languages-atlas).
8 UNESCO (whc.unesco.org) classifies "world heritage" into *natural heritage*; *tangible cultural heritage*, which includes objects (from buildings and monuments to books, works of art, and artifacts); and *intangible cultural heritage*, which includes "practices, representations, expressions, knowledge, skills" (such as song, music, drama, dance, rituals, festivals or artisanal traditions) transmitted from generation to generation, though only insofar as they are "compatible with existing international human rights instruments, as well as with the requirements of mutual respect among communities, groups and individuals, and of sustainable development" (www.unesco.org/culture/ich/en/convention).
9 We use "crisis of life" because a good example of the combination of utilitarian pragmatism, sense of crisis, and conviction about inherent value is offered by the Crisis of Life video project (www.crisisoflife.net). There, scientists and activists talk about ways to stop the biodiversity crisis to ensure the survival of all living beings, including humans. While the explicit arguments concern mostly the utility and sustainability of ecosystems, the images (other than those of the interviewees) exclusively show charismatic organisms and landscapes, and the accompanying music is either sentimental or alarming and ominous.
10 Chakrabarty (2014, 9, 21) reports that paleoclimatologists "see climatic tipping points and species extinction as perfectly repeatable phenomena" and demonstrate that, while the current phase of warming is anthropogenic, "it is only contingently so" since it happens elsewhere in the universe, and happened and will happen again on Earth without the human species. Chakrabarty (ibid., 9) insists that the climate crisis

"requires us to move back and forth between thinking on these different time scales all at once." But since he also judiciously notes that the "'global' of globalization literature . . . cannot be thought without humans directly" (ibid., 22), it remains unclear how thinking across the time scales helps humans to face their current predicament.

11 www.dannen.com/decision/45–07–17.html.
12 http://ist-socrates.berkeley.edu/~schwrtz/SftP/Towards.html.
13 http://www.iucnredlist.org/technical-documents/categories-and-criteria.
14 http://www.iucn.org/news_homepage/news_by_date/?13487/Lemurs-of-Madagascar-three-year-conservation-plan-launched.
15 http://www.unesco.org/new/en/culture/themes/endangered-languages/language-vitality/.
16 Glut or merit, and just to provide a recent example, in November 2014, the UN Committee for the Safeguarding of the Intangible Cultural Heritage inscribed on its list of Intangible Cultural Heritage in Need of Urgent Safeguarding a dance from Kenya, a male-child cleansing ceremony from Uganda, and an oral tradition from Venezuela. See full lists in http://www.unesco.org/culture/ich/index.php?lg=en&pg=00559.
17 Convention on Biological Diversity, article 2 (www.cbd.int/convention/text/). The Convention was opened for signature in June 1992.
18 www.futerra.co.uk/downloads/Branding_Biodiversity.pdf.
19 The literature on this topic is immense. Among recent publications in English, good entry points can be found in Goldie (2010) and Bagnoli (2011). The history and anthropology of emotions have been on the rise since the 1980s, and various disciplines, from biology to geography to cultural studies, have also undergone their "affective turn."
20 Gottlieb's definition of "religion" overlaps with spirituality and belief systems that make room for the transcendental and the supernatural. For a balanced discussion of more theologically rigorous ways of "greening" religion, see Garreau (2010); and Milton (2002) for an attempt at joining emotion, religion and environmental issues. Founded in 1996, *Ecotheology: Journal of Religion, Nature and the Environment* became in 2007 the *Journal for the Study of Religion, Nature and Culture*.
21 We speak of "conflicts over environmental issues" rather than of "environmental conflict" or "environmentally induced conflict." The latter notions, which connect human-induced environmental scarcity or degradation with interstate or intercommunal conflicts (and therefore do not refer to conflicts over the control or distribution of non-renewable resources), have been criticized on several grounds (Hagmann 2005). One of them is the neglect of what "issues" is meant to underline: the role of people's motivations, perceptions, meanings and emotions. Although climate change has been frequently approached as a security matter, there is no evidence of its links with armed conflict (Gleditsch 2012), nor consensus on "what kinds of environmental changes have what kinds of influences on what kinds of conflict or cooperation" (Bernauer, Böhmelt and Koubi 2012, 6; see also Koubi et al. 2012).

Bibliography

All Internet links in the text, the notes and the bibliography were accessed and checked on 8 December 2014.

Adams, W. M., 2004. *Against Extinction: The Story of Conservation*. Earthscan, London.
Austin, P. K. and Sallabank, J., 2011. Introduction. In: Austin, P. K. and Sallabank, J., eds., *The Cambridge Handbook of Endangered Languages*. Cambridge University Press, New York.
Bäckstrand, K., 2003. Civic Science for Sustainability: Reframing the Role of Experts, Policy-Makers and Citizens in Environmental Governance. *Global Environmental Politics*, 3 (4), 24–41.

Bagnoli, C., ed., 2011. *Morality and the Emotions*. Oxford University Press, New York.

Banking Against Doomsday. *The Economist*, 10 March 2012 (www.economist.com/node/21549931).

Barrow, M. V. Jr., 2009. *Nature's Ghost: Confronting Extinction from the Age of Jefferson to the Age of Ecology*. University of Chicago Press, Chicago.

Beckford, C. L., Jacobs, C., Williams, N. and Nahdee, R., 2010. Aboriginal Environmental Wisdom, Stewardship, and Sustainability: Lessons from the Walpole Island First nations, Ontario, Canada. *Journal of Environmental Education*, 41 (4), 239–248.

Beer, G., 2009. Darwin and the Uses of Extinction. *Victorian Studies*, 51 (2), 321–331.

Benedictus, L., 2009. Should pandas be left to face extinction? [Interviews with Chris Packham and Mark Wright]. *The Guardian*, 23 September (www.theguardian.com/environment/2009/sep/23/panda-extinction-chris-packham).

Berkes, F., 2012. *Sacred Ecology* [originally subtitled *Traditional Ecological Knowledge and Resource Management*], 3rd ed. Routledge, New York.

Berliner, D., 2010. Perdre l'esprit du lieu. Les politiques de l'Unesco à Luang Prabang (RDP Lao). *Terrain*, 55, 90–105.

Bernauer, T., Böhmelt, T. and Koubi, V., 2012. Environmental Changes and Violent Conflict. *Environmental Research Letters*, 7, 1–8.

Boitani, L., 2000. *Action Plan for the Conservation of Wolves in Europe (Canis lupus)*. Council of Europe, Strasbourg.

Bowker, G. C., 2000. Biodiversity Datadiversity. *Social Studies of Science*, 30, 643–683.

Bowman, A., 2010. Are we armed only with peer-reviewed science? The scientization of politics in the radical environmental movement. In: Skrimshire, S., ed., *Future Ethics: Climate Change and Apocalyptic Imagination*. Continuum, London, 173–196.

Boym, S., 2001. *The Future of Nostalgia*. Basic Books, New York.

Brantlinger, P., 2003. *Dark Vanishing: Discourse on the Extinction of Primitive Races, 1800–1930*. Cornell University Press, Ithaca.

Brennan, A. and Lo, Y.-S., 2011. Environmental Ethics. In: Zalta, E. N., ed., *The Stanford Encyclopedia of Philosophy*, Fall 2011 ed. (plato.stanford.edu/archives/fall2011/entries/ethics-environmental).

Brosius, J. P., 1997. Endangered Forest, Endangered People: Environmentalist Representations of Indigenous Knowledge. *Human Ecology*, 25 (1), 47–69.

Brosius, J. P., 1999. Analyses and Interventions. Anthropological Engagements with Environmentalism. *Current Anthropology*, 40 (3), 277–309.

Brosius, J. P. and Hitchner, S. L., 2010. Cultural diversity and conservation. *International Social Science Journal*, 61 (199), 141–168.

Buell, F., 2010. A Short History of Environmental Apocalypse. In: Skrimshire, S., ed., *Future Ethics: Climate Change and Apocalyptic Imagination*. Continuum, London, 13–36.

Buijs, A. and Lawrence, A., 2013. Emotional conflicts in rational forestry: Towards a research agenda for understanding emotions in environmental conflicts. *Forest Policy and Economics*, 33, 104–111.

Cadden, J., 2004. Trouble in Earthly Paradise: The Regime of Nature in Late Medieval Christian Culture. In: Daston, L. and Vidal, F., eds., *The Moral Authority of Nature*. Chicago University Press, Chicago, 207–231.

Cadden, J., 2013. *Nothing Natural Is Shameful: Sodomy and Science in Late Medieval Europe*. University of Pennsylvania Press, Philadelphia.

Cameron, D., 2007. Language endangerment and verbal hygiene: History, morality and politics. In: Duchêne, A. and Heller, M., eds., *Discourses of Endangerment: Ideology and Interest in the Defense of Languages*. Continuum, London, 268–285.

Carrus, G., Passafaro, P. and Bonnes, M., 2008. Emotions, habits and rational choices in ecological behaviours: The case of recycling and use of public transportation. *Journal of Environmental Psychology*, 28, 51–62.

Casado de Otaola, S., 2010. *Naturaleza patria: Ciencia y sentimiento de la naturaleza en la España del regeneracionismo*. Fundación Jorge Juan/Marcial Pons, Madrid.

Cass, N. and Walker, G., 2009. Emotion and rationality: The characterisation and evaluation of opposition to renewable energy projects. *Emotion, Space and Society*, 2, 62–69.

Caston, D., 2013. Biocultural Stewardship: A Framework for Engaging Indigenous Cultures. *Minding Nature*, 6 (3) (www.humansandnature.org/biocultural-stewardship – a-framework-for-engaging-indigenous-cultures-article-158.php).

Catlin, G., 1841. *The Manners, Customs, and Condition of the North American Indians*. Published by the author, London.

Chakrabarty, D., 2014. Climate and Capital: On Conjoined Histories. *Critical Inquiry*, 41 (1), 1–23.

Clayton, S. D. and Saunders, C. D., 2012. Introduction: Environmental and Conservation Psychology. In: Clayton, S. D., ed., *The Oxford Handbook of Environmental and Conservation Psychology*. Oxford University Press, New York, 1–8.

Clifford, J., 2013. *Returns: Becoming Indigenous in the Twenty-First Century*. Harvard University Press, Cambridge, MA.

Cocks, M., 2006. Biocultural Diversity: Moving beyond the Realm of 'Indigenous' and 'Local' People. *Human Ecology*, 34 (2), 185–200.

Crichton., M., 2003. Environmentalism as Religion. In: *Three Speeches By Michael Crichton* (scienceandpublicpolicy.org/commentaries_essays/crichton_three_speeches.html).

Cronon, W., 1996a. The Trouble with Wilderness; or, Getting Back to the Wrong Nature. In: Cronon, W., ed., 1995. *Uncommon Ground: Rethinking the Human Place in Nature*. W. W. Norton, New York, 69–90.

Cronon, W., 1996b. Introduction: In Search of Nature. In: Cronon, W., ed., *Uncommon Ground: Rethinking the Human Place in Nature*. W. W. Norton, New York, 23–68.

Daston, L., 2012. The Sciences of the Archive. *Osiris*, 27, 156–187.

Daston, L. and Vidal, F., eds., 2004. *The Moral Authority of Nature*. University of Chicago Press, Chicago.

Davies, N. and Schwitzer, N., 2013. Lemur Conservation Status Review: An Overview of the Lemur Red-Listing Results 2012. In: Schwitzer, C. et al., eds., *Lemurs of Madagascar: A Strategy for Their Conservation 2013–2016*. International Union for Conservation of Nature, Gland, 13–33.

Delord, J., 2007. The nature of extinction. *Studies in History and Philosophy of Biological and Medical Sciences*, 38, 656–667.

Descola. P., 2008. Who Owns Nature? (www.booksandideas.net/Who-owns-nature.html?lang=fr).

Dirzo, R., Young, H. S., Galetti, M., Ceballos, G., Isaac, N.J.B. and Collen, B., 2014. Defaunation in the Anthropocene. *Science*, 25 July, 345 (6195), 401–406.

Dobrin, L. M. and Berson, J., 2011. Speakers and language documentation. In: Austin, P. K. and Sallabank, J., eds., 2011. *The Cambridge Handbook of Endangered Languages*. Cambridge University Press, Cambridge, 189–209.

Dove, M. R., 2006. Indigenous People and Environmental Politics. *Annual Review of Anthropology*, 35, 191–208.

Dowie, M., 2009. *Conservation Refugees: The Hundred-Year Conflict between Global Conservation and Native Peoples*. MIT Press, Cambridge, MA.

Dunlap, T.R., 2004. *Faith in Nature: Environmentalism as Religious Quest*. University of Washington Press, Seattle.

Dyson, F., 2008. The Question of Global Warming. *The New York Review of Books*, 12 June (www.nybooks.com/articles/archives/2008/jun/12/the-question-of-global-warming).

Economou, G.D., 2002 [1972]. *The Goddess Natura in Medieval Literature*. University of Notre Dame Press, Notre Dame.

ESA, Endangered Species Act, 1973 (www.nmfs.noaa.gov/pr/laws/esa/text.htm).

Eser, U., 2009. Ethical perspectives on the preservation of biocultural diversity. *Die Bodenkultur*, 60, 9–14.

Fabre, D., ed., 2013. *Émotions patrimoniales*. Éditions de la Maison des sciences de l'homme, Paris.

Forest Peoples Programme, 2014a. *The Palangka Raya Declaration on Deforestation and the Rights of Forest Peoples* (www.forestpeoples.org/topics/climate-forests/news/2014/03/palangka-raya-declaration-deforestation-and-rights-forest-people).

Forest Peoples Programme, 2014b. *Securing Forests Securing Rights: Report of the International Workshop on Deforestation and the Rights of Forest Peoples* (www.forestpeoples.org/topics/rights-land-natural-resources/publication/2014/securing-forests-securing-rights-report-intern).

Frey, B.S. and Steiner, L., 2011. World Heritage List: does it make sense? *International Journal of Cultural Policy*, 17 (5), 555–573.

Garreau, J., 2010. Environmentalism as Religion. *The New Atlantis. A Journal of Technology & Society*, Summer (www.thenewatlantis.com/publications/environmentalism-as-religion).

Gill, S.D., 1987. *Mother Earth: An American Story*. University of Chicago Press, Chicago.

Gissibl, B., Höhler, S. and Kupper, P., 2012. Towards a Global History of National Parks. In: Gissibl, B., Höhler, S. and Kupper, P., eds., *Civilizing Nature: National Parks in Global Historical Perspective*. Berghahn, New York, 1–27.

Gleditsch, N.P., 2012. Whither the weather? Climate change and conflict. *Journal of Peace Research*, 49 (1), 3–9.

Goldie, P., ed., 2010. *The Oxford Handbook of Philosophy of Emotion*. Oxford University Press, New York.

Gorenflo, L.J., Romaine, S., Mittermeier, R.A. and Walker-Painemilla, K., 2012. Co-occurrence of linguistic and biological diversity in biodiversity hotspots and high biodiversity wilderness areas. *PNAS*, 109 (21), 8032–8037.

Gottlieb, R.S., 2004. *This Sacred Earth: Religion, Nature, Environment*, 2nd ed. Routledge, New York.

Gottlieb, R.S., 2006. *A Greener Faith: Religious Environmentalism and Our Planet's Future*. Oxford University Press, New York.

Gqola, P.D., 2005. A Question of Semantics? On *Not* Calling People "Endangered." In: Huggan, G. and Klasen, S., eds., *Perspectives on Endangerment*. Georg Olms Verlag, Hildesheim, 51–59.

Grant, C., 2014. *Music Endangerment: How Language Maintenance Can Help*. Oxford University Press, New York.

Grove, R.H., 1992. Origins of Western Environmentalism. *Scientific American*, July, 22–27.

Gruber, J.W., 1970. Ethnographic Salvage and the Shaping of Anthropology. *American Anthropologist*, New Series 72, 1289–1299.

Gutmann, A., 1994. Introduction. In: Gutmann, A., ed., *Multiculturalism: Examining The Politics of Recognition*. Princeton University Press, Princeton, 3–24.

Hagmann, T., 2005. Confronting the Concept of Environmentally Induced Conflict. *Peace, Conflict and Development*, 6, 1–22.

Hamblin, J.D., 2013. *Arming Mother Nature: The Birth of Catastrophic Environmentalism*. Oxford University Press, New York.

Hames, R., 2007. The Ecologically Noble Savage Debate. *Annual Review of Anthropology*, 36, 177–190.

Harmon, D., 2001. On the meaning and moral imperative of diversity. In: Maffi 2001b, 53–70.

Harrison, D. and Hitchcock, M., eds., 2005. *The Politics of World Heritage: Negotiating Tourism and Conservation*. Channel View Publications, Clevedon.

Harrison, R., 2012. *Heritage: Critical Approaches*. Routledge, New York.

Heazle, M., 2006. *Scientific Uncertainty and the Politics of Whaling*. University of Washington Press, Seattle.

Heinich, N., 2009. *La fabrique du patrimoine. De la cathédrale à la petite cuillère*. Maison des Sciences de l'Homme, Paris.

Heinich, N., 2010–2011. The Making of Cultural Heritage. *The Nordic Journal of Aesthetics*, 40–41, 119–128.

Heinich, N., 2012. Les émotions patrimoniales: de l'affect a l'axiologie. *Social Anthropology/ Anthropologie Sociale*, 20, 19–33.

Heise, U.K., 2010. Lost Dogs, Last Birds, and Listed Species: Cultures of Extinction. *Configurations*, 18, 49–72.

Heller, M. and Duchêne, A., 2007. Discourses of endangerment: Sociolinguistics, globalization and social order. In: Duchêne, A. and Heller, M., eds., *Discourses of Endangerment: Ideology and Interest in the Defense of Languages*. Continuum, London, 1–13.

Heller-Roazen, D., 2005. *Echolalias: On the Forgetting of Language*. MIT Press, Cambridge, MA.

Heyd, D., 2010. Cultural Diversity and Biodiversity: a tempting analogy. *Critical Review of International Social and Political Philosophy*, 13 (1), 159–175.

Hughes, J.D., 2009. *An Environmental History of the World: Humankind's Changing Role in the Community of Life*, 2nd ed. Routledge, New York.

Hughes, J.D., 2014. *Environmental Problems of the Greeks and Romans: Ecology in the Ancient Mediterranean*, 2nd ed. Johns Hopkins University Press, Baltimore.

Jones, R., 2013. *The Medieval Natural World*. Routledge, New York.

Kals, E. and Müller, M.M., 2012. Emotions and Environment. In: Clayton, S., ed., *The Oxford Handbook of Environmental and Conservation Psychology*. Oxford University Press, Oxford, 128–147.

Kearns, L., 2004. The Context of Eco-theology. In: Jones, G., ed., *The Blackwell Companion to Modern Theology*. Blackwell, New York, 466–484.

Kincaid, H., Dupré, J. and Wylie, A., eds., 2007. *Value-Free Science? Ideals and Illusions*. Oxford University Press, New York.

Kirshenblatt-Gimblett, B., 2006. World Heritage and Cultural Economics. In: Karp, I., Kratz, C., Szwaja, L. and Ybarra-Frausto, T. with Buntinx, G., Kirshenblatt-Gimblett, B. and Rassool, C., eds., *Museum Frictions: Public Cultures/Global Transformations*. Duke University Press, Durham, 161–202.

Kitcher, P., 2010. The Climate Change Debates. *Science*, 328, June 4, 1230–1234.

Kolbert, E., 2014. *The Sixth Extinction: An Unnatural History*. Henry Holt, New York.

Koubi, V., Bernauer, T., Kalbhenn, A. and Spilker, G., 2012. Climate variability, economic growth, and civil conflict. *Journal of Peace Research*, 49 (1), 113–127.

Krech, S., 1999. *The Ecological Indian: Myth and History*. W.W. Norton, New York.

Krznaric, R., 2010. Empathy and Climate Change: Proposals for a Revolution of Human Relationships. In: Skrimshire, S., ed., *Future Ethics: Climate Change and Apocalyptic Imagination*. Continuum, London, 153–172.

Kuper, A., 2003. The Return of the Native. *Current Anthropology*, 44, 389–402.

Kupper, P., 2014. *Creating Wilderness: A Transnational History of the Swiss National Park*. G. Weiss, trans. Berghahn, New York.

Labadi, S., 2010. Introduction: investing in cultural diversity. *International Social Science Journal*, 61 (199), 5–13.

Leroy, N., 2007. Réglementation et ressources naturelles: l'exemple de la forêt en Comtat Venaissin. *Médiévales*, 53, 81–92.

Leys, R., 2011. The Turn to Affect: A Critique. *Critical Inquiry*, 37, 434–472.

Locher, F. and Fressoz, J.-B., 2012. Modernity's Frail Climate: A Climate History of Environmental Reflexivity. *Critical Inquiry*, 38, 579–598.

Loh, J. and Harmon, D., 2005. A global index of biocultural diversity. *Ecological Indicators*, 5, 231–241.

Lorimer, J., 2006. Nonhuman charisma: which species trigger our emotions and why? *ECOS*, 27 (1), 20–27.

Lorimer, J., 2007. Nonhuman charisma. *Environment and Planning D: Society and Space*, 25, 911–932.

Lowenthal, D., 1990. Awareness of Human Impacts: Changing Attitudes and Emphases. In: Turner, B. L. et al., ed., *The Earth As Transformed by Human Action: Global and Regional Changes in the Biosphere over the Past 300 Years*. Cambridge University Press, New York, 121–135.

Lowenthal, D., 1998. *The Heritage Crusade and the Spoils of History*. Cambridge University Press, New York.

Maffi, L., 2001a. Introduction: On the Interdependence of Biological and Cultural Diversity. In: Maffi 2001b, 1–50.

Maffi, L., ed., 2001b. *On Biocultural Diversity: Linking Language, Knowledge, and the Environment*. Smithsonian Institution Press, Washington.

Maffi, L., 2005. Linguistic, cultural, and biological diversity. *Annual Review of Anthropology*, 34, 599–617.

Maffi, L., 2010. What is Biocultural Diversity? In: Maffi, L. and Woodley, E., eds., *Biocultural Diversity Conservation: A Global Sourcebook*. Earthscan, London, 3–11.

Maier, D.S., 2012. *What's So Good About Biodiversity? A Call for Better Reasoning About Nature's Value*. Springer, New York.

McNeill, J.R. and Mauldin, E.S., eds., 2012. *A Companion to Global Environmental History*. Blackwell, Oxford.

Metrick, A. and Weitzman, M.L., 1996. Patterns of Behavior in Endangered Species Preservation. *Land Economics*, 72 (1), 1–16.

Milton, K., 2002. *Loving Nature: Towards an Ecology of Emotion*. Routledge, London.

Mittermeier, R.A. et al., 2008. Lemur diversity in Madagascar. *International Journal of Primatology*, 29 (6), 1607–1656.

Mowaljarlai, D., Vinnicombe, P., Ward, G.K. and Chippindale, C., 1988. Repainting of images in Australia and the maintenance of Aboriginal culture. *Antiquity*, 62, 690–696.

Mufwene, S.S., 2005. Globalization and the Myth of Killer Languages: What's Really Going On? In: Huggan, G. and Klasen, S., eds., *Perspectives on Endangerment*. Georg Olms Verlag, Hildesheim, 19–48.

Nadasdy, P., 2005. Transcending the Debate over the Ecologically Noble Indian: Indigenous Peoples and Environmentalism. *Ethnohistory*, 52 (2), 291–331.

O'Connor, S., Barham, A. and Woolagoodja, D., 2008. Painting and repainting in the West Kimberley. *Australian Aboriginal Studies*, issue 1, 22–38.

Oosthoek, J. and Gills, B.K., eds., 2008. *The Globalization of Environmental Crisis*. Routledge, New York.

Paddle, R.N., 2000. *The Last Tasmanian Tiger: The History and Extinction of the Thylacine*. Cambridge University Press, New York.

Pádua, J.A., 2002. *Um sopro de destruição: pensamento político e crítica ambiental no Brasil escravista, 1786–1888*. Jorge Zahar, Rio de Janeiro.

Poulot, D., 1997. *Musée, nation, patrimoine (1789–1815)*. Gallimard, Paris.

Powers, J., 2014. Naomi Klein on *This Changes Everything*, Her New Book About Climate Change. *Vogue*, 26 August (www.vogue.com/1009011/naomi-klein-this-changes-everything-climate-change).

Radkau, J., 2014. *The Age of Ecology*. P. Camiller, trans. Polity, Cambridge.

Rashkow, E.D., 2008. *The Nature of Endangerment: Histories of Hunting, Wildlife, and Forest Societies in Western and Central India, 1857–1947*. Doctoral Dissertation, School of Oriental and African Studies, University of London.

Redford, K.H., 1991. The Ecologically Noble Savage. *Cultural Survival Quarterly*, 15 (1), 46–48.

Reis, G. and Roth, W.M., 2009. A Feeling for the Environment: Emotion Talk in/for the Pedagogy of Public Environmental Education. *Journal of Environmental Education*, 41 (2), 71–87.

Rizzo, I. and Mignosa, A., eds., 2013. *Handbook on the Economics of Cultural Heritage*. Edward Elgar Publishing, Cheltenham.

Rozzi, R., 2012. Biocultural Ethics: Recovering the Vital Links between the Inhabitants, Their Habits, and Habitats. *Environmental Ethics*, 34, 27–50.

Sandler, R., 2013. The Ethics of Reviving Long Extinct Species. *Conservation Biology*, 28 (2), 354–360.

Savater, F., 2009. Lamento por Babel. *El País*, 26 May (elpais.com/diario/2009/05/26/cultura/1243288807_850215.html).

Schnapp, A., 2013. Conservation of Objects and Monuments and the Sense of Past During the Greco-Roman Era. In: Schnapp, A. with von Falkenhausen, L., Miller, P.N. and Murray, T., eds., *World Antiquarianism: Comparative Perspectives*. Getty Research Institute, Los Angeles, 159–175.

Schuster, J.M., 2004. Making a list: Information as a tool of historic preservation. In: Ginsburgh, V.A., ed., *Economics of Art and Culture*. Elsevier, Amsterdam.

Scruton, R., 2007. A Righter Shade of Green. *The American Conservative*, 16 July (www.theamericanconservative.com/articles/a-righter-shade-of-green).

Siebert, C., 2011. Food Ark. *National Geographic*, July (ngm.nationalgeographic.com/2011/07/food-ark/siebert-text).

Silverman, S., 1995. Introduction. In: Silverman, S. and Parezo, N., eds., *Preserving the Anthropological Record*, 2nd ed. [1st ed. 1992]. Wenner-Gren Foundation for Anthropological Research, New York (copar.org/par).

Smith, L., 2006. *Uses of Heritage*. Routledge, New York.

Sober, E., 1986. Philosophical Problems for Environmentalism [excerpts]. In: Gruen, L. and Jamieson, D., eds., 1994. *Reflecting on Nature: Readings in Environmental History*. Oxford University Press, New York, 345–362.

Sodikoff, G.M., ed., 2011. *The Anthropology of Extinction: Essays on Culture and Species Death*. Indiana University Press, Bloomington.

Stocking, G.W. Jr., ed., 1989. *Romantic Motives: Essays on Anthropological Sensibility*. University of Wisconsin Press, Madison.

Stoczkowski, W., 2009. UNESCO's doctrine of human diversity: A secular soteriology? *Anthropology Today*, 25 (3), 7–11.

Sullivan, R., 2002. *A Whale Hunt: How a Native-American Village Did What No One Thought It Could*. Scribner, New York.

Takacs, D., 1996. *The Idea of Biodiversity. Philosophies of Paradise*. Johns Hopkins University Press, Baltimore.

Taylor, C., 1994 [1992]. The Politics of Recognition. In: Gutmann, A., ed., *Multiculturalism: Examining The Politics of Recognition*. Princeton University Press, Princeton, 25–73.

Todd, J., 1986. *Sensibility: An Introduction*. Methuen, London.

Toledo, V.M. et al., 2002. Mesoamerican Ethnoecology: A Review of the State of the Art. In: Stepp, J.R., Wyndham, F.S. and Zarger, R.K., eds., *Ethnobiology and Biocultural Diversity*. University of Georgia Press, Athens, GA, 561–574.

Tsing, A.L., 2005. *Friction: An Ethnography of Global Connection*. Princeton University Press, Princeton.

Turin, M., 2011. Born Archival: The Ebb and Flow of Digital Documents from the Field. *History and Anthropology*, 22 (4), 445–460.

Turner, S.S., 2007. Open-Ended Stories: Extinction Narratives in Genome Time. *Literature and Medicine*, 26 (1), 55–82.

UNESCO, 2001. Universal Declaration on Cultural Diversity (portal.unesco.org/en/ev.php-URL_ID=13179&URL_DO=DO_TOPIC&URL_SECTION=201.html).

UNESCO, 2008. *Links Between Biological and Cultural Diversity. Report of the International Workshop Organized by UNESCO With Support from The Christensen Fund* (Paris, 26–28 September 2007). UNESCO, Paris.

Vining, J. and Ebreo, A., 2002. Emerging Theoretical and Methodological Perspectives on Conservation Behavior. In: Bechtel, R.B. and Churchman, A., eds., *Handbook of Environmental Psychology*. Wiley, New York, 541–558.

Walker, W. et al., 2014. Forest carbon in Amazonia: the unrecognized contribution of indigenous territories and protected natural areas. *Carbon Management*, doi: 10.1080/17583004.2014.990680.

West, P., Igoe, J. and Brockington, D., 2006. Parks and Peoples: The Social Impact of Protected Areas. *Annual Review of Anthropology*, 35, 251–277.

Whyte, K.P., 2013. On the role of traditional ecological knowledge as a collaborative concept: a philosophical study. *Ecological Processes*, 2:7, doi:10.1186/2192-1709-2-7.

Wilcove, D.S., 2010. Endangered species management: the US experience. In: Sodhi, N.S. and Ehrlich, P.R., eds., *Conservation Biology for All*. Oxford University Press, New York, 220–261.

Williams, B., 1995. Must a Concern for the Environment be Centred on Human Beings? In: Williams, B., *Making Sense of Humanity and Other Philosophical Papers, 1982–1993*. Cambridge University Press, Cambridge, 233–240.

Worrell, R. and Appleby, M.C., 2000. Stewardship of Natural Resources: Definition, Ethical and Practical Aspects. *Journal of Agricultural and Environmental Ethics*, 12, 263–277.

Yoder, A. and Welch, C., 2012. Lemurs Are The Most Threatened Mammals In The World: IUCN Red-List Workshop 2012 (lemur.duke.edu/lemurs-are-the-most-threatened-mammals-in-the-world-2012-iucn-red-list-workshop-cont).

Part I

Affects and values

Species, cultures, and languages as constitutive elements of culture have long been at the heart of preservation projects. Particularly since the 1990s, in theory if not always in practice, those projects tend to be animated by the so-called "biocultural axiom" – the assumption that the diversity of life is essentially *biocultural*, that protecting biodiversity requires protecting human cultures, and vice versa. At the same time that it is depicted as a feature of the world, biocultural diversity has emerged as a value; it has become the essential normative good to be defended for itself, for the sake of cultural and environmental justice, and more generally for humankind's survival. Without the notion of biodiversity, extinction was part of the balanced order of nature; once it was perceived as diminishing biodiversity, it became a threat (David Sepkoski's chapter). Insofar as biodiversity has the rank of value, extinction represents the anti-value par excellence. No matter what cost-benefit analyses might show, proposing to let the panda die counts as an eminently immoral suggestion.

In all spheres, from scientific research to policy-making to grassroots activism, the processes whereby biocultural diversity rose to be such a global value have been imbued with affect. Emotions do not merely energize and drive action: they convey meanings, shape perceptions and sustain values. In the endangerment regime, metaphorical slippages are a powerful mechanism for bringing that about. Thus, the striking growth of projects to document and protect endangered languages is nourished not only by the conviction that languages embody worldviews and that if a language disappears, so does access to that worldview, but also by the application to languages of organicist metaphors. Languages are not living organisms, but claiming that they are under threat and could die like endangered species contributes to bringing them into the emotional, ethical and cognitive orbit of biocultural diversity at the expense of considering the social and political conditions that drive speakers of a language to give it up and adopt another (Shaylih Muehlmann's chapter).

Feelings of loss permeated émigré Claude Lévi-Strauss' emotion at discovering in a New York second-hand bookshop a collection of the *Annual Reports of the Bureau of American Ethnology*. The volumes represented, he later wrote, "most of what will remain known about the American Indian." He imagined a future when "the last native culture will have disappeared from the Earth and our only interlocutor will be the electronic computer." And in old age, he produced the photographic memoir *Saudades do Brasil* – a nostalgic homage to the Amazonian Indian, a particular ethnographic experience, and a certain kind of anthropological data. As Lévi-Strauss understood, the volumes found in New York and the negatives developed and printed decades after the pictures were taken are the second-order embodiments of lost cultures: neither tools nor artifacts, but verbal and iconographic depictions of those material objects. The threat to cultures, so anxiously felt by the main actors of "salvage anthropology," thus transmutes into a threat to data – a process of "second-order endangerment" (Rebecca Lemov's chapter) that accelerates with the rapid obsolescence of electronic supports and the expanding galaxy of "big data," and generates worries that are no less distressing than those of earlier ethnographers.

1 "Languages die like rivers"

Entangled endangerments in the Colorado Delta

Shaylih Muehlmann

Languages die like rivers.
Words wrapped around your tongue today . . .
Shall be faded hieroglyphics
Ten thousand years from now.
           – Carl Sandburg (1916)

The poem by Carl Sandburg that opens this chapter tackles the main theme examined in the pages that follow: the analogy between the death of languages and that of rivers. Specifically, I will analyze the ways the relationship between nature and language has been theorized in the context of their endangerment, drawing from my ethnographic research in the delta of the Colorado River in northern Mexico.

Contemporary understandings of language change are usually framed through biological metaphors. The very notion of "language death" implies that a language has a "life" and a "lifespan," which is not an accurate description of how languages exist and evolve but a highly questionable and contested comparison between the life of biological organisms and the historical formation of languages. Additionally, since the 1990s within the language documentation field in North America, many scholars and activists have made a much more literal connection between languages and organisms or species, and also with "ecosystems" and environments. This proliferation of metaphorical affinities between languages and the natural world, as well as the attempts to posit the empirical connections between these domains, demands a more careful analysis of the relationship between the idea of "language death" and the demise of species, rivers or other natural habitats.

For a variety of reasons, the language-species analogy has generated a great deal of intellectual discomfort. In part, this is because it exhibits blatant inaccuracies, as there are many important ways that language and nature are not alike. Returning to Sandburg's poem, for instance, we may note that in the most obvious sense, rivers do not technically die because they are not biologically alive to begin with. The same criticism has been made about the idea that languages "die," as they are not living creatures in a biological sense either (Crawford 1995). The metaphor of death nevertheless turns out to be powerful in both cases because languages and rivers are both essential, though in very different ways, to human life.

Precisely because the comparison, if taken literally, raises serious objections, the phrase "languages die like rivers" resonates with what I want to explore in a number of more specific respects. This chapter draws on ethnographic research at the end of the Colorado River among some of the communities this ecosystem supports. Over the past century, the Colorado Delta, facing intensive agricultural development and drought, has become one of the most contested ecosystems in North America. Most years, the river does not reach the sea, but runs dry right across the international border with Mexico. The ecological changes on the river have been entangled with dramatic linguistic and social changes for the indigenous people who have lived by the river for centuries. As indigenous languages have been increasingly displaced by English and Spanish, the dual "endangerment" of languages and the river has blended environmental and linguistic narratives in a single discourse of crisis and impending extinction.

I conducted multi-year research among a group of Cucapá people who live at the end of the Colorado and are considered an "endangered language community" because they are shifting from their native language, Cucapá, to Spanish.[1] The Cucapá people have relied on fishing as their primary means of subsistence for centuries, but as a result of the 1944 water treaty between the United States and Mexico, ninety percent of the water in the Colorado is diverted before it reaches Mexico. The remaining ten percent that crosses the border is increasingly being directed to the manufacturing industry on the border. Additionally, in 1997, the Mexican federal government created a biosphere reserve in the only part of the Cucapá traditional fishing grounds that was still viable, right at the mouth of the river, where some groundwater meets the sea. This reserve effectively criminalizes fishing in the area, and the Cucapá people have been unsuccessful in legal disputes to regain fishing rights in the protected zone (Muehlmann 2013).

These ecological and economic changes have also coincided with a series of more cultural changes in the community as the majority of children are no longer learning Cucapá but are growing up fluent in Spanish instead. The Cucapá language has already reached a stage of advanced obsolescence.[2] During my first period of fieldwork in 2005–2006, only a handful of elders were identified by residents as active Cucapá speakers. The elders fluent in Cucapá at that time were between 60 and 80 years old, although there were various degrees of receptive competence among some who were slightly younger. In the other places where Cucapá people live (which include several villages in Mexico and the reservation in Somerton, Arizona), there was a similar age-based distribution of speakers. This sociolinguistic situation has remained relatively stable over the last decade, although two prominent elders fluent in Cucapá have since passed away.

The connections between the sociolinguistic and environmental changes this group has experienced since the early 20th century have been couched in the language of extinction cited above. Environmentalists and local fishermen have often construed the Cucapá people as one more endangered species wiling away to extinction in this severely threatened ecosystem. In this case, the metaphor is a way of incorporating the Cucapá people's often inconvenient presence into conservation programs (Muehlmann 2013). This has also led to the allegation that the "extinction" of the Colorado River has caused the "extinction" of the Cucapá culture. Such a formulation rests quite uneasily with most Cucapá people, who perceive their identities as still very much alive and resilient. However, the idea that with the end of the river, everything changed for the community resonates strongly among many of its members.

Don Madaleno, the traditional chief of the Cucapá at the time of my fieldwork and one of the elders fluent in the language, spoke poignantly about the relationship between his language and the river. He said, "When the Colorado River ended, everything ended." While it is clearly not a coincidence that the "end" of the river has coincided so closely with the "end" of the Cucapá language, the relationship between these two concurrent losses is not transparent. Like a great deal of language-environment correlations, the actual relationships that hold these two phenomena in mutual occurrence is more complicated than it may appear on the surface.

Such narratives of joint endangerment are not unique to this region. The connections between the imminent extinctions of the world's biological and cultural diversity have become a rallying cry among language and environmental advocates. In what follows, I will contextualize my ethnographic example within the growing body of work that has explored the language/nature connections in the context of endangerment paradigms. I will argue that the discourses that emerged in the 1980s and 1990s, drawing together the domains of biological and linguistic endangerment, had the effect of renewing earlier concerns about environmental determinism among sociocultural and linguistic anthropologists, while raising a new interest among linguists and evolutionary biologists in pursuing encompassing frameworks that have often been criticized for being reductionist. I will also explore the ways that the surge in discussion about the juncture of environmental and linguistic domains is an opportunity to more carefully examine, from an ethnographic standpoint, the political processes that I will argue ultimately mediate apparently causal relationships between language inequalities and other forms of social and economic marginalization.

The entangling of environmental and linguistic endangerment

The narratives of impending extinction that have characterized endangerment discourses over the last century have been shaped by the unique way that the environmental movement emerged in the Western world, particularly at the end of the 20th century. The idea that human beings are altering and destroying the environment to the extent that our own survival and well being is threatened has strongly influenced ideas about linguistic and cultural endangerment. Since the 1990s, certain features of environmental discourses on extinction have been incorporated into discourses about endangered languages, particularly in campaigns to "save" some languages from "extinction." In a 2005 review of the literature on language endangerment, Luisa Maffi, co-founder in 1996 of the NGO

Terralingua, which promotes the notion of "biocultural diversity" and works to sustain it, reviews how parallels were established between the ideas of biocultural or biolinguistic diversity and the better known phenomenon of biodiversity in order to rally linguists and others around the issue of language endangerment (2005, 602).

Therefore, language advocates appealed to the legitimacy of ecological discourses by drawing on the paradigm of biodiversity conservation that had gained such success in the environmental movement (see Conklin and Graham 1995; Hecht and Cockburn 1990). While these authors have highlighted the rhetorical appeal of discourses of diversity for framing language in the context of the environmental movement, the merging of these discourses is not just strategic. For example, Maffi herself, among many others, have promoted empirical programs of investigation to research the connections between linguistic, biological and cultural diversity understood as joint material processes. I will explore the way these empirical efforts have developed later on in this chapter. For now, however, it is important to note that the interest in the relationship between biological and linguistic diversity did not begin merely as a co-opting of environmental discourses about endangerment but rather it emerged at a specific historical moment and was part of larger ideological shifts in the understanding of language in relation to minority movements, the nation-state, culture and changes in the global economy.

The notion of linguistic diversity and in particular "biolinguistic" or "biocultural" diversity gained prominence at a historical juncture when language ideologies appeared to disconnect from nation-state ideologies (Muehlmann 2007). Since the rise of European nation-states in the nineteenth century, discourses among language advocates in the context of language policy emphasized constitutional rights for linguistic minorities within specific nations, and within structures of regulation in state-run institutions. Several authors have explored the way that discussions of language changed as a result of the nation-state's apparent loss of political primacy (see Duchêne and Heller 2007; Patrick 2003). The perception of expanding global interconnection problematized the discourse on multilingualism that appealed to the legitimacy of the nation-state to argue for minority rights (Duchêne 2008). As a result, it shifted the discussion of language diversity from a framework connected to national identities to a discussion of language diversity as connected to universal values of linguistic science or biodiversity. Ultimately, this brought questions of biological diversity and linguistic diversity into the same fold, highlighting the ways they were threatened by the same phenomenon: that is, the expansion of global capitalism.

These kinds of comparisons came to characterize the literature on language endangerment in the 1990s, and they also dominated campaign strategies among organizations – both government and non-governmental – which borrowed elements from contemporary environmental rhetoric of biodiversity in their program development and promotion (Muehlmann and Duchêne 2007). Therefore, the entanglement of distinct endangerment discourses has had an impact on both popular scholarship and language policy.

The slippage between literal and metaphorical connections

The analogy between social change and the life history of the organism has an early place in western European philosophy and has fueled a recurrent tendency to see living organisms as the "prototype of all dynamic wholes and, consequently, to attribute a cyclical development to individuals and institutions alike" (Nisbet 1969, 70). Conceptions of language change have been no exception.

Assumptions about the nature of language are partially inspired by the idea that there are general principles that apply similarly to languages and biological species and also motivated by the observation that similar processes of extinction are occurring with both. The idea is that languages, and their endangerment, are simply *comparable to* the endangerment of biological species without a necessary empirical connection specifically developed. This means there is often remarkable ambiguity as to whether the links between language and species are made figuratively or literally. Roslyn M. Frank (2008, 6) notes that it is common to see authors use the metaphor of language death figuratively in one instance and then literally in another. In this section, I will explore how the slippage of metaphorical and empirical connections between biological and linguistic endangerment create an underlying tension in the literature that has focused on these themes.

This kind of slippage is evident in a much-cited article published in the journal *Nature* in 2003 by zoologist William Sutherland, entitled "Parallel Extinction Risk and Global Distribution of Languages and Species." Sutherland argues that, by the criteria of biological species extinction risk, languages appear "more threatened than birds or mammals" (2003, 275). More specifically, he claims that rare languages are more likely to become endangered than common ones. He explains that this is so because areas with high language diversity also have high bird and mammal diversity, and all three show similar relationships to area, latitude, area of forest and, for languages and birds, maximum altitude. Sutherland admits that although similar factors explain the diversity of languages and biodiversity, the factors explaining extinction risk for each are different: endangerment of birds and mammals increases with human density because of habitat loss, whereas the threat to languages does not. Sutherland thus recognizes the shortcomings of the analogy, but his conclusion obscures the figurative character of his initial argument, since it states that when standard measures of species risk are applied to languages, languages come out even more endangered than birds and animals.

Along with the wide public dissemination of such analogies, there has also been considerable debate about the accuracy and utility of comparing languages and biological organisms. For instance, it has been pointed out that unlike biological organisms, languages have no genetic material and thus carry no mechanisms for natural selection (Crawford 1995). Therefore, their prospects of survival are not determined by any intrinsic capacity for adaptation but by social forces alone. Such differential processes in the history of a language cannot be accounted for in the language as organism metaphor. Moreover, conceiving of language loss as a Darwinian process may have political connotations, since it suggests that some languages are inherently more "fit" than others (Crawford 1995). While contemporary linguists do not tend to make such arguments, the idea that some languages are superior to others continues to be a popular opinion among the general public and, consequently, affects language policy.

One example of how this belief is playing out in contemporary language policy in the United States are the English Only policies that, especially since the 1980s, have demanded the legal protection of English and the restriction of other languages. Often fueled by anti-minority politics, these policies have been instituted because of the belief that English unites the nation, is superior to immigrant languages, and that there should therefore be a requirement of immigrants to learn English (Crawford 1992). These policy efforts have focused on restricting or prohibiting the use of languages other than English in public spaces.[3]

While the parallel drawn between endangered species and endangered languages has been successful in raising public awareness of language loss, it has not uniformly evoked the type of attention intended, at times generating hostility toward the language preservation project, rather than support. For example, in May 2003, the science writer David Berreby published in the *New York Times* an article entitled "Fading Species and Dying Tongues: When the Two Part Ways." The article is a critical appraisal of the concept of "endangered languages" and the revitalization movement, which has focused on programs and policies aimed at "revitalizing" (i.e. re-establishing language fluency in a community of endangered language speakers). But the article is specifically a reaction to Sutherland's 2003 article in *Nature*. Berreby criticizes the comparison between animals and languages, quoting the question posed by linguist Michael Krauss (1992), "Should we mourn the loss of Eyak or Ubykh less than the loss of the panda or the California condor?" Berreby (2003) points out, "Ubykh, a language of Turkey, is a human creation. The panda is not; it is our neighbor, not our invention." His article is both a critique of the biological metaphor as applied to language and an allegation that its discursive utility resides in the promotion of academic self-interest. He writes, "It is no surprise that linguists and activists promote maintaining spoken languages to study. Just as the Poultry and Egg Council wants us to eat eggs, linguists want languages to study" (Berreby 2003).

Unsurprisingly, Berreby's article prompted controversy and response by many linguists, including letters to the *New York Times* (Garrett 2003; Hinton 2003; Ostler 2003; Pianfetti 2003), as well as much discussion on Internet forums and blogs. Numerous reactions acknowledged the problems with the biological metaphor but took issue with the various ways that Berreby allegedly misrepresented the language revitalization movement.

In *The Ecology of Language Evolution*, University of Chicago linguist Salikoko S. Mufwene, who specializes in the development of "creoles," explores several additional shortcomings of the metaphor – a major one being that it does not capture variation within a language. He also suggests that the concept of "parasitic species" is more adequate because a language does not exist without speakers just as parasites cannot exist without hosts.[4] Ultimately, however, the critiques to which Mufwene and others are responding (and indeed formulating) with regard to the inexactitudes of the metaphor, which many linguists have acknowledged, are not what have inspired dismay. Many of the responses are motivated not simply by the inaccuracies of the biological metaphor but by the perception that the revitalization movement is prioritizing languages over people and animals. Sutherland's article, for example, is representative of an increasing trend in the narratives of language advocacy, in which the comparison between language and species is subsumed by a powerful rhetoric that conflates the two. This creates a more insidious confusion, since speaking metaphorically about "language endangerment" as if languages were the same as biological species obscures the relationship that might exist between a language and its natural environment, including the living beings therein.

The literal connections between linguistic diversity and biodiversity

Another strain of the endangerment literature has established much more literal links between language and environmental endangerment than the rhetorical and heuristic approaches reviewed thus far. Several authors have called specific attention to what they say are remarkable overlaps in the global distribution of languages and biodiversity, and have thus prompted speculation about the empirical links potentially responsible for this correspondence (Nettle 1998). While the geographical correlation between areas of high biological and linguistic diversity has been much cited, the reasons for it are not fully understood and have been widely misattributed.

In general, the global distribution of species and languages has prompted investigation into the possible existence of common threats. For example, David Harmon (1996), Terralingua co-founder and a principal investigator in its Index of Linguistic Diversity, argued that 10 of the top 12 "megadiversity" spots in the world, as defined by the World Conservation Union, also figure among the top 25 most linguistically diverse areas, and suggests that "biogeographic" factors might account for the distribution. He explains that land masses with a variety of terrains, climates and ecosystems, or with geophysical barriers (such as island territories), might foster a higher number of both species and languages through presumably more isolated or self-sufficient habitats. He therefore hypothesizes a process of co-evolution between small human groups and local ecosystems.

Other authors, however, have emphasized that the picture is far more complex. For example, Peter Mühlhäusler (1996), a specialist of endangered languages in the Asia/Pacific region, demonstrated that linguistic distinctiveness can develop in the absence of isolation. He showed that high concentrations of linguistically distinct communities have often coexisted in the same areas and communicated through complex networks of multilingualism. The existence of these cases in the same areas highlights the importance of sociocultural factors in the formation and definition of linguistic areas. Studies have demonstrated the importance of factors such as the size of an economy in determining language diversity. Large scale economic systems, for example, bring about linguistic spread and lower diversity (Nichols 1990). Empirical inconsistencies, as well as the high level of disagreement and complexity associated with the correlation between linguistic and biological diversity, are an indication of the nascent stage that research on this topic is at, and may suggest that the correlation does not hold systematically (for example, it has not been found in Central and South America; Manne 2003).

Some authors argue that the problem lies in the very framework that seeks to establish such connections, encapsulated in the terms "biocultural" or "biolinguistic diversity." This critique has come from anthropology and integrationist linguistics respectively, with each tradition taking a different line of attack. Maffi (2005) points out that the idea of correlations between the culture, languages and geographical zones has long been unpopular among anthropologists because it seems to connote environmental determinism. Early on, cultural anthropologist Alfred L. Kroeber (1928) explored correlations between cultural and natural, though such an exploration never developed into a programmatic line of research. According to Maffi, early linguists had already reacted critically to Charles Darwin's (1859; 1871) likening of languages to natural organisms. In fact, philosopher Robert Richards has documented how, initially, Darwin's loose comparisons between languages and organisms caught on like "fire" before they were more critically evaluated (Richards 2002, 25).

British anthropologist Tim Ingold (2007) discusses biologists' attempts to integrate the study of cultural evolution with the science of evolutionary biology in a single framework. He argues that social and cultural anthropologists have largely withdrawn from those attempts, meeting them with dismissal at best and often with outright hostility. As we have seen, such frameworks tend to treat language and culture as traits that can be understood comparatively and analyzed as emerging through a process of cultural or linguistic evolution. Ingold argues that these approaches have met with hostility because they are an "affront to the millions of intelligent human beings for whom traditions are real and important but who are not, on that account, trait-bearing cultural clones whose only role in life is to express . . . information that has been transmitted to them from previous generations" (Ingold 2007, 14).

Ingold also points out that resisting comprehensive reductionist frameworks for understanding biology and culture together is not a new stance for anthropology. Franz Boas wrote that "any attempt to explain cultural form on a purely biological basis is doomed to failure" (Boas 1940, 165). Arguments against such a framework have been important in resisting some of the more extreme forms of determinism, especially those emerging since the 1960s in debates about the alleged racial basis of intelligence and aggression (Chagnon 1968; Sahlins 1976) or about the apparent biological basis of gender (Butler 2004; Martin 1987).

In the context of language endangerment, the idea that correlations exist between areas of linguistic and biocultural diversity as a result of joint evolutionary processes may recall forms of environmental determinism that came to prominence in the late 19th century and early 20th century in geography and anthropology. The fundamental argument was that aspects of physical geography influenced the psychology of individuals and defined the culture of the society those individuals formed. Yale geographer Ellsworth Huntington's (1922) theory that "climatic energy" determined human accomplishments was particularly crude, but it was nonetheless widespread. In his view, "civilization" could thrive only where allowed by temperate climates such as those of Western Europe and northeastern North America. Such views have been roundly rejected since then, both for their inability to accommodate historical agency and, perhaps more so, for their uses to justify racism and imperialism.

While environmental determinism was an early stumbling block for investigative frameworks attempting to encompass biological and cultural change, contemporary theorists concerned with endangerment have raised other objections. The attempts at measuring cultures and languages to standardize and compare them to biodiversity have been particularly problematic. In the 1990s, David Harmon, already mentioned, took linguistic diversity as a major indicator of cultural diversity, and the loss of "language richness" as a proxy for the loss of "cultural richness." More recently, he developed a slightly more nuanced algorithm involving three "cultural indicators" of diversity: language, ethnicity and religion. With his collaborators, Harmon (2002) has elaborated an Index of Biocultural Diversity intended to measure trends in "biocultural diversity" by country and "traditional knowledge," the global state of which is also often seen as being correlated to linguistic diversity (Maffi 2005).

One problem with the measurement of diversity, particularly through such mechanisms as an "index," is that every *difference* is construed as *diversity*. Ingold (2000) argues that a key consequence of such a model, which he calls "genealogical," is that it compares individual items in terms of qualities they possess by virtue of an essential nature, regardless of their position in relation to one another in the world. As Ingold emphasizes, such comparisons assume that the world is intrinsically divided up into discrete units. Linguists, likewise, generally recognize that languages are not discreet units, and that "counting" them therefore involves a great deal of arbitrary construction of boundaries. Yet scholars interested in language obsolescence generally tolerate and reproduce estimates about "the number of languages in the world," despite the fact that such figures are essentially fictions as there is no way to accurately count languages.

Language archives are one of the more extreme examples of the itemizing and objectifying logic evident in efforts to index languages. A number of national funding bodies, such as the US National Science Foundation and the German Volkswagen Stiftung, and the Hans Rausing Endangered Language Project at the School of Oriental and African Studies, have set up archives to serve as repositories for data on endangered languages. UNESCO's "Atlas of the World's Languages in Danger" is perhaps the most emblematic of these archiving attempts both because of its scope and its status as a visual emblem of the endangered language movement. According to the Atlas' most recent editor, Christopher Moseley, the primary aim of this Atlas is to provide indicators on the status of linguistic diversity and the numbers of speakers of endangered languages. In the online version of UNESCO's Atlas, languages are shown by points on a map and one can click on a language to view a series of facts about it such as its "main" name (and any alternative names) and its International Organization for Standardization (ISO) code. The ISO code is a single three-letter code assigned to each language in the world that has been recognized as "a separate language."[5]

Moseley (2012) explains in his brief history of the Atlas that it was originally inspired by the IUCN Red List of Threatened Species, instigated by the International Union for Conservation of Nature (IUCN). This Red List was created as a response to 1992 Convention on Biological Diversity (United Nations), signed at the Earth Summit in Rio de Janeiro. The IUCN's Red List uses degrees of endangerment of species very similar to those used in the UNESCO Atlas. Therefore, UNESCO's endangered language atlas draws on both the sense of urgency inspired by the specter of endangered animals worldwide as well as on the legitimacy of the "museum effect" conjured by the archive, which imposes an order of hierarchy that makes it look natural (Dias 1994).

The emphasis in such archives on documenting languages "recognized as separate" and the avoidance of documenting dialects (Moseley 2012, 5) is that this framework supports the assumption that "variation" is less valuable than "diversity" (Ingold 2000), and that the variation between languages is more valuable than the variation "within." Rather than prompting a re-evaluation of the arbitrary segmentation of languages in the endangerment paradigm, such observations have led experts focusing on other levels of language variation to appeal to endangerment to call attention to their chosen language units. For example, in their article "Moribund Dialects and the Endangerment Canon: The Case of Ocracoke Brogue," Walt Wolfram and Natalie Schilling-Estes (1995) describe how the "canon" has neglected the dialects of "safe" languages. In their view, such neglect derives from the questionable assumption that inter-language variation is more valuable than intra-language diversity, which in turn presupposes that the systematic boundaries between languages are "obvious and discreet" (1995, 697). For the same reasons, there have recently been similar appeals for more attention to creoles and pidgins (Garrett 2012), as well as to vernaculars and supervernaculars (Blommaert 2011).

Because of the persistence of such problematic arguments in the endangerment literature, critics have charged that the idea that languages exist as separate, identifiable units actually rests at the heart of mainstream linguistics (Harris 1981) and provides the foundation for claims about language diversity and death (Orman 2013). Among others, pioneer of integrationism Roy Harris has argued that the belief that languages exist as internally structured systems and invariant units has underwritten all of "orthodox linguistics." The theoretical foundation of "integrationism" is this critique of "mainstream" linguistics (often referred to in this literature as "segregationism," or any approach that assumes that communication is independent of its users or contexts). The major intervention of integrationist linguistics is that almost all conceptualizations of linguistic diversity and their associated socio-political discourses are based on fundamentally dubious notions of language. As Jon Orman writes, "In all cases, a measure of diversity is arrived through an enumeration of holistic units of some kind, be it individual language systems which essentially reduce to language or (dia)lect names or, in more technically sophisticated accounts . . . particular typological or 'genetic' features" (Orman 2013, 3). The criticized methods assume that it is possible to establish objective criteria for determining the sameness or difference of linguistic phenomena.

For integrationists, then, what mainstream approaches see as linguistic diversity is really a form of metalinguistic diversity, a diversity of folk classifications, based upon a theoretically questionable notion of what languages are. Thus, given that the notion of "a language" is itself a social construct, it would be more accurate to consider "language death" as a meta-metalinguistic concept (Orman 2013).

From a more anthropological perspective, the very idea of language as some-thing that can be studied as a "non metalinguistic" phenomenon, as something that remains real outside of our own perceptions of it, is also problematic. "A definition of language," Raymond Williams wrote, "is always, implicitly or explic-itly, a definition of human beings in the world" (1977, 21). And as work on language ideology has long stressed, all views of language are mediated by social experiences, and the potency of "language" derives precisely from its being bound up with ordinary people's perceptions, experiences and history. Anthropologists (e.g. Hill 2002) have emphasized that the languages we count in the world today are the products of the rise of nation-states and the colonial regimes they imposed on the world. In short, thought of as "languages" and "dialects," languages and their dialects are very recent historical artifacts. Additionally, the boundaries that have been drawn between languages are not based on their internal consistency or homogeneity, but rather on the historical circumstances that allowed them to be identified, documented and standardized.

Nonetheless, this view of language that takes "metalinguistic" conceptions of languages and their boundaries seriously, as well as the perceptions people have of their language's demise, is extremely important for understanding the role of language in wider social contexts. It matters how people experience the links between language practices and identity, or those between language changes and other transformations, for example in environmental or economic domains. As we shall now see, the ways those links are embodied and the reasons why they are important are profoundly interconnected with power relationships and unequal distribution of resources. It is at that level that the links between language death and environmental degradation become clearer.

The role of language, power and resources in endangerment

While it is clear that the specific processes that result in language "death" and "diversity" are varied and far from definitively established, there is consensus that the risk of obsolescence results from asymmetrical relations of power. As Craw-ford (1998, 8) writes, "Language death does not happen in privileged communi-ties. It happens to the dispossessed and the disempowered." Across the various strains of the literature and promotional material focused on language endanger-ment, obsolescent languages are referred to as "marginalized," "disempowered," "sub-dominate," "oppressed," "disenfranchised" or "underprivileged." However, this vocabulary of inequality is employed in a very loose way, often disconnected from the specific local and historical circumstances to which it refers.

The literature on language death usually construes power as an unequal distribution and control of resources. For example, authors often link the rapid decrease in the number of languages over the past centuries with European colonization and Western economic expansion (Krauss 1992; Maffi 2001; Nettle and Romaine 2000). A fairly representative portrayal of these dynamics is traced by Daniel Nettle and Suzanne Romaine (2000), who focus on colonization and its effects on language change. They claim that the rapid loss of language diversity has occurred primarily in the last thousand years and has resulted from the emergence of massive differentials in the control societies have over the environment. These differentials run in favor of Eurasia against the other continents. Further, they argue that changes in economic lifestyles since the invention of agriculture have enabled populations that are technologically, militarily and/or economically more powerful to dominate others and "impose" their languages on them. Accordingly, in their view, the present decrease in the number of languages prolongs an older political trend.

Globalization thus appears as a less overtly political and more economic form of colonization. Nettle and Romaine (2000) highlight the role of political economy in the expansion of some languages, for example English in the present-day United Kingdom and Tok Pisin in Papua New Guinea. They claim that the scale of expansion of a language is correlative with whether a language is used in all sectors of an economic system or restricted only to some. The authors also highlight how economic inequalities among populations have forced many of their vernaculars to the condition of "peripherality" (129). They give the example of the Celtic languages in the United Kingdom since the development of Old English. In summary, Nettle and Romaine claim that the marginalization of minority groups is the main cause of "language death." This marginalization, they argue, is organized by differential access to resources, which is in turn mediated by participation in economic activity.

Mühlhäusler (1996), in contrast, argues that the causal relationship between economic marginalization and linguistic marginalization may actually work the other way around. In *Language Change and Linguistic Imperialism in the Pacific Region*, he describes how in many parts of the Pacific, there are long chains of interrelated dialects and languages with no clear internal boundaries. He explains that despite this continuum, linguists have "identified" and documented languages wherever the center of economic and communicational activities has been located, which would often be those places that linguists and missionaries settled. As a result, the documented varieties have become the "languages" representative of their related dialect chain (6). He argues that similar situations have arisen elsewhere where language varieties in colonial centers have become the de facto centers of linguistic systems labeled as "languages."

Mühlhäusler emphasizes that this process of labeling languages and linguistic groupings has several adverse side effects. The most important is that it assigns unequal status to different forms of language, which initiates the formation of social and economic inequalities. Therefore, he argues that the processes of labeling languages and linguistic groupings assign a different status to varieties of language, which "paves the way for social and economic differentials" (1996, 6). Whereas Nettle and Romaine view economic inequality as the cause of language death, Mühlhäusler interprets the marginalization of language varieties as the cause of economic inequality. While these views are not mutually exclusive, their different emphases again point to the complexities of the causal connections between colonial power dynamics and language marginalization.

Mufwene (2001) has pointed out that both accounts of the link between economic and linguistic marginalization obscure the effect of different colonization styles, including settlement, exploitation and trade colonization. He argues that these styles do not bear equally on language vitality. Trade colonization, which typically led to either settlement or exploitation, was cautiously based on egalitarian relations with the host population. In such cases, language shift did not result from the simple imposition of a language through the domination of one group over another. Mufwene thus argues that we should develop more thorough knowledge of the circumstances by which languages become endangered before dramatic generalizations are made (376).

Despite the complexities that these discussions reveal about the potential causal connections between linguistic and economic and environmental marginalization, their relationship to each other is often presented as a straightforward rationale for language revitalization. The statistical dictum that "correlation does not imply causality" is relevant here since, in their most simplified form, rhetoric on language preservation implies that language endangerment is *the cause* of the wider endangerment crisis (Muehlmann 2007).

It is common for language minority advocates to rely heavily on economic and political inequalities to justify claims about the need to protect their languages (cf. Crawford 1998; Crystal 2000; Hinton and Hale 2001; Nettle and Romaine 2000). This line of argument simply takes it for granted that language endangerment is the cause, not the effect of other forms of inequality and injustice. Furthermore, any argument that language preservation would work to reverse these inequalities is largely absent from the literature. The issue of whether the focus on language is tantamount to addressing an epiphenomenon rather than the root of the problem is not explored. In a critical vein, Mufwene (2002) notes that sometimes the problem of language endangerment "boils down to a choice between saving speakers from their economic predicament and saving a language" (377). He thus frames language shift as a survival strategy and urges linguists to bring the interests of speakers to the center of the debate.

Linguistic anthropologist Monica Heller (2002) has advanced these debates by shifting the attention to language as a process implicated in the construction of inequality. She argues that the concern over language death ultimately indicates the "importance of understanding language as a social practice which plays a central role in processes of categorization which serve to regulate access to the production and distribution of material and symbolic resources" (2002, 17). Heller argues that understanding language in this way places the discussion of language change in the context of the social construction of inequality, and that doing so helps avoid the discursive traps that reproduce the homogenizing and segmenting of languages and identities.

Therefore, one way of understanding the connections between what we apprehend as language death and something like the end of a river is to look at how both are mediated by power relations and economic marginalization. In order to more fully explore this connection, I now turn to the joint "death" of the Cucapá language and the Colorado River in northern Mexico.

"When the Colorado River ended, everything ended"

Before massive dams were built upstream in the mid-20th century, the delta of the Colorado River in Mexico was an expanse of wetlands covering more than 2 million acres. The delta provided subsistence for the residents of the region and habitat for wildlife and marine fisheries in the Gulf of California. It was also an important flyway for birds flying north from Central America and south from Canada. Now, as a result of the over-development of the watershed in the United States, the Colorado River has one of the highest rates of species extinction and endangerment on the continent (Bergman 2002, 29). It is within this environmental and geographical context that the Cucapá language is also obsolescing.

From the perspective of some Cucapá people, these processes have been inseparable from each other. As mentioned, Don Madaleno made this clear when he said, "When the Colorado River ended, everything ended." Unlike the kinds of connections made in the academic literature, Don Madaleno's remark was based on his long-term experience of a changing linguistic and natural landscape, and it concerned how life has changed more generally in the Colorado Delta. Don Madaleno was born on February 16th, 1934, one year before the completion of the Hoover Dam, the first of the large dams on the river. Over the course of his life, which ended in 2009, he witnessed the rest of the Colorado being cut-off above the lower part of the delta where his village is located. There is also a causal implication in Don Madaleno's remark, for he saw the lack of water in the Colorado River as resulting in less fish, fewer trees and fewer animals. He would rarely go a day without reminiscing about the piles of fish in the river and the clouds of ducks in the sky that still existed when he was growing up on the banks of the river. But how the decline of the Cucapá language has been connected with these changes is a less obvious matter.

The lack of water in the river has been one factor among the many social and economic transformations that have led to a shift to Spanish as a dominant language. But this was mediated by other processes that were largely political and economic. The "end" of the Colorado River in Cucapá lands profoundly eroded the former role of fishing and hunting as the main source of livelihood. Until the early 20th century, the Cucapá followed semi-nomadic subsistence patterns coordinated with the river's stages through the year. Their main source of food was fish but they also hunted rabbits and deer and planted corn, beans and pumpkins (Alvarez de Williams 1974; Gifford 1933; Sánchez 2000).

The loss of these traditional subsistence activities forced a rapid integration into the encroaching economic system of the border region; the Cucapá people began working in factories, as farmhands, in construction, building roads and selling scrap metal. And these are jobs that demand speaking Spanish, not Cucapá. Therefore, the link made by Don Madaleno was, on closer inspection, not so much about the river and the language as about economic and political factors closely related to power dynamics and the distribution of resources. When pressed about the connection further, he said the reasons for the Cucapá language's obsolescence had to do with political turmoil. "With all these fights, the fight for the land, for the water, and to fish, we don't have time to focus on teaching the children how to speak Cucapá."

The notion that when languages disappear they take their associated "cultures" with them has been most powerfully conjured in relation to indigenous languages and cultures, which are often the focus of discussions on endangered language. This has been a relevant issue for Cucapá fishing activists because some government officials have claimed that they are not sufficiently "indigenous" to qualify for preferential fishing rights since there is a low level of indigenous language fluency in the community.

While more subtle links have been made between language loss and the loss of "immaterial" resources such as solidarity and prestige (cf. Gal 1979; Hill and Hill 1977; Kulick 1992), many linguists have supported strong claims that identities or cultures are explicitly linked and mapped onto each other, such that when languages are lost, cultures are as well (Harrison 2007; Nettle and Romaine 2000; Woodbury 1993). Campaigns to "save" endangered languages have been connected to efforts to rescue cultural heritage, knowledge and practices (Crystal 2000; Maffi 2001; Skutnabb-Kangas 2000). These ideas about the language-culture link have been met with some discomfort among many Cucapá people, especially in a political climate in which the relationship between livelihood (in this case, fishing) and cultural identity has become much more relevant for local people than a link with their indigenous language, which most Cucapá people no longer speak.

Most elders, like Don Madaleno, understand the failed transmission of Cucapá to younger generations as a result of a political and economic situation in which parents were too busy struggling to subsist or too involved in conflict with the government to teach their children. Adriana, Don Madaleno's daughter, who has since the 1990s been an activist on behalf of Cucapá fisher people, has often expressed a similar view. One afternoon, for instance, while waiting in a court in Mexicali to receive a fishing permission, she commented to several journalists, "We should be at home teaching our language and traditions to our children. Instead, we're here yet another year arguing with lawyers about the right to feed them."

Other, more general, economic factors have been important in the shift to Spanish as well. When today's elders were entering the Mexican workforce in the 1950s and 1960s, there was a powerful disincentive to speak Cucapá because indigenous people faced discrimination. With the introduction of universal primary education in Spanish in the 1940s and 1950s, there was also a strong disciplinary move in many parts of Mexico to suppress indigenous languages. For example, Cruz, the father in the family with whom I stayed during my fieldwork, only went to grade school for several years, but some of his most vivid memories involve being punished for speaking Cucapá. By the 1980s, while children faced less violent reactions to learning or speaking Cucapá, this language was still stigmatized through structural means. For example, Felix, a fisherman in his late 30s, explained that people do not teach their children Cucapá because "they see more benefits speaking Spanish or English than speaking Cucapá." He explained that to get hired working on a farm or in tourist camps, you do not need Cucapá. So he did not see any benefit in teaching his children a language that will not help them "advance" in the world.

While these larger conditions have had an important role in the shift to Spanish as a dominant language, locals also point to other factors. Doña Katiana, a woman in her mid 60s, attributed the loss of Cucapá to new patterns of use, corresponding to the processes of language change that Dell H. Hymes (1961) called "internal." She explained that, looking back, she realized that it was not possible for her children to learn to speak Cucapá because of the traditional social organization of the community, or what she called "the old ways." She remembered that it was customary for the elders to engage in everyday activities separately from the younger generations. Her grandparents and parents would eat together and the children would then eat separately. When she had her own children, 11 in total, she followed this custom. As Spanish became the dominant language, these exclusive spaces where the elders spoke would have provided the only opportunities for the young people to learn Cucapá. However, because children were excluded from these spaces of linguistic socialization in Cucapá, they did not learn the language.

Therefore, one insight that can be drawn from considering the literature on language and environmental change in the context of the "end" of the Colorado River is that the processes that result in the obsolescence of a language are profoundly situated in specific historical, political and ecological circumstances. The extent to which these circumstances can be abstracted and applied to other linguistic/environmental scenarios in general appears to be limited. However, an ethnographic approach to language loss and ecological change does allow us to unravel some of the very situated ways that language use and ideas about language are related to the access to certain types of resources – in this case, fishing rights and access to the Colorado River.

Conclusions

In this chapter, I have argued that the particular endangerment sensibility that emerged in the 1980s and 1990s, which drew together the domains of biological and linguistic endangerment, had the effect of reviving earlier objections to environmental determinism among sociocultural and linguistic anthropologists, while raising a new enthusiasm among linguists and evolutionary biologists. The ways that the endangerment paradigm has incorporated cultural and linguistic domains also calls critical attention to the contemporary appeal and acceptance of essentialized portrayals of cultural and linguistic "traits" as quantifiable entities that can be tracked, imperiled and "saved."

While "language death" is always a misfortune for linguists, the scope and significance of this problem for the speakers of endangered languages has yet to be fully explored. What is clear is that languages play a central role in their speakers' lives, not simply as abstract systems, but in some cases as a central organizing element of everyday life and often as a key category in the distribution of access to resources, including environmental ones.

Views about the links between the river's failure to reach the delta in Mexico and the Cucapá language's obsolescence among Cucapá people underscored the importance of an integrated understanding of how processes of linguistic, economic and environmental marginalization overlap and reinforce each other. But this is not the kind of integrative vision that collapses these processes into a single conceptual resource construed vaguely as "diversity." Rather, the case study of the Colorado Delta highlights the fact that processes of linguistic and environmental change cannot be conflated or taken as causally linked in any simplified or generalizable manner. What language obsolescence and environmental degradation chiefly share is their deep involvement in the creation and structure of social inequality. For those interested and engaged in projects dealing with the preservation of natural and cultural heritages, this implies the need to approach their efforts in ways that, above all, are sensitive to the relations of inequality that manifest themselves in anxieties about endangerment and extinction.

Notes

1 The Cucapá language is of the Yuman family and is genetically quite close to the nearby Kiliwa and Maricopa languages. James Crawford produced a grammar (1966) and a dictionary (1989) for Cucapá that was based on research he did in the 1940s.

2 "Language obsolescence" is the more technical term used in the literature to refer to the process whereby speakers of a given language decrease, eventually resulting in no fluent speakers of the variety. There is a great deal of imprecision in the various terms used to denote language obsolescence, including for example: "language endangerment," "language death," "language loss," and "language extinction." Often these terms are used interchangeably to refer to the process by which languages fail to be learned by a younger generation at a society level. "Language attrition" is usually used to refer to the loss of proficiency in a language on an individual level and "language loss" is a more overarching term (referring to the level of an individual or a society).

3 The residential school system implemented in the United States and Canada in the late 19th century, which removed and isolated children from the influence of their families and traditions in order to assimilate them into dominant culture, is one of the most forceful examples of such language ideologies put into implementation in public policy. The objectives of this system were based on the assumption Native American and Canadian cultures and languages were inferior to English (Jaine 1995; Kroskrity and Field 2009; Miller 1996).

4 Mufwene wishes to refine rather than reject the language as organism analogy, and extends it to the concept of a "language ecology," which he and others have elaborated

as an analytical tool (Haugen 1972; Mufwene 2001; Mühlhäusler 1996). Such exten-sion is motivated by the observation that language death often results from changes in the "ecology" that include the society using the endangered language as one of its codes. In this context, "ecological" has been defined in terms of "the cultural construal of a system of resources, accessible through local technologies, on which the well being of a community depends" (Hill 1996, 6).

5 These codes were not devised by UNESCO but were originally devised by SIL Inter-national (2009), one of the first and still most active research agencies on endangered languages, which also maintains a database, Ethnologue, kept up to date by the con-tributions of linguists globally. SIL is an international Christian non-governmental organization that has worked with language communities worldwide in language docu-mentation efforts. SIL began in the 1930s when the documentation of Indo-European languages was still in its infancy with the goals of introducing literacy and the transla-tion of the Bible in these languages (Quakenbush 2007).

Bibliography

Alvarez de Williams, A., 1974. Los Cucapa Del Delta Del Rio Colorado. *Calafia*, 2 (5), 40–47.

Bergman, C., 2002. *Red Delta: Fighting for Life at the End of the Colorado River*. Fulcrum Publishing, Golden, CO.

Berreby, D., 2003. Fading Species and Dying Tongues: When the Two Part Ways. *New York Times*, 27 May, F3.

Blommaert, J., 2011. Supervernaculars and Their Dialects. *Working Papers in Urban Lan-guage and Literacies*, 81. King's College London, London.

Boas, F., 1940. *Race, Language and Culture*. Free Press, New York.

Butler, J., 2004. *Undoing Gender*. Taylor and Francis, New York.

Chagnon, N. A., 1968. *Yanomamo, the Fierce People*. Holt, Rinehart and Winston, New York.

Conklin, E. and Graham, L., 1995. The Shifting Middle Ground: Amazonian Indians and Eco-Politics. *American Anthropologist*, 97 (4), 695–710.

Crawford, J. M., 1966. *The Cocopa Language*. Thesis. UMI Dissertation Services Berkeley, University of California, Berkeley.

Crawford, J., 1989. *Cocopa Dictionary*. University of California Press, Berkeley.

Crawford, J., 1992. *Hold Your Tongue: Bilingualism and the Politics Of "English Only."* Addison-Wesley, Reading, MA.

Crawford, J., 1995. Endangered Native American Languages: What is to Be Done and Why? *The Bilingual Research Journal*, 19 (1), 17–38.

Crawford, J., 1998. *Endangered Native American Languages: What is to Be Done and Why?* (http://www.ncela.us/files/rcd/BE021828/Endangered_Native_American.pdf). Accessed 29 June 2014.

Crystal, D., 2000. *Language Death*. Cambridge University Press, Cambridge.

Darwin, C., 1859. *On the Origin of Species by Means of Natural Selection*. Murray Press, London.

Darwin, C., 1871. *The Descent of Man, and Selection in Relation to Sex*, 2 vols. Murray Press, London.

Dias, N., 1994. Looking at Objects: Memory, Knowledge in Nineteenth-century Ethno-graphic Displays. In: Robertson, G., ed., *Travellers' Tales: Narratives of Home and Dis-placement*. Routledge, London, 164–176.

Duchêne, A., 2008. *Ideologies Across Nations: the Construction of Linguistic Minorities at the United Nations*. Mouton de Gruyter, New York.

Duchêne, A. and Heller, M., eds., 2007. *Discourses of Endangerment: Interest and Ideology in the Defence of Language.* Continuum, International Publishing Groups, New York.

Frank, R.M., 2008. The Language-organism-species Analogy: A Complex Adaptive Systems Approach to Shifting Perspective on 'Language'. In: Frank, R.M., Dirven, R., Ziemke, T. and Bernardez, E., eds., *Body, Language and Mind: Volume 2: Sociocultural Situatedness.* Mouton de Gruyter, Berlin, 215–262.

Gal, S., 1979. *Language Shift: Social Determinants of Linguistic Change in Bilingual Austria.* Academic Press, New York.

Garrett, A., 2003. Letter to the Editor. *New York Times,* 4 June, 5.

Garrett, P., 2012. Dying Young. In: Sodikoff, G.M., ed., *The Anthropology of Extinction.* Indiana University Press, Bloomington, 143–164.

Gifford, E.W., 1933. The Cocopa. *University of California Publications in American Archaeology and Ethnology,* 31 (2), 257–333.

Harmon, D., 1996. Losing Species, Losing Languages: Connections between Biological and Linguistic Diversity. *Southwest Journal of Linguistics,* 15, 89–108.

Harmon, D., 2002. *In Light of Our Differences: How Diversity in Nature and Culture Makes us Human.* Smithsonian Institution Press, Washington.

Harris, R., 1981. *The Language Myth.* Duckworth, London.

Harrison, D., 2007. *When Languages Die: The Extinction of the World's Languages and the Erosion of Human Knowledge.* Oxford University Press, New York.

Haugen, E.I., 1972. The Ecology of Language. In: Anwar, S., ed., *The Ecology of Language: Essays by E. Haugen.* Stanford University Press, Stanford.

Hecht, S. and Cockburn, A., 1990. *The Fate of the Forest: Developers, Destroyers and Defenders of the Amazon.* Verso, New York.

Heller, M., 2002. Language, Education and Citizenship in the Post-national Era: Notes from the Front. *The School Field: International Journal of Theory and Research in Education,* 13 (7), 15–31.

Hill, J., 1996. *Languages on the Land: Toward an Anthropological Dialectology* [Lecture]. The David Skomp Distinguished Lectures in Anthropology, Indiana University, 21 March.

Hill, J., 2002. "Expert Rhetorics" in Advocacy for Endangered Languages: Who Is Listening, and What Do They Hear? *Journal of Linguistic Anthropology,* 12 (2), 119–133.

Hill, J. and Hill, K., 1977. Language Death and Relexification in Tlaxcalan Nahuatl. *International Journal of the Sociology of Language,* 12, 55–67.

Hinton, L., 2003. Letter to the Editor. *New York Times,* 3, 4.

Hinton, L. and Hale, K., 2001. *The Green Book of Language Revitalization in Practice.* Academic Press, San Diego.

Huntington, E., 1922. *Civilization and Climate.* Yale University Press, New Haven.

Hymes, D.H., 1961. Functions of Speech: An Evolutionary Approach. In: Gruber, F., ed., *Anthropology and Education.* University of Pennsylvania, Philadelphia, 55–83.

Ingold, T., 2000. *The Perception of the Environment.* Routledge, New York.

Ingold, T., 2007. The Trouble with Evolutionary Biology. *Anthropology Today,* 23, 13–17.

Jaine, L., 1995. *Residential Schools: The Stolen Years.* University Extension Press, Extension Division, University of Saskatchewan, Saskatoon.

Krauss, M., 1992. The World's Languages in Crisis. *Language,* 68 (1), 4–10.

Kroeber, A.L., 1928. Native Culture of the Southwest. *University of California Publications in American Archaeology and Ethnology,* 23, 375–398.

Kroskrity, P.V. and Field, M.C., 2009. *Native American Language Ideologies: Beliefs, Practices, and Struggles in Indian Country.* University of Arizona Press, Tucson.

Kulick, D., 1992. *Language Shift and Cultural Reproduction.* Cambridge University Press, Cambridge.

Maffi, L., ed., 2001. *On Biocultural Diversity: Linking Language, Knowledge, and the Environment.* Smithsonian Institution Press, Washington.

Maffi, L., 2005. Linguistic, Cultural and Biological Diversity. *Annual Review of Anthropology,* 29, 599–617.

Manne, L.L., 2003. Nothing has yet lasted forever: Current and Threatened Levels of Biological Diversity. *Evolutionary Ecology Research,* 5, 517–527.

Martin, E., 1987. *The Woman in the Body: A Cultural Analysis of Reproduction.* Beacon Press, Boston.

Miller, J.R., 1996. *Shingwauk's Vision: History of Native Residential Schools.* University of Toronto Press, Toronto.

Moseley, C., 2012. *The UNESCO Atlas of the World's Languages in Danger: Context and Process.* World Oral Literature Project.

Muehlmann, S., 2007. Defending Diversity: Staking Out a Common, Global Interest. In Duchêne, Alexandre and Monica Heller eds., *Discourses of Endangerment: Interest and Ideology in the Defence of Language.* Continuum, New York, 14–34.

Muehlmann, S., 2013. *Where the River Ends: Contested Indigeneity at the End of the Colorado River.* Duke University Press, Durham.

Muehlmann, S. and Duchêne, A., 2007. Beyond the Nation-State: International Agencies as New Sites of Discourses on Bilingualism. In: Heller, M., ed., *Bilingualism: A Social Approach.* Palgrave Macmillan, New York, 96–110.

Mufwene, S.S., 2001. *The Ecology of Language Evolution.* Cambridge University Press, Cambridge.

Mufwene, S.S., 2002. Colonization, Globalization and the Plight of 'Weak' Languages. *Linguistics,* 38, 375–395.

Mühlhäusler, P., 1996. *Language Change and Linguistic Imperialism in the Pacific Region.* Routledge, London.

Nettle, D., 1998. Explaining Global Patterns of Language Diversity. *Journal of Anthropological Archaeology,* 17 (4), 354–374.

Nettle, D. and Romaine, S., 2000. *Vanishing Voices.* Oxford University Press, Oxford.

Nichols, J., 1990. Linguistic Diversity and the First Settlement of the New World. *Language,* 66 (3), 475–521.

Nisbet, R.A., 1969. *Social Change and History: Aspects of the Western Theory of Development.* Oxford University Press, London.

Orman, J., 2013. Linguistic Diversity and Language Loss: A View from Integrational Linguistics. *Language Sciences,* 1–11.

Ostler, R., 2003. Letter to the Editor. *New York Times,* 4 June, 2.

Patrick, D., 2003. *Language, Politics, and Social Interaction in an Inuit Community,* 8. Walter de Gruyter.

Pianfetti, A., 2003. Letter to the Editor. *New York Times,* 4 June, 4.

Quakenbush, S., 2007. Chapter 4: SIL International and Endangered Austronesian Languages. In: Rau, D.V. and Florey, M., eds., *Documenting and Revitalizing Austronesian Languages.* University of Hawaii Press, Honolulu, 42–65.

Richards, R.J., 2002. The Linguistic Creation of Man: Charles Darwin, August Schleicher, Ernst Haeckel, and the Missing Link in 19th-Century Evolutionary Theory. In: Doerres, M., ed., *Experimenting in Tongues: Studies in Science and Language.* Stanford University Press, Stanford, 21–48.

Sahlins, M., 1976. *The Use and Abuse of Biology: An Anthropological Critique of Sociobiology.* University of Michigan Press, Ann Arbor.

Sánchez, Ogás, Y., 2000. *A La Orilla Del Río Colorado.* Colonia Pro-Hogar, Mexicali.

Sandburg, C., 1916. *Chicago Poems*. Henry Holt and Company, New York.

SIL International, 2009. *Ethnologue: Languages of the World*, 16th ed. SIL International, Dallas.

Skutnabb-Kangas, T., 2000. *Linguistic Genocide in Education or Worldwide Diversity and Human Rights?* Routledge, New York.

Sutherland, W., 2003. Parallel Extinction Risk and Global Distribution of Languages and Species. *Nature*, 423, 276–279.

Williams, R., 1977. *Marxism and Literature*. Oxford University Press, Oxford.

Wolfram, W. and Schilling-Estes, N., 1995. Moribund Dialects and the Endangerment Canon: The Case of Ocracoke Brogue. *Language*, 71 (4), 696–721.

Woodbury, A. C., 1993. *A Defense of the Proposition, 'When a language dies, a culture dies'*. University of Texas, 101–129.

2 Extinction, diversity, and endangerment

David Sepkoski

The passage of the Endangered Species Act (ESA) in 1973 is viewed as a watershed moment for the modern biological conservation movement. It is generally interpreted as the culmination of initiatives extending back to the beginning of the 20th century by biologists, politicians, and concerned members of the public to foster awareness of and provide a legal basis for the protection of endangered species (Barrow 2009). Another watershed moment came 13 years later, in 1986, with the meeting in Washington, DC of the National Forum on BioDiversity, an event spearheaded by biologists Walter Rosen and E.O. Wilson that effectively initiated public and political awareness of the modern biodiversity crisis (Takacs 1996). Together, these two events are central to the establishment of what this volume is describing as the "endangerment sensibility," which may be defined as "a network of concepts, values and practices dealing with objects considered threatened by extinction and evanescence, and with the techniques aimed at preserving them" that has come to have a central place in contemporary scientific, cultural, and political discourse (Vidal and Dias this volume).

At the heart of the endangerment sensibility are two central assumptions: first, that some entity or entities are threatened with potential loss, and second, that those entities can and should be protected because of their inherent (i.e. non-instrumental) worth. Under the endangerment sensibility, the threat to be protected against is extinction, and what is to be preserved is not any single entity, but rather diversity itself. An important consequence, then, of the emergence of the endangerment sensibility has been the simultaneous appearance of a discourse in which diversity – whether biological or cultural – is seen to have an inherent, normative value. In this sense, events such as the passage of the ESA or the beginnings of biodiversity awareness can be seen as part of a larger scientific and cultural phenomenon, and a history that is longer and more complex, than simply the rise of conservation biology in the United States. Indeed, the biodiversity initiative itself has, in recent years, contributed to a larger discourse about the value of diversity and the links between biological and cultural diversity.

Both the endangerment sensibility and the broader discourse of diversity share a fairly recent historical emergence that can be located, quite precisely, in the late 1970s to mid-1980s. It is at that time that a new set of public and political discourses focusing on a wide array of endangered entities – species, ecosystems, languages, artifacts, and cultures – were initiated, along with the development of the sense that diversity itself is a value to be protected and celebrated in Western societies. But why did these discourses appear at this precise historical moment? While the answer is complex and multifaceted, I argue that a central causal factor was the development of a new biological understanding of the phenomenon of extinction, which both recast the nature of the threat of endangerment to biological entities, and also underlay a revaluing of the biological concept of diversity. This shift in biological understanding profoundly influenced the wider social, political, and cultural awareness of extinction and diversity that has shaped the endangerment sensibility of the early 21st century.

We can only understand how these discourses became linked at the end of the 20th century – and indeed why the endangerment sensibility itself came to exist when it did – by historicizing the reciprocal development of ideas about extinction and diversity in biology and culture over the past 200 years. Broadly speaking, this history reveals a transition from a scientific belief that extinction was a slow, gradual, and inevitable process driven by natural competition between individuals that was the direct consequence of natural selection (a "Darwinian view"), to one in which extinctions were understood to be sometimes catastrophic, geologically sudden events that could extinguish multiple taxonomic groups in a single blow, and in which the Darwinian logic of natural selection was not always predictive of survival. A central feature was a new biological interest in mass extinctions, which has led to what some observers have described as a "new catastrophism" in paleontology and evolutionary biology (Leakey and Lewin 1995, 229; Raup 1991, xiv). Extinction came to be defined not, as biologists since the early 19th century had believed, as a process that contributes to an endlessly renewing natural equilibrium, but rather as catastrophic and potentially permanent loss to standing biological diversity, with potentially profound ecological and evolutionary consequences. This new understanding of extinction had emerged by the early 1980s, when it immediately became part of the rhetoric of the modern biodiversity crisis. The current crisis is now commonly compared to earlier mass extinction events in Earth's history, and is often labeled the "sixth extinction" (e.g. Benton 2003; Jablonski 2004; Leakey and Lewin 1995; Myers 1979, ix; Wilson 1992, 32).

An additional argument of this chapter is that biological concepts of extinction and diversity are linked, and that biological understanding of extinction has always both reflected and reinforced broader cultural perceptions and valuations of diversity and endangerment. The central components of the extinction discourse that prevailed during the era of Darwin and beyond were that extinction 1) is slow and gradual; 2) is reciprocally balanced by the replenishment of new species (nature's economy); and 3) is in some sense progressive – that is, by reflecting the "fair" outcome of natural competition, it contributes to the robustness of living ecosystems by weeding out "unfit" individuals or species. Under this view, diversity is an inherent and self-renewing property of the "economy of nature," and thus requires no special protection or independent valuation. As this chapter will show, this particular extinction discourse was also implicated in a cultural and political ideology – especially in Britain and the US – that supported imperialism and downplayed the value of protecting species and peoples. However, by the 1970s, a radically new extinction discourse emerged that held that extinction 1) can be sudden and catastrophic; 2) can have permanent ecological and evolutionary consequences for diversity; and 3) is in some cases agnostic toward selectivity – in other words, extinctions do not necessarily target the "weakest" or least "fit" organisms and species. In this much more contingent and unpredictable view of the history of life, diversity came to be seen as a precious and potentially irreplaceable resource, an idea that has found great resonance in discourses surrounding the preservation of endangered linguistic and cultural resources as well.

Extinction and the balance of nature in the 19th century

Before the first decades of the 19th century, most naturalists believed that extinction was impossible. During the 17th and 18th centuries, this belief often took on theological overtones, relating to the notion of plentitude and the conviction that God created nothing in vain, and also reflected an assumption that nature is self-renewing or "autogenic" (Barrow 2009, Ch. 1; Pietarinen 2004). In his influential essay *The Oeconomy of Nature*, for example, Linnaeus had remarked that in order "to perpetuate the established course of nature in a continued series, the divine wisdom has thought fit, that all living things should contribute and lend a helping hand to preserve every species; and . . . the death and destruction of any one thing should always be subservient to the restitution of another" (Linnaeus 1762, 40).[1] Likewise, when, in his *Notes on the State of Virginia*, Thomas Jefferson famously denied that the wooly mammoth was extinct, he did so on the basis that "such is the balance of nature, that no instance can be produced, of her having permitted any one race of her animals to become extinct; of her having formed any link in her great work so weak as to be broken" (Jefferson 1801, 77).

Even after the 1820s, when the reality of extinctions had become accepted, notions of plentitude and natural balance continued to influence scientific understanding of extinction. While natural philosophers had speculated about the extinction of species before, it is generally accepted that the first great theorizer of extinction was Georges Cuvier, whose studies of fossil deposits in the Paris basin served as the basis for a theory of Earth history in which a series of great "revolutions" or catastrophic mass extinctions have punctuated the history of life (Rudwick 1997). What was most unpalatable to the existing European scientific outlook about Cuvier's theory was not the possibility that God might be removed from the equation (indeed, Cuvier did not propose this), but rather its implicit challenge to the balance of nature. In fact, though Darwin's evolutionary ideas would eventually be far more upsetting to traditional theology, they nonetheless preserved an essential commitment to equilibrium and balance in nature that was conspicuously absent in Cuvier (Cuddington and Ruse 2004, 101).

Most simply put, the dominant view among naturalists during the 19th century – especially in Britain – was that extinction took place in such a manner that nature's balance was never upset, and that the total number of species in existence at any one time (and hence the diversity of life) remained constant throughout the life's history. In consequence, Cuvier's theory was rejected as speculative "catastrophism," and any discussion of sudden, catastrophic "mass extinctions" was pushed to the lunatic fringe for more than a century after the publication of *Origin of Species* in 1859 (Sepkoski 2012). The responsibility for this cannot be laid solely at Darwin's door: the exclusion of catastrophic mass extinction from serious biological consideration rested on a set of assumptions that went deeper than the empirical facts of biology. Writing in 1804, the English physician and geologist James Parkinson (and first identifier of "Parkinson's disease") argued that accepting the fact of extinction need not disturb the belief that nature exists in a balance, since "that plan, which prevents the failure of a genus, or species, from disturbing the general arrangement, and oeconomy of the system, must manifest as great a display of wisdom and power, as could any fancied chain of beings, in which the loss of a single link would prove the destruction of the whole" (Parkinson 1804, 468). In other words, Parkinson argued, nature could maintain an equilibrium even in the face of the extinction of the occasional species – or even genus – provided that the loss was made good with the creation of a new species somewhere else, thus preserving the divine rationality of "those laws, by which the regulation of the oeconomy of creation was decreed."

However, the most important influence on the idea that diversity remains in a stable natural equilibrium was the Scottish geologist Charles Lyell, whose philosophy of "uniformitarianism" had a profound effect on Darwin and on the subsequent development of paleontology and evolutionary thought. In his foundational three-volume work *Principles of Geology* (1830–33), Lyell helped naturalize the process of extinction as a regular phenomenon, while also driving a stake through the heart of Cuvier's catastrophic theory of mass extinction (Rudwick 2005). In the second volume of *Principles*, Lyell extended his geological theory that the physical environment of the Earth has fluctuated very gradually around a stable mean into a set of consequences for the development of life. In the first place, he argued, since "species are subject to incessant vicissitudes . . . it will follow that the successive destruction of species must now be part of the regular and constant order of nature" (Lyell 1830–33, vol. 2, 141). While Lyell at this point did not believe in evolution, he nonetheless granted that slow environmental change would force organisms to adapt, and that in some cases, individuals or even species would fail to do so, and become extinct. However, he continued, the process of extinction would always be balanced by the appearance of new species to replace them, and vice-versa: "the addition of any new species, or the *permanent* numerical increase of one previously established, must always be attended either by the local extermination or the numerical decrease of some other species" (Lyell 1830–33, vol. 2, 142). This process, Lyell suspected, was the result of "some regular and constant law" (Lyell 1830–33, vol. 2, 128), and he even went so far as to speculate that the "Author of Nature" had "ordained that the fluctuations of animate and inanimate creation should be in perfect harmony with each other" (Lyell 1830–33, vol. 2, 159). In short, while Lyell envisioned an Earth that was subject to constant fluctuation and variation, he nonetheless held firm in the belief that "the successive extinction of terrestrial and aquatic species . . . [is] part of the economy of our system" (Lyell 1830–33, vol. 2, 168).

Thirty years later, Darwin essentially adopted Lyell's position wholesale. In the *Origin*, Darwin treated the relationship between extinction and speciation as a dynamic equilibrium, and argued that the total number of living species remained stable over time (Cuddington and Ruse 2004, 101–102). However, whereas Lyell had adapted the principle of equilibrium to an essentially static chain of being, Darwin made it central to his theory of evolution, as a logical consequence of the central mechanism of natural selection. If natural selection is the principle that favors those individuals – and, ultimately, species – who are best suited to survival and reproduction, then extinction is simply the fate of those who cannot successfully compete. Again and again in the *Origin*, Darwin reinforced this point, explaining that "it inevitably follows, that as new species in the course of time are formed through natural selection, others will become rarer and rarer, and finally extinct" (Darwin 1859, 110), and that since a species is "maintained by having some advantage over those with which it comes into competition . . . the consequent extinction of less-favoured forms almost inevitably follows" (Darwin 1859, 320). Because of the Malthusian principle of limited resources and fierce competition, natural selection is essentially a zero-sum game where the number of winners will always be balanced by an equal number of losers. Far from being mysterious or anathema, extinction is an essential process for keeping nature in a healthy balance. However, since natural selection is a process that acts only on individuals, extinction will proceed only slowly and gradually, as a lengthy period of time will be required for species to become rare through loss of individual members and, ultimately, go extinct. Darwin also took care to remind us that while "on the theory of natural selection the extinction of old forms and the production of new and improved forms are intimately connected together," nonetheless, "the old notion of all the inhabitants of the earth having been swept away at successive periods by catastrophes, is very generally given up" (Darwin 1859, 317).

In part because it strongly endorsed the Lyellian view of equilibrium in nature, Darwin's view of extinction became the standard position in the century after the *Origin*. I argue that this testifies as much to the enduring cultural popularity of ideas of balance and economy as it does to any decisive scientific evidence. Indeed, as we will see shortly, during the later part of the 20th century, many paleontologists have pointed out that much of the evidence in the fossil record actually contradicts Darwin's interpretation. However, while the long-standing belief that the Earth or Mother Nature takes care of herself has been persistent, it cannot merely be the case that the Darwinian/Lyellian view of extinction makes us feel good about nature. One suspects that there are other cultural or ideological factors that supported – and may have even influenced – this particular biological formulation of extinction.

Extinction discourse and the (non)value of diversity

At this point, it is worth briefly reviewing the tenets of the dominant 19th century view of extinction, which Darwin strongly endorsed: 1) extinction is a regular, law-abiding, and natural process; 2) extinction is driven primarily by competition – individuals or species that become extinct have failed to remain adapted to their environments, or to compete for resources, and therefore "deserve" to die; 3) extinction is inevitable – it is the logical consequence of natural selection; 4) extinction tends to be equally balanced by the appearance of new species (speciation), thus maintaining the "economy of nature"; and 5) the number of taxa (i.e. diversity) in the world therefore exists in dynamic equilibrium. It is worth noting that some naturalists admitted occasional exceptions to these basic rules: for example, Lyell and other authors gave fairly extensive attention to human-caused extinction, which was observed could produce more rapid or widespread extermination than the "natural" variety (Lyell 1830–33, vol. 2, 147–149), and Darwin acknowledged rare episodes in the fossil record – for example, the disappearance of the ammonites – that appear to have been "wonderfully sudden" (Darwin 1859, 318). But, for the most part, the received view was that extinction was gradual, inevitable, and "fair." In fact, one view that developed during the late 19th century was the theory of "racial senescence," which essentially built on Lyell's observation that "species, as well as individuals, are mortal" (Lyell 1830–33, vol. 2, 130), to propose that species had natural life-spans and could essentially die of "old age."

There are a number of ways one could connect views about biology to prominent values in Victorian society, and, indeed, many authors have done so (e.g. Desmond 1989; Ruse 1979). A case in point is the phenomenon known as "social Darwinism," which is doubly a misnomer: first, because many of the tenets associated with social Darwinism hew closer to the ideas of Herbert Spencer than to Darwin's (Hofstadter 1955; Ruse 1999); and second, for the more basic reason that ideas of competition and struggle existed in Western social thought long before Darwin (e.g. in Malthus and Adam Smith). In this section, however, I focus only on the biological topic of extinction, which I connect to a single cultural/political issue: the extinction of groups of human beings.

Darwin's view, as has been pointed out ever since 1859, appears to endorse a ruthlessly competitive view of nature, and his view of extinction seems to naturalize extinction as an inevitable and perhaps even progressive force. Darwin's and Lyell's views about extinction were part of a much larger 19th century discourse related to British and European imperial expansion, and in particular to questions about the justification for exploiting and eradicating the native peoples, flora, and fauna encountered during colonization. Patrick Brantlinger has identified what he calls an "extinction discourse" at this time, which he argues "is a specific branch of the dual ideologies of imperialism and racism" (Brantlinger 2003, 1). According to Brantlinger, British and American expansion was often underwritten by an explicit belief that it was justifiable to subjugate and even exterminate so-called "savage" tribes because such "races" were doomed anyway by the inexorable logic of biology (Brantlinger 2003, 6–7). This attitude also clearly implicates Victorian biological and anthropological theories of race, which I will not discuss here, many of which (e.g. polygenism) developed independently from or even at odds with the Darwinian view (Bowler 1986; Desmond and Moore 2009). What I want to focus on is the extent to which biological ideas about extinction informed this aspect of Victorian ideology, and to what extent this "extinction discourse" can be broadened out to more general cultural attitudes about the protection of diversity. The message, to put it bluntly, was that if extinction is inevitable and even beneficial, why should society care if it happens?[2]

Both Lyell and Darwin explicitly discussed the extinction of human races as a natural phenomenon, which they and other authors reasoned at least implicitly to mitigate the responsibility of Europeans for the negative consequences of colonial expansion. In *Principles*, Lyell wrote, for example, about the inevitability of widespread (non-human) extinction in the wake of colonialism: "We must at once be convinced, that the annihilation of species has already been effected, and will continue to go on hereafter, in certain regions, in a still more rapid ratio, as the colonies of highly-civilized nations spread themselves over unoccupied lands" (Lyell 1830–33, vol. 2, 156). Yet he saw this as little cause for regret, arguing that "if we wield the sword of extermination as we advance, we have no reason to repine at the havoc committed, nor to fancy, with the Scottish poet, that 'we violate the social union of nature.'" Why? Because extinction is part of the natural order of nature: "we have only to reflect, that in thus obtaining possession of the earth by conquest . . . we exercise no exclusive prerogative. Every species which has spread itself from a small point over a wide area, must, in like manner, have marked its progress, by the diminution, or the entire extirpation, of some other" (Lyell 1830–33, vol. 2, 156). Lyell made it clear that this explanation applied equally to "the extirpation of savage tribes of men by the advancing colony of some civilized nation." While he did pause to express regret for this fact, he offered no apology, since, as he was quick to note, this was the natural and *inevitable* course of nature: "few future events are more certain than the speedy extermination of the Indians of North America and the savages of New Holland in the course of a few centuries, when these tribes will be remembered only in poetry and tradition" (Lyell 1830–33, vol. 2, 175).

It is difficult to exaggerate how prevalent this view was, especially in Britain and the US, not only among those who stood directly to benefit from imperial expansion, but also with scientists and other "disinterested" observers. As Brantlinger puts it, this extinction discourse was "a powerful nexus of ideas that has been hegemonic for countless European explorers, colonists, writers, artists, officials, missionaries, humanitarians, and anthropologists" (Brantlinger 2003, 190). Whether or not it was seen as desirable, cultural and racial extinction was, almost universally, regarded as inevitable. In *Descent of Man* (1871), Darwin acknowledged that "from the remotest of times successful tribes have supplanted other tribes," a fact that he attributed chiefly to the same natural competition that existed among "lower" organisms, observing that "at the present day civilized nations are everywhere supplanting barbarous nations . . . and they succeed mainly, though not exclusively, through their arts, which are the product of the intellect" (Darwin 1871, 160). Just as extinction is the inevitable corollary of natural selection in a state of nature, among human groups, "extinction follows chiefly from the competition of tribe with tribe, and race with race," with the result that "when civilized nations come into contact with barbarians the struggle is short" (Darwin 1871, 238).

Darwin and Lyell were hardly alone among Victorian men of science in holding this view, nor did Darwin's theory of evolution via natural selection produce it out of nowhere. Darwin first entertained these ideas while aboard the *HMS Beagle* during the early 1830s (and well before his evolutionary ideas were fully developed), remarking in his account of that voyage that "wherever the European has trod, death seems to pursue the aboriginal," and observing that "the varieties of man seem to act on each other in the same way as different species of animals – the stronger always extirpating the weaker" (Darwin 1909, 459). Well before Darwin's evolutionary ideas were published, James Cowles Pritchard (sometimes spelled Prichard) had written an essay in the *Edinburgh New Philosophical Journal*, "On the Extinction of Human Races," where he argued that human extinction occurs naturally when tribes or races of people are placed in natural competition. Pritchard's essay is somewhat remarkable for the degree of guilt it expresses over the "whole races [that] have become extinct during the few centuries which have elapsed since the modern system of colonization have commenced," and for its resolve "to take up seriously the consideration, whether any thing can be done effectually to prevent the extermination of the aboriginal tribes" (Pritchard 1840, 168, 170). Nonetheless, the language of the essay is fatalistic throughout, referring to the expansion of Europeans as "the harbinger of extermination to the native tribes," whose "allotted time of their destruction is at hand," and the best he offers is an early version of salvage ethnography: "to obtain much more extensive information than we now possess of their physical and moral characters" before extinction takes place (Pritchard 1840, 169). Likewise, in his *Natural History of the Human Species* (1851), Charles Hamilton Smith discussed the inevitability of human extinction through competition as regrettable but inevitable, stating that "it would be revolting to believe that the less gifted tribes were predestined to perish beneath the conquering and all-absorbing covetousness of European civilization, without an enormous load of responsibility resting on the perpetrators"; nonetheless, "their fate appears to be sealed in many quarters, and seems, by a preordained law, to be an effect of more mysterious import than human reason can grasp" (Smith 1851, 207).[3]

Darwin's theory of natural selection, then, certainly did not create this extinction discourse, but it apparently made the "preordained law" responsible for inevitable extinction seem considerably less mysterious. In an 1864 essay titled "The Origin of Human Races and the Antiquity of Man Deduced from the Theory of Natural Selection," Darwin's close friend and supporter Alfred Russell Wallace used natural selection to essentially justify European imperial expansion. After first clarifying that the struggle for existence applies equally to human individuals and "tribes" as to "lower" organisms, Wallace argued that

> [i]t is the same great law of '*the preservation of favored races in the struggle for life*,' which leads to the inevitable extinction of all those low and mentally undeveloped populations with which Europeans come in contact. . . . The intellectual and moral, as well as the physical qualities of the European are superior . . . [and] enable him when in contact with the savage man, to conquer in the struggle for existence, and to increase at his expense, just as

the more favourable increase at the expense of the less favorable varieties in the animal and vegetable kingdoms, just as the weeds of Europe overrun North America and Australia, extinguishing native production by the inherent vigour of their organisation, and by their greater capacity for existence and multiplication.

(Wallace 1864, clxiv–clxv)

Wallace concluded that, therefore, "it must inevitably follow that the higher – the more intellectual and moral – must displace the lower and more degraded races; and the power of 'natural selection,' still acting on his mental organisation, must ever lead to the more perfect adaptation of man's higher faculties to the conditions of surrounding nature, and to the exigencies of the social state" (Wallace 1864, clxix). This process, Wallace concluded, would bring about the inevitable improvement of the human race, ultimately producing a utopian society he described as "as bright a paradise as ever haunted the dreams of seer or poet" (Wallace 1864, clxx).

One could produce a litany of such statements, describing not only the expansion of European nations to the Americas, Africa, and Oceana, but also affairs closer to home, such as the Irish famine where, for example, Charles Edward Trevelyan argued that "Supreme Wisdom has educed permanent good out of transient evil" in reducing the Irish population by "natural" means (Trevelyan 1848, 1). Many similar statements were equally approving. In his travelogue *Greater Britain*, Charles Dilke opined that "[n]atural selection is being conducted by nature in New Zealand on a grander scale than any we have contemplated, for the object of it here is man," and enthused about "a struggle which at once crushes and starves out of life every weakly plant, man, or insect, and fortifies the race by continual buffetings" (Dilke 1869, 392). In his *The Uncivilized Races of Men in All Countries of the World*, John George Wood explained that while "we cannot but regret that entire races of men, possessing many fine qualities, should be thus passing away," it is nonetheless "impossible not to perceive that they are but following the order of the world, the lower race preparing a home for the higher," noting that such races "occupied precisely the same relative position toward the human race as do the lion, tiger, and leopard toward the lower animals, and suffered in consequence from the same law of extinction" (Wood 1870, 790). And this is to say nothing of the writings of Herbert Spencer, who more than any Victorian intellectual applied the struggle for existence to human social organization. Spencer wrote about the inevitability of human extinction in many places; while he regarded actual physical extermination of "inferior races" to have "injurious effects on the moral natures" of advanced societies, he nonetheless endorsed a nonviolent "purifying process" by which "a competition of societies during which the best, physically, emotionally, and intellectually, spread most, and leave the least capable to disappear gradually, from failing to leave a sufficiently-numerous posterity" (Spencer 1972, 173–174).

The crux of the matter is that, in Victorian society and beyond, extinction was considered both an inevitable and a progressive process, whether applied to humans or to "lower" organisms. This view came from biology, but it is inseparable from a broader set of cultural and political attitudes about race and social progress. It certainly did not promote the active protection of threatened peoples or organisms, nor did it celebrate intrinsic biological or cultural differ-ence. "Diversity" was not an independent value at this time in biology or culture because it had not been identified as something necessary for biological or cul-tural stability. Extinction was understood to happen constantly (albeit slowly), without any consequences to the health of ecosystems or civilizations. In fact, if anything, extinction was a positive good: by removing the unfit, it acted for the betterment of species or "races." From the perspective of biologists, there was little sense that when species or cultures disappeared, some valuable resource was being lost; rather, through the law-abiding process of natural selection, nature was constantly improving its stock.

The new catastrophism and a science of diversity

I now move my narrative ahead to the later 20th century, conscious that in so doing, I am ignoring an important history of biological conservation efforts and ecological awareness-raising going back to the late 19th century. That history is parallel to the one I am describing, and dovetails with my story when we reach the beginnings of the biodiversity movement, but it would require much greater space to discuss. The development I want to focus on here – which has received scant attention in the history of conservation biology – is the emergence of a new set of ideas about extinction that were put forward during the 1970s and 1980s. These ideas, which emerged primarily from paleontology and effectively rehabili-tated the notion of catastrophic mass extinctions, were a crucial ingredient in a new appreciation of the threat posed by extinction and a new positive valuing of biological diversity that are combined in the biodiversity movement. This "new catastrophism" was explicitly grounded in better empirical understanding of the patterns of mass extinction in the fossil record, but it also undoubtedly was both influenced by and contributed to a broader culture of anxiety that prevailed dur-ing the Cold War era, centered around perceived threats of nuclear Armageddon, political uncertainty, and environmental disaster (Badash 2009; Hamblin 2013).

Along with the study of mass extinctions, the second half of the 20th century saw a new biological approach to the study of diversity. It is worth noting that the word "diversity" has a slightly different meaning as a colloquial term than in its precise scientific definition. Some authors tend to conflate the two meanings, or miss the very specific way in which biologists and ecologists now employ the term. For example, in some definitions, biological diversity is misunderstood as being essentially reducible to taxonomy: one observer has commented for example that biological diversity, "both as a vernacular and a scientific concept, is about the classification of perceptible things and phenomena, especially species" (Oksanen 2004, 2). The biological study of diversity may indeed rely heavily on classifi-cation, and it is fair to say that what was understood as "diversity" up through Darwin's day might fit the above description.[4] But, as paleontologist Peter Ward points out, "diversity is also a highly technical term with a rigid mathematical definition; it encompasses not only the number of species, but also the relative abundance of the various coexisting species" (Ward 2000, 27). In other words, when biologists now talk about diversity, they do not simply refer to the total number of taxonomic entities that exist at a particular time and place, but also invoke a fairly complex understanding of the relationships and interdependence of these taxa with one another and with their physical environments. I am arguing that this new technical understanding of diversity was in part a product of a new understanding of extinction; that it was highly influential on the establishment of the biodiversity movement; and that, consciously or unconsciously, formulations of cultural diversity now often invoke this more complex "ecological" sense.

In this "highly technical sense," the study of diversity is a relatively recent scientific field (Heyd 2010, 160). Based on a very informal survey, I found that between 1897 (the first year records were available) and 1959, the term appears in the title of articles in the biological sciences archived by JSTOR only 18 times. But between 1960 and 1969 (the decade when much of the discipline of theoretical ecology was formalized), this number jumps to 113, and each of the following decades charts an exponential rise in its use: during the 1970s, 381 times; the 1980s, 595 times; the 1990s (the beginning of the era of biodiversity consciousness), 1,494 times; and the 2000s, 2,171 times.[5] This rise tracks very closely to a Google n-gram graph of the appearance of the term in the text of books during the 20th century; it also – not coincidentally, I argue – tracks the pattern of an n-gram for the term "mass extinction" (Figure 2.1). As I illustrate below, this growing scientific interest in biological diversity would, by the 1980s, connect with broader cultural valuations of diversity.

An important ingredient in the new scientific study of diversity was the development of theoretical ecology during the 1950s and 1960s by scientists like G. Evelyn Hutchinson, Robert A. MacArthur, and E.O. Wilson, who developed mathematical models for understanding species abundance, migration, and population limits (Dritschilo 2008; Kingsland 1985). These models describe the influence of geographical area, immigration and emigration, and finite resources on ecological diversity, demonstrating that such factors tend to produce a stable equilibrium in diversity (Sepkoski 2012, Ch. 4). In particular, MacArthur and Wilson's "species-area effect" (that species richness increases exponentially with habitable area) and their equilibrial model of island biogeography (which predicts that in island populations, the number of species immigrating will reach an equilibrium with the number leaving or becoming extinct) had a great influence on the development of paleontological approaches to global and historical patterns of extinction and diversity, and figure into how estimates of current biodiversity are measured (Sepkoski 2012).

What this new ecological understanding of diversity particularly pointed to was the need to better understand the phenomenon of extinction, which is a vitally important factor in diversification. Writing to a colleague in 1979, the paleontologist and extinction pioneer David Raup commented that "I am becoming more and more convinced that the key gap in our thinking for the

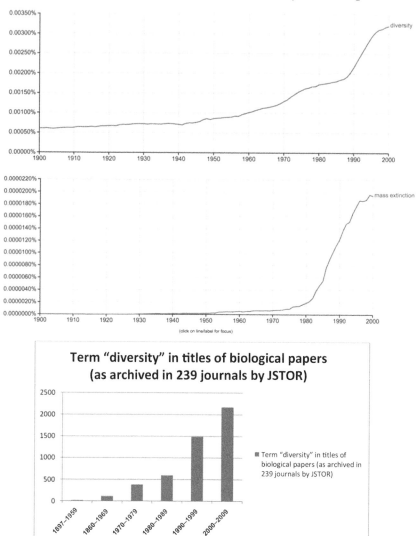

Figure 2.1 Comparison of Google n-grams to analysis of JSTOR articles with the term "diversity" in the titles.

last 125 years is the nature of extinction" (David Raup to Thomas J.M. Schopf, 28 January 1979). What he meant was that paleontology – and biology more generally – had no adequate theory for the causes and consequences of extinction. Here, and many other times in his career, Raup laid the blame at Darwin's doorstep: by focusing exclusively on natural selection and competitive replacement as the cause of extinction, Darwin's view effectively presented a tautology with little explanatory value, where "the only evidence we have for the inferiority of victims of extinction is the fact of their extinction" (Raup 1991, 17). This recalls many of the 19th century justifications for allowing human extinctions as well: Smith, for example, had argued that since only imperfect races can become extinct, the prospective extinction of the Native Americans was "proof" that the race was imperfect (Smith 1851, 276). It also, Raup argued, failed to agree with the fossil record. As he wrote,

> [i]f we take neo-darwinian theory at face value, the fossil record makes no sense. That is, if we have (a) adaptation through natural selection and/or species selection and (b) extinction through competitive replacement or displacement, then we ought to see a variety of features in the fossil record that we do not such as: (a) clear evidence of progress, (b) decrease in evolutionary rates (both morphologic and taxonomic), (c) possibly a decrease in diversity.
> (Raup to Schopf, 28 January 1979)

Raup's explanation for the persistence of the Darwinian model was in part that scientists had overly relied on a simplistic assumption that all extinction is selective, or as Stephen Jay Gould put it, assumed an "overly Darwinian world of adaptation, gradual change, and improvement" (Gould 1991, xv; Raup 1986, 50). But Raup also suggested that cultural values may have played a role as well, noting that Darwin's view "describes a fair game where goodness triumphs in the end" and that it therefore "fits well with the traditional Calvinist view that many of us grew up with" (Raup 1986, 52).

What the fossil record did show was a complex pattern of steep rises and sharp plummets in levels of diversity over the history of life. Beginning in the 1970s, paleontologists began serious efforts to inventory the entire marine fossil record, essentially by counting the number of fossil taxa (genera and families, primarily) that had ever been discovered, and analyzing the resulting data (Sepkoski 2012). What this analysis showed was that there is a general trend toward increasing diversification, beginning during the Cambrian explosion 500 million years ago and continuing to the present. This pattern is, however, influenced by two important factors: in the first place, at particular times in the history of life, global assemblages of organisms appear to have reached a stable equilibrium – where the rates of their extinctions and speciations were in balance – which often persisted for many tens of millions of years, before experiencing a slow decline, only to be replaced by a new global assemblage that began an exponential increase in diversification before reaching its own equilibrium plateau. This was the finding of the paleontologist Jack Sepkoski, who in a series of papers proposed that three successive "evolutionary faunas" had characterized the marine diversity of the last 500 million years (Sepkoski 1984). In some respects, this conclusion seems to agree with Darwin's view, which held that the number of species at any one time will tend toward a dynamic equilibrium. However, in another crucial respect, Sepkoski's model differed from Darwin's: each successive fauna diversified to reach a significantly *higher* diversity plateau than the previous one, and the current, "modern" fauna of the last 100 million years has not yet ceased its exponential diversification. In other words, while diversity appears to tend toward equilibrium for lengthy periods of time, diversity has not been stable over the entire history of life, and in general, the pattern is one of increasing diversity (number of taxa) over time (Sepkoski et al. 1981).

The second critical departure from Darwin's interpretation that this analysis produced was a new appreciation for the evolutionary significance and consequences of mass extinction. While paleontologists had begun rehabilitating the study of mass extinctions since the early 1960s (Sepkoski 2012, Chs. 2–3), the faunal analysis conducted by Sepkoski (and Raup, who was his colleague at the University of Chicago) showed that major, catastrophic mass extinctions had played a key role in perturbing the evolutionary system at least five times during the Phanerozoic eon. These mass extinctions were episodes that, typically, lasted no more than a few million years, but that produced drops in standing diversity of anywhere from 50 to 95 percent. In 1984, Raup and Sepkoski published a paper arguing that there have been five such extinction events over the past 500 million years and, remarkably, that these extinctions (along with more minor though still "mass" extinctions) appear to follow a regular periodicity, occurring roughly every 26 million years (Raup and Sepkoski 1984). The major evolutionary interpretation they advanced was that these events could not be explained as the product of natural selection alone; they were catastrophic episodes that effectively "reshuffled the deck" for evolution, wiping out formerly long-standing taxa (like the dinosaurs) and ushering new ones to evolutionary prominence (such as the mammals). If the significance of these mass extinctions was to be credited, this presented an entirely new view of extinction: while normal or "background" extinctions probably occurred slowly and constantly, as Darwin had held, a significant mechanism in the history of life and of diversification were events that appeared to follow no Darwinian rules of selectivity, in which entire taxa disappeared through no "fault" of their own.

The Raup-Sepkoski extinction work happened to coincide with another major event in paleontology: the discovery, by a team led by geologist Walter Alvarez, of circumstantial evidence that a massive meteorite impact had occurred 65 million years ago, at the exact moment when the dinosaurs disappeared from the fossil record. This, potentially, was the kind of non-selective trigger implied by the Raup-Sepkoski work, and together these findings created a sensation in the scientific community and the popular media. Major magazines and newspapers, from *Time* and *Newsweek* to the *New York Times* and the *Washington Post*, gave the new impact-extinction theories front page billing, and many paleontologists were thrust into the international spotlight. It is not difficult to understand why this work would have caused a stir: the dinosaurs have always been the most charismatic and popular prehistoric creatures, and their demise had remained an enigma for over a century. Another factor was the era of Cold War anxieties of nuclear annihilation and environmental catastrophe. If the dinosaurs could go, the idea went, then so could we humans. In fact, the model of "nuclear winter" that frightened the public during the mid-1980s was actually developed from climate models produced to estimate the atmospheric effects of the massive asteroid that likely struck 65 million years ago, making the juxtaposition of the fates of humanity and the dinosaurs more than merely metaphorical (Badash 2009; Ehrlich et al. 1983). As newspaper columnist Ellen Goodman opined in 1984, "I wonder whether every era gets the dinosaur story it deserves," noting that "the scientists of the 19th century – a time full of belief in progress – saw evolution as part of the planet's plan of self-improvement . . . Those who lived in a competitive economy valued the 'natural' competition of species. The best man won." But, she continued, "surely we are now more sensitive to cosmic catastrophe, to accident," adding, "in that sense, the latest dinosaur theory fits us uncomfortably well. 'Our' dinosaurs died together in some meteoric winter, the victims of a global catastrophe. As humans, we fear a similar shared fate." However, as Goodman observed in closing her column, "the difference is that their world was hit by a giant asteroid while we – the large-brained, adaptable creatures who inherited the earth – may produce our own extinction" (Goodman 1984).

One upshot of this extinction work was the creation of a cottage industry in paleontological studies of mass extinction, and the legitimation of a new catastrophism. Another was that extinction was essentially redefined in terms of diversity: mass extinctions are recognized in the fossil record, explicitly, as those periods when diversity drops significantly in a short amount of time. Raup, Sepkoski, and other authors continued during the 1990s and beyond to explore the dynamics and evolutionary consequences of major mass extinctions, honing in on the potential challenges they presented to the older, Darwinian view. Raup has most succinctly reduced the problem to a question of whether extinction is caused by "bad genes or bad luck," or, as he has put it, whether "the evolution of life [is] a fair game, as the survival-of-the-fittest doctrine so strongly implies?" (Raup 1991, xi). While Raup and others acknowledge that background extinction tends to follow selective logic, episodes of mass extinction introduce a new set of rules. It is not exactly the case that selection plays no role in mass extinction: when evolutionary/environmental catastrophes occur – such as an asteroid impact – a certain number of species will undoubtedly be killed immediately. For those organisms, extinction has been purely "bad luck." But even in the aftermath, when selection begins again, the rules that formerly governed adaptive success may have been completely upended, to the extent that previously dominant species will suddenly find themselves poorly adapted for their new circumstances. In those cases, the unfortunate organisms may indeed have the wrong genes for survival, but it is still "bad luck" to possess those genes. Raup has termed this process "wanton extinction" or "nonconstructive selection," and paleontologist David Jablonski has expanded the idea into a very detailed understanding of how extinction and recovery function in mass extinction "regimes."

A major lesson from all of this is that it is extremely difficult to predict winners and losers following mass extinctions, although so-called "weedy" species (e.g. cockroaches) tend to persist at the expense of more specialized organisms (Jablonski 2004, 174). It is also the case that mass extinctions impact diversity more than just by temporarily reducing it. Jablonski has found that the clades (evolutionary lineages) that survive mass extinctions tend to be more homoge-neous in their constituent species. This can have a narrowing effect on diversity over time: if life reached a point of maximal difference (or "disparity," mean-ing major genetic and morphological dissimilarity between higher taxa) early on during the Cambrian explosion, and if each subsequent extinction event has stochastically removed some number of those disparate higher taxa (orders or phyla), then evolution has less and less genetic and morphological difference to "work with" as time goes on. This means that even if diversity – as measured by the sheer number of species alive – has increased, it has become a more homoge-neous kind of diversity, since those species are clustered within fewer and fewer higher taxa. This was the argument that Gould famously made in his book *Won-derful Life*, and it was both inspired and confirmed by the work of his fellow pale-ontologists on extinction (Gould 1989). And while the history of life has shown us that diversity does indeed rebound from extinction events, the time required can often be on the order of tens of millions of years – a fact of dubious comfort to any organisms who experience the extinction itself (Jablonski 2004, 174).

Extinction, biodiversity, and the value of diversity

It was at the height of the scientific and public interest in mass extinctions that, in 1986, the biodiversity movement formally began. There were certainly earlier contributing factors – a long history of conservation efforts focused on preserv-ing individual endangered species, for example (Barrow 2009; Farnham 2007).[6] But there was something genuinely new about the way the major proponents of biodiversity – people like Wilson and Norman Myers – mobilized interest in pro-tecting not one or a few individual species or habitats, but the entire diverse global ecosystem itself. Biodiversity, in other words, helped make diversity a normative value. The reasons for this are many and complex, but I will point to a few: in the first instance, ecologists began during the mid-20th century to better appreci-ate the fragility and interconnectedness of ecosystems (Barrow 2009, 205). One could not focus on just the big, "charismatic" vertebrates and expect success; the insects and even microbes matter, too, if one wanted to maintain healthy habi-tats. Second, this occasioned a transition to a less romantic, and more utilitar-ian, environmentalist ethos than the one that existed in the late 19th and early 20th centuries (Takacs 1996, 22). Conservation arguments increasingly tended to promote the economic, biomedical, and even ethical reasons for preserving all life, rather than those related to aesthetics and recreation. The biodiversity movement would follow this trend. Third, and quite simply, the pace of human depletion of the natural environment got a lot faster. Rainforest destruction, environmental pollution, sprawl, and a host of other problems had been acceler-ating since the demographic expansion of Western societies in the 1950s, mak-ing their consequences more and more apparent. Fourth, arguments began more frequently to be focused, from the 1960s and onward, on the danger of unfore-seen consequences. While the utilitarian value of most species was unknown, the rapid pace of discovery in the pharmaceutical and other industries suggested that previously unknown or humble organisms might have great worth. Likewise, as the laws of ecological relationships were better understood, it occurred to many that irreparable harm might be done to fragile ecosystems before it was even realized. Diversity itself, in other words, became conceptualized as a vital resource.

And finally, biologists interested in conservation efforts became aware of the new science of extinction and its consequences, which gave them both a sense of the scope of the current crisis, and also tools and data with which to predict its consequences. As Wilson put it in *The Diversity of Life*, "the laws of biological diversity are written in the equations of speciation and extinction" (Wilson 1992, 220). Paleontological studies of mass extinction are cited in nearly every scientific statement about biodiversity, and comparing the current crisis to extinctions in the geologic past is one of the most prominent and effective tools in biodiversity's rhetorical arsenal. It was paleontologists like Raup, for example, who calculated the "normal" rate of extinction – often cited as about 2–3 species per year – against which current trends could be juxtaposed. Paleontological studies of mass extinction also gave biodiversity proponents a set of arguments about the potential consequences – both for ecological recovery and in evolutionary terms – of allowing a "sixth extinction" to proceed unchecked. And extinction studies have helped silence the appeals to nature's ability to endlessly renew itself that characterized an earlier era of thinking. The fact that mass extinctions can and do occur, and that they have dramatic short and long-term consequences for diversity, has contributed a greatly enhanced sense of impending danger that was absent in earlier conservation rhetoric.

Paleontologists themselves have also been prominent contributors to the discourse surrounding the biodiversity crisis. When the first conference on biodiversity was held in 1986, Raup presented a paper on "Diversity Crises in the Geological Past," which appeared as the fifth chapter in the massive published proceedings (Wilson 1988a). The arguments Raup made became familiar staples in subsequent biodiversity appeals: that the paleontological data on extinction "can place our knowledge of present-day extinctions in the larger time context of the global evolution of life" (Raup 1988, 51), and that "without consideration of the time perspective available from the geological record, a full evaluation of the contemporary extinction problem may prove as difficult as would be the case . . . if an epidemiologist were to treat an infectious disease without medical records" (Raup 1988, 57). Indeed, Wilson's own introduction to the volume invoked paleontology directly, comparing the current crisis to "the great natural catastrophes at the end of the Paleozoic and Mesozoic eras," and warning that "the modern episode exceeds anything in the geologic past" (Wilson 1988b, 11–12). These arguments were later taken up by many other paleontologists: for example, Sepkoski warned in 1997 that "it may indeed be possible that we are on the brink of the greatest of all mass extinctions," and rallied his colleagues with the argument that "we are the only scientists who have ever seen biodiversity crises to their end . . . and have some idea of what happens in their aftermath" (Sepkoski 1997, 536). Michael Benton, who has produced important analyses of past extinction and diversity, similarly concluded that "extinction events in the fossil record . . . can give indications of what might, or might not, happen in the future" (Benton 2003, 8), and pointed out that "comparison of present crises with documented ancient examples at least allows scientists and policy makers to work with real facts and figures" (Benton 2003, 16). And Jablonski has forcefully marshaled his own work on extinction recovery to sound a warning about the present crisis, noting that "in order to understand the dynamics of biodiversity . . . we need to understand extinctions and their complex aftermath" (Jablonski 2004, 174), and reminding us that "whatever the exact magnitude of present-day diversity losses, rebounds in the fossil record suggest that they will not be recouped in the next thousand years" (Jablonski 1991, 755).

One of the most striking and important contributions of the paleontological work to biodiversity discourse has been the erosion of the time-honored faith that nature is capable of endlessly renewing itself through competition. This fact has not been lost even on detractors of the conservation movement: in a 1979 op-ed piece provocatively titled "Give me that Old-Time Darwin," a public relations consultant named Sam Witchel criticized efforts to preserve endangered species like the much-maligned snail darter, lamenting "now come the ecologists with their New Truth that completely contradicts everything Darwin revealed," and proposing that "what seems to be missing from the ecologists' dogma is the recognition that all forms of life, including man, must fight for survival" (Witchel 1979). In response to such claims, Gould penned a 1990 essay in which he reminded "some apologists for development [who] have argued that extinction at any scale . . . poses no biological worry but, on the contrary, must be viewed as part of an inevitable natural order," and that "capacity for recovery at geological scales has no bearing whatever upon the meaning of extinction today" (Gould 1990). Wilson took on this trope even more directly, commenting that data on diversity loss "should give pause to anyone who believes that what *Homo sapiens* destroys, Nature will redeem," at least "within any length of time that has meaning for contemporary humanity" (Wilson 1992, 31). And Jablonski reiterated this point a decade later, noting that while "it might seem . . . that catastrophe and renewal go hand-in-hand, so that there really is nothing to worry about because Mother Nature will right herself eventually . . . a key message of the fossil record is the great disparity between the time scales of extinction and recovery" (Jablonski 2004, 174).

As David Takacs has pointed out, during the 1980s and 1990s, the biodiversity movement brought about a new way of seeing, and valuing, natural diversity that embodied not only scientists' interpretations of empirical evidence, but also their "political, emotional, aesthetic, ethical, and spiritual feelings" (Takacs 1996, 1). Biological diversity came to be seen, by scientists, policy makers, and the general public, not just as important for ecological survival or medical and economic development, but as something "good" in itself. This shift occurred at the same time that many Western societies began to identify other kinds of diversity – cultural or linguistic, for example – as an inherent normative good. While I do not argue that the influence of biodiversity on wider valuations of cultural diversity was strictly causal, appreciation of biological and cultural diversity are nonetheless linked. As Takacs again observes, "as biologists link themselves with the forces promoting the multicultural ethic that has made normative and political headway in our society, different kinds of diversity thus become symbiotically and metaphysically linked in inherent 'goodness'" (Takacs 1996, 43). He is not alone in drawing this conclusion. David Heyd argues that "cultural diversity is but one manifestation of the culture of diversity, the other being biodiversity," and he contends that "it is by no means a coincidence that despite the different origins of the two movements, their evolution took place more or less simultaneously" (Heyd 2010, 160). Both manifestations of this culture of diversity, Timothy Farnham explains, are linked by a shared conception of diversity as a resource: "It was assumed that by maintaining diversity in the natural world we could keep all our options open. . . . In this way, we have come to believe that with greater diversity – whether cultural or biological – comes greater value" (Farnham 2007, 7).

One of the clearest examples of the overlap between valuations of biological and cultural diversity is in the rhetoric used by the United Nations and UNESCO to promote these ideals. A few years after the initial biodiversity conference was held in Washington, an "Earth Summit" was held in Rio de Janeiro where 150 nations signed the United Nations' "Convention on Biological Diversity," which explicitly called attention to "the intrinsic value of biological diversity" (UN 1992, 1). A decade later, UNESCO produced the "Universal Declaration on Cultural Diversity," which framed cultural diversity in the same language of "resource" in which biological diversity was being presented: "The Declaration aims both to preserve cultural diversity as a living, and thus renewable treasure that must not be perceived as being unchanging but as a process guaranteeing the survival of humanity" (UNESCO 2002). It went on to make the analogy between both forms of diversity explicit, stating in Article 1 that "as a source of exchange, innovation and creativity, cultural diversity is as necessary for humankind as biodiversity is for nature" (UNESCO 2002).

This sense that cultural and biological diversity are not merely similar, but are actually manifestations of the same phenomenon, can be seen in the emergence of a new term – "biocultural diversity" – at around the same time. The term seems to have been first coined at a 1996 conference held in Berkeley that was jointly sponsored by UNESCO, the World Wildlife Fund, and the newly founded organization Terralingua (which is devoted to preservation of endangered languages) on "Endangered Languages, Endangered Knowledge, Endangered Environments" (Maffi 2001b). Papers from the conference were published in a volume titled *On Biocultural Diversity*, and reflect the way in which key elements of the biological understanding of extinction and diversity influenced discussions of cultural and linguistic endangerment. As conference organizer Luisa Maffi puts it, species, ecosystems, and cultural and linguistic groups "are facing comparable threats of radical diversity loss," highlighting the sense in which extinction is now being conceptualized as the loss of diversity (Maffi 2001a, 3). Maffi's paper focuses on "a threat to linguistic diversity that may be far greater in magnitude than the threat facing biodiversity," characterizing the potential loss of endangered languages as "an extinction crisis" of "unprecedented" proportions (Maffi 2001a, 5). In terms of the strategies for response to this linguistic extinction crisis, Maffi argues that "issues of linguistic and cultural diversity conservation may be formulated in the same terms as for biodiversity conservation: as a matter of 'keeping options alive' and of preventing 'monocultures of the mind'" (Maffi 2001a; Smith 2001, 38). One of the perceived threats inherent in the biodiversity crisis is the homogenizing effect of extinction – the realization that homogeneous "monocultures" imbalance the ecosystem and place remaining species at greater risk – and Maffi clearly has that biological analogy in mind. Other chapters in the volume draw on similar analogies between biological and cultural notions of diversity and endangerment. For example, one author characterizes cultural diversity as "the variation in culturally heritable information and its distribution across cultural lineages," making an explicit comparison between cultural and genetic diversity, and even going so far as to discuss phylogenetic branching, "drift," and isolation as factors in linguistic diversification (Smith 2001, 96–97). Another notes the similar role played by "colonizing cultures" in the reduction of both biological and cultural diversity: just as colonizing societies reduce biological diversity by replacing native flora and fauna with fewer, high-yield imported species of plants and animals, so globalization reduces linguistic diversity by imposing languages on colonized peoples. Thus, he argues, "the destruction of biodiversity and linguistic diversity have the same cause" (Wollock 2001, 250–251).

This conflation of biological and cultural diversity is nowhere more evident than in a UNESCO booklet published in 2003 titled *Sharing a World of Difference: The Earth's Linguistic, Cultural, and Biological Diversity*. This document defines biocultural diversity as "interlinkages between linguistic, cultural, and biological diversity," and asserts that "the diversity of life on Earth is formed not only by the variety of plant and animal species and ecosystems found in nature (biodiversity), but also by the variety of cultures and languages in human society (cultural and linguistic diversity)" (Skutnabb-Kangas et al. 2003, 9). The booklet begins by presenting some of the now-familiar arguments for preserving biodiversity, such as the "unforeseen consequences" of damage to the "delicate relationships" in the ecosystem, and the endangerment of "potential for adaptation" by reducing genetic diversity, arguing that "diversity is the basic condition of the natural world" (Skutnabb-Kangas et al. 2003, 9–10). "However," the document continues, "diversity is not only a characteristic of the natural world. The idea of 'diversity of life' goes beyond biodiversity. It includes cultural and linguistic diversity found among human societies" (Skutnabb-Kangas et al. 2003, 18). This cultural diversity can be thought of "as the totality of the 'cultural and linguistic richness' present within the human species," a quantity analogous to species and genetic richness in biology, and the world's 6–7,000 languages are "the total 'pool of ideas'" represented in human culture, all of which are threatened by a "linguistic and cultural extinction crisis" (Skutnabb-Kangas et al. 2003, 19, 28–29). But the conclusion the booklet reaches goes beyond mere analogical relationship: "biological diversity and linguistic diversity are not separate aspects of the diversity of life, but rather intimately related, and indeed, mutually supporting ones," and "the extinction crises that are affecting these manifestations of the diversity of life may be converging also" (Skutnabb-Kangas et al. 2003, 35). The central message is that, like biological diversity, cultural diversity is a resource for ensuring a healthy cultural "ecosystem" that, if lost, will be lost forever.[7]

Conclusion

The concepts of extinction, endangerment, and diversity are thus linked in a shared history extending back more than 200 years. What this chapter has argued is that transformations in biologists' conceptions of the nature of extinction have influenced the way Western societies have understood biological and cultural diversity, and the urgency those societies have felt regarding the protection of that diversity. There are many ways in which analysis of this history could be broadened to consider a deeper array of scientific and cultural manifestations of these connections, but even from a rather cursory overview, a basic pattern is clear: during the 19th century, at a time when naturalists understood nature to be an essentially endlessly renewable resource, diversity was taken for granted, and extinction was not perceived as a threat to the economy of nature, and diversity *per se* did not have normative value. Rather, extinction was understood to be nature's way of strengthening and improving itself by weeding out the unfit, and competition was celebrated as the source of natural progress. This view, of course, supported Victorian ideologies of social progress and imperial expansion, and justified a lack of concern about the inevitable victims of progress (combined with, at most, romantic nostalgia for cultures that passed away). When competition is natural, extinction is inevitable, and not to be resisted.

We now live in a society where cultural and biological diversity are considered to be precious resources, and where threats to those resources are perceived from all directions. We fundamentally value diversity – as an inherent normative good – in a way that previous Western societies did not. This is partially due, I have argued, to a new extinction discourse in which biologists 1) understood extinction as a potentially catastrophic and irreversible process, 2) characterized extinction explicitly in terms of its effect on diversity, and 3) no longer conceptualized survival as a "fair game," with extinction penalizing only those individuals and species that "deserve" it. These ideas developed first in the context of ecology and paleontology, but have ramified outwards to perceptions of cultural and linguistic diversity, and have become central to cultural valuations of diversity itself. There is obviously an important sense in which scientists have themselves been influenced by changing cultural norms; it is quite likely that the new understanding of extinction was made more acceptable by a cultural and political context in which nuclear proliferation and environmental catastrophe were looming specters (Hamblin 2013). After all, Cuvier's view of extinction, which was not terribly different from the current paleontological view, was considered anathema by the scientific establishment of Victorian Britain for reasons that extended well beyond the empirical. Westerners now think of the natural world – and our own place in it – as inherently tenuous and insecure, in part because we now appreciate that the history of life offers few guarantees or reassur-ances. The things we value most must actively be protected, and the very activi-ties that have given our species domination over the Earth can accelerate our own demise. After all, the dinosaurs had 135 million years of ascendancy before their catastrophe wiped the slate clean; we humans have had barely 10,000. We may not actually destroy the Earth, but, as Gould reminds us, "on geological scales, our planet will take good care of itself and let time clear the impact of any human malfeasance" (Gould 1990).

Notes

1 This 1762 publication is a translation of *Specimen academicum de oeconomia naturae*, which was defended as a thesis by Linnaeus' student Isaac J. Biberg in 1749. This was one of some 186 essays, largely written by Linnaeus, defended by his students for their doctoral degrees (Egerton 2007).
2 This is not to say, however, that no European authors expressed concern or regret about vanishing cultures. Nonetheless, even those active in "aboriginal protection" societ-ies (such as James Cowles Pritchard, discussed below) nearly always acknowledged the inevitability of cultural extinction as a result of European contact (Stocking 1987, 48–50, 273).
3 On 19th century anthropologists' response to the impending crisis of the extinction of human races, see also Gänger this volume.
4 Darwin did not use the term in a way that reflected even the basic definition of diversity as species richness. The word does appear as a noun (as opposed to adjec-tival forms like "diverse" or "diversified") some 18 times in the text of the first edi-tion of the *Origin*, but in every case, it is used merely as a synonym for "variety." For Darwin, diversity was essentially a comparative term, meant to indicate an amount or degree of difference between the features of organisms; it does not convey a sense of ecological interdependence, nor is it usually presented as a broader phenomenon that is threatened or in need of preservation. In fact, in the two instances in the *Origin* where Darwin invoked "diversity" in a somewhat broader sense, it was presented as an example of how "beautiful" or "harmonious" the balance of nature is (Darwin 1859, 74, 169).
5 There are a variety of additional factors that could explain this pattern: for example, the increase in the number of biological journals during this period, or the increase in the number of papers published each year by biologists. My figures are absolute numbers, not percentages for all biological papers published, which would likely tell a somewhat different story. However, the mere fact that an increasing number of biologists were writing papers about "diversity" over this period is significant in itself.
6 Public interest in pre-historic mass extinctions was spurred by a number of books and magazine and newspaper articles, beginning in the mid-1980s (Sepkoski 2012, Ch. 9).
7 The views of Maffi and other members of the Terralingua foundation have been criti-cized by some authors on the grounds that they rely on a problematic equation between species and languages. See Vidal and Dias and Muehlmann this volume.

Bibliography

Badash, L., 2009. *A Nuclear Winter's Tale: Science and Politics in the 1980s, Transformations: Studies in the History of Science and Technology*. MIT Press, Cambridge, 317–388.
Barrow, M. V., 2009. *Nature's Ghosts: Confronting Extinction from the Age of Jefferson to the Age of Ecology*. University of Chicago Press, Chicago.

Benton, M.J., 2003. *When Life Nearly Died: The Greatest Mass Extinction of All Time*. Thames and Hudson, London/New York.

Bowler, P.J., 1986. *Theories of Human Evolution: A Century of Debate, 1844–1944*. Blackwell, Oxford.

Brantlinger, P., 2003. *Dark Vanishings: Discourse on the Extinction of Primitive Races, 1800–1930*. Cornell University Press, Ithaca, NY.

Cuddington, K. and Ruse, M., 2004. Biodiversity, Darwin, and the Fossil Record. In: Oksanen, M. and Pietarinen, J., eds., *Philosophy and Biodiversity*. Cambridge University Press, Cambridge, 101–118.

Darwin, C., 1859. *On the Origin of Species*. J. Murray, London.

Darwin, C., 1871. *The Descent of Man, and Selection in Relation to Sex*. John Murray Albemarle Street, London.

Darwin, C., 1909. *The Voyage of the Beagle*. Harvard Classics. P. F. Collier & Son, New York.

Desmond, A., 1989. *The Politics of Evolution: Morphology, Medicine, and Reform in Radical London*. University of Chicago Press, Chicago.

Desmond, A. and Moore, J., 2009. *Darwin's Sacred Cause: How a Hatred of Slavery Shaped Darwin's Views on Human Evolution*. Houghton Mifflin Harcourt, Boston.

Dilke, C. W., 1869. *Greater Britain: a Record of English-Speaking Countries during 1866 and 1867*, 4th ed. Macmillan and Co, London.

Dritschilo, W., 2008. Bringing Statistical Methods to Community and Evolutionary Ecology: Daniel S. Simberloff. In: Harman, O. and Dietrich, M., eds., *Rebels, Mavericks, and Heretics in Biology*. Yale University Press, New Haven, 356–371.

Egerton, F. N., 2007. Linnaeus and the Economy of Nature. *Bulletin of the Ecological Society of America*, January, 72–88.

Ehrlich, P. R., Harte, J., Harwell, M. A., Raven, P. H., Sagan, C., Woodwell, G. M., . . . Teal, J. M., 1983. Long-Term Biological Consequences of Nuclear War. *Science*, 222 (4630), 1293–1300.

Farnham, T.J., 2007. *Saving Nature's Legacy: Origins of the Idea of Biological Diversity*. Yale University Press, New Haven.

Goodman, E., 1984. Musings of a Dinosaur Groupie. *The Washington Post*, 3 January, A17.

Gould, S.J., 1989. *Wonderful Life: The Burgess Shale and the Nature of History*, 1st ed. W. W. Norton, New York.

Gould, S.J., 1990. The Golden Rule: A Proper Scale for Our Environmental Crisis. *Natural History*, 99 (9), 24–28.

Gould, S.J., 1991. Introduction. In: Raup, D. M., ed., *Extinction: Bad Genes or Bad Luck*. W.W. Norton, New York, xiii–xvii.

Hamblin, J.D., 2013. *Arming Mother Nature: The Birth of Catastrophic Environmentalism*. Oxford University Press, New York.

Heyd, D., 2010. Cultural Diversity and Biodiversity: A Tempting Analogy. *Critical Review of International Social and Political Philosophy*, 13 (1), 159–179.

Hofstadter, R., 1955. *Social Darwinism in American Thought*, Rev. ed. Beacon Press, Boston.

Jablonski, D., 1991. Extinctions: A Paleontological Perspective. *Science*, 253 (5021), 754–757.

Jablonski, D., 2004. The Evolutionary Role of Mass Extinctions: Disaster, Recovery and Something Inbetween. In: Taylor, P. D., ed., *Extinctions in the History of Life*. Cambridge University Press, New York, 151–177.

Jefferson, T., 1801. *Notes on the State of Virginia*, 8th American ed. Boston.

Kingsland, S. E., 1985. *Modeling Nature: Episodes in the History of Population Ecology, Science and its Conceptual Foundations*. University of Chicago Press, Chicago.

Leakey, R. and Lewin, R., 1995. *The Sixth Extinction: Patterns of Life and the Future of Humankind*. Doubleday, New York.

Linnaeus, C., 1762. The Oeconomy of Nature. *Miscellaneous Tracts Relating to Husbandry and Physick*. J. Dodsley, London, 39–129.

Lyell, C., 1830–33. *Principles of Geology: Being an Attempt to Explain the Former Changes of the Earth's Surface, by Reference to Causes Now in Operation*. J. Murray, London.

Maffi, L., 2001a. Introduction: On the Interdependence of Biological and Cultural Diversity. In: Maffi, L., ed., *On Biocultural Diversity*. Smithsonian Institution Press, Washington, DC, 1–52.

Maffi, L., ed. 2001b. *On Biocultural Diversity*. Smithsonian Institution Press, Washington, DC.

Myers, N., 1979. *The Sinking Ark: A New Look at the Problem of Disappearing Species*. Pergamon Press, New York.

Oksanen, M., 2004. Biodiversity Considered Philosophically: An Introduction. In: Pietarinen, J. and Oksanen, M., eds., *Philosophy and Biodiversity*. Cambridge University Press, Cambridge, 1–26.

Parkinson, J., 1804. *An Examination of the Mineralized Remains of the Vegetables and Animals of the Antediluvian World*. C. Whittingham, London.

Pietarinen, J., 2004. Plato on Diversity and Stability in Nature. In: Oksanen, M. and Pietarinen, J., eds., *Philosophy and Biodiversity*. Cambridge University Press, Cambridge, 85–100.

Pritchard, J.C., 1840. On the Extinction of Human Races. *The Edinburgh New Philosophical Journal*, 28, 166–170.

Raup, D.M., 1986. *The Nemesis Affair: A Story of the Death of Dinosaurs and the Ways of Science*. W.W. Norton & Co, New York.

Raup, D.M., 1988. Diversity Crises in the Geological Past. In: Wilson, E.O., ed., *Biodiversity*. National Academy Press, Washington, DC, 51–57.

Raup, D.M., 1991. *Extinction: Bad Genes or Bad Luck?* W.W. Norton & Co, New York.

Raup, D.M. and Sepkoski, J.J. Jr., 1984. Periodicity of extinctions in the geologic past. *Proceedings of the National Academy of Sciences of the United States of America*, 81 (3), 801–805.

Rudwick, M.J.S., 1997. *Georges Cuvier, Fossil Bones, and Geological Catastrophes: New Translations & Interpretations of the Primary Texts*. University of Chicago Press, Chicago.

Rudwick, M.J.S., 2005. *Lyell and Darwin, Geologists: Studies in the Earth Sciences in the Age of Reform*. Burlington VT, Aldershot, Hampshire, Great Britain.

Ruse, M., 1979. *The Darwinian Revolution: Science Red in Tooth and Claw*. The University of Chicago Press, Chicago.

Ruse, M., 1999. *Mystery of Mysteries: Is Evolution a Social Construction?* Harvard University Press, Cambridge.

Sepkoski, D., 2012. *Rereading the Fossil Record: The Growth of Paleobiology as an Evolutionary Discipline*. University of Chicago Press, Chicago.

Sepkoski, J. J. Jr., 1984. A Kinetic Model of Phanerozoic Taxonomic Diversity. III. Post-Paleozoic Families and Mass Extinctions. *Paleobiology*, 10 (2), 246–267.

Sepkoski, J.J. Jr., 1997. Biodiversity; Past, Present, and Future. *Journal of Paleontology*, 71 (4), 533–539.

Sepkoski, J.J. Jr., Bambach, R.K., Raup, D.M. and Valentine, J.W., 1981. Phanerozoic Marine Diversity and the Fossil Record. *Nature (London)*, 293 (5832), 435–437.

Skutnabb-Kangas, T., Maffi, L. and Harmon, D., 2003. *Sharing a World of Difference: The Earth's Linguistic, Cultural, and Biological Diversity*. UNESCO – Terralingua – World Wide Fund for Nature, Paris.

Smith, C.H., 1851. *The Natural History of the Human Species: Its Typical Forms, Primeval Distribution, Filiations, and Migrations*. Gould and Lincoln, Boston.

Smith, E.A., 2001. On the Coevolution of Cultural, Linguistic, and Biological Diversity. In: Maffi, L. ed. *On Biocultural Diversity*. Smithsonian Institution Press, Washington, DC, 95–117.

Spencer, H., 1972. *Herbert Spencer on Social Evolution: Selected Writings. Edited and with an Introduction by J.D.Y. Peel*. University of Chicago Press, Chicago, London.

Stocking, G.W. Jr., 1987. *Victorian Anthropology*. The Free Press, New York.

Takacs, D., 1996. *The Idea of Biodiversity: Philosophies of Paradise*. John Hopkins University Press, Baltimore.

Trevelyan, C.E., 1848. *The Irish Crisis*. Longman Brown Green & Longmans, London.

UN, 1992. *Convention on Biological Diversity*.

UNESCO, 2002. *UNESCO Universal Declaration on Cultural Diversity*, Paris.

Wallace, A.R., 1864. The Origin of Human Races and the Antiquity of Man Deduced from the Theory of Natural Selection. *Journal of the Anthropological Society of London*, 2, clviii–clxxxvii.

Ward, P.D., 2000. *Rivers in Time: The Search for Clues to Earth's Mass Extinctions*. Columbia University Press, New York.

Wilson, E.O., ed., 1988a. *Biodiversity*. National Academy Press, Washington, DC.

Wilson, E.O., 1988b. The Current State of Biological Diversity. In: Wilson, E.O., ed., *Biodiversity*. National Academy Press, Washington, DC, 3–18.

Wilson, E.O., 1992. *The Diversity of Life*. Belknap Press, Cambridge, MA.

Witchel, S., 1979. Give Me That Old-Time Darwin. *New York Times*, 3 May, A23.

Wollock, J., 2001. Linguistic Diversity and Biodiversity: Some Implications for the Language Sciences. In: Maffi, L., ed., *On Biocultural Diversity*. Smithsonian Institution Press, Washington, DC, 248–264.

Wood, J.G., 1870. *Uncivilized Races of Men in All Countries of the World: Being a Comprehensive Account of their Manners and Customs, and of their Physical, Social, Mental, Moral and Religious Characteristics*. J.B. Burr & Co, Hartford, CT.

3   Anthropological data in danger, c. 1941–1965

Rebecca Lemov

1

In 1966, the same year he published *La Pensée Sauvage*, the French celebrity-anthropologist Claude Lévi-Strauss addressed a bicentennial celebration commemorating the birth of James Smithson, the founder-donor of what would become the world's largest museum complex, the Smithsonian. The anthropologist took the opportunity to perform a sweeping assessment, and his talk subsequently appeared in the form of a brief shot-across-the-bow missive in the pages of *Current Anthropology*. No less formidable a title than "Anthropology: Its Achievements and Future" bespoke the scale of the challenge Lévi-Strauss took on. Manifesto, missive, or broadside: whatever its function may have been, its tone was curious, almost Borgesian. Setting the scene, he recalled his early days in New York in 1941, when, poking around in a used book store somewhere – he did not specify where or which, only that it was a bookstore that "specialized in secondhand government publications" – he stumbled across a selection of most of the original forty-eight *Annual Reports* of the Bureau of American Ethnology, as published by the Smithsonian Institution beginning in 1881.

The cache included Mallery's *Pictographs*, Matthews' *Mountain Chant*, Fewkes' *Hopi Katcinas*, and such treasure-trove collections as Stevenson's *Zuni Indians*, Boas' *Tsimshian Mythology*, Roth's *Guiana Indians*, and Curtin and Hewitt's *Seneca Legends*. It also included less renowned documents on ancient shell art of the Americas, animal carvings from mounds of the Mississippi Valley, and "Cessions of land by Indian tribes to the United States." The price was good: $2 or $3 apiece. Yet the anthropologist was short of funds, and "$3 represented all I had to spend on food for the same number of days." In fact, though he did not mention it in his commemorative talk, he was a just-arrived refugee from Vichy France, fleeing the anti-Semitic laws that had forbidden him to teach and would soon have deported him along with other French Jews unlucky enough to have no escape. Landing in New York by the good graces of the New School for Social Research, as part of an effort to rescue European scholars and artists under threat from totalitarian regimes, the scholar completed a peripatetic journey from the Maginot Line to Marseilles to Martinique that finally ended with a narrow escape via a Swedish banana boat through San Juan, Puerto Rico (Wilcken 2011, 120). This partly explains his lack of funds that day in the bookstore – and perhaps too his impressionable state of mind, as he recalled it.

His feelings verged on wonder at encountering the ethnographies: "I can hardly describe my emotion at this find. That these sacrosanct volumes, in their original green and gold bindings, representing most of what will remain known about the American Indian, could actually be bought and privately owned was something I had never dreamed of" (Lévi-Strauss 1966, 124; cf. same quotation in Wilcken, who adds the textual detail about the original bindings, 121 fn 39). Calculating into the sticker price the remarkable allure of these volumes and the psychological state into which their chance encounter had precipitated him, he resolved to purchase at least one of these "marvelous publications" that very day, despite the fact that food money was in short supply. Over time, he scraped together enough to acquire forty-seven of the forty-eight-book series. Along with the several hundred *Bulletins* and *Miscellaneous Publications* also from the Bureau, they constituted a repository beyond monetary value: "invaluable materials" (124).

In this tale as told, strong passions struggle for balance. The anthropologist expressed a shadowed form of wonder in which the emotional high of acquisition vied with an emotional low of impending loss. The cultures he sought to study were disappearing faster than the half-life of radioactive substances, and within a century, or sooner, Lévi-Strauss wrote, these endangered entities would live only in memory. By then, "the last native culture will have disappeared from the Earth and our only interlocutor will be the electronic computer." Soon would come the day when all true native cultures would be gone and their ragged denizens would remain to inhabit the bedraggled tropics, out of date. As for anthropologists, they would long since have been absorbed in communicating and calculating with computers. Technological mediation would be unavoidable. You would no longer be able to *go and see* for yourself, for example, to the Mato Grosso as described in *Tristes Tropiques*. Seeing for yourself – or *autopsia*, an early modern term denoting "firsthand experience" – was a value at the heart of scientific observation and empiricism itself for half a millennium (on the history of scientific observation, see Daston and Lunbeck 2011, 3; on autopsia, see Pomata 2011, 65–66). Yet now it was, at least in ethnography it seemed, a losing proposition. Not for much longer would it be a viable way to proceed. Only a hundred years were left, Lévi-Strauss predicted, for direct study before the computer took over central functions of human rationality, for surely that is what he meant by postulating it as "our only interlocutor."

Already the huge fieldwork projects of Big Social Science were amassing cross-disciplinary teams of techno-modern field scientists, shaping them out of the dross of Malinowski-style "from the door of my tent" observers with their diaries and complexes. The Six Cultures study, for example, began in 1954 as the fruit of a three-way institutional collaboration among Harvard, Yale, and Cornell Universities, funded by the Ford Foundation's Behavioral Science division. It functioned to create massive stockpiles of data in the face of the "inadequacy of available ethnographic materials" about subjects such as childhood and infancy around the world (LeVine 2010, quoting John Whiting 1994). However, there was another route to compensation: instead of remaining a "science on the wane," anthropology could recover if it revived its own neglected data. The sacred green-and-gold volumes would then represent, Lévi-Strauss crowed, "a living inspiration for the scientific task ahead of us."

Here, a vital historical question relevant to the question of endangerment arises: what was rapidly disappearing and where precisely did the emergency lie? Was it the groups of people or the documentation of those groups of people that were soon to be past? As seen here, Lévi-Strauss *focused on documentation, on data*. Heavy-hearted, he prospectively mourned the imperiled Bureau of Ethnology collection; what he felt about the Hopi, Zuni, or Tsimshian themselves was less clear. He did not seem as directly concerned with the wane of people's traditional life ways as much as the fragility of the documents describing and preserving them. Perhaps this was related to the fact that his escape to New York took place burdened by a now legendary trunk full of notes, file cards, travel diaries, maps, and sketches from his own eight weeks of fieldwork (Lévi-Strauss 1973). He brought these things with him and took pains to protect them, even as colonial officials threatened him en route and the FBI demanded to inspect them on his arrival. In this way, he carved out a positive project in the face of a personal and professional crisis.

Rather than reflect on the fact that Lévi-Strauss' one hundred year countdown to ethnography's end is now more than half over, I would like to explore a peculiar "structure of feeling" embedded in the operations of his future-oriented, past-preserving prognostication. It points to what can be called "second-order endangerment." This is a form of the endangerment sensibility with a difference. It is a set of ideas and practices that, instead of applying to first-order phenomena, such as pelicans or roseate spoonbills, or rare languages or "primitive" cultures, address themselves to the second-order phenomena of their respective data collections. Such collections may be found in shipping trunks full of notes, Bureau of Ethnology volumes, pre-digital data banks of dreams, or an electronic database storing field observations. These containers and the data they held, it was urgently felt, were endangered. But first, it is important to clarify what is meant by a structure of feeling. Drawing from Raymond Williams' use of the term, in speaking about a structure of feeling that I will associate with second-order endangerment (specifically, the endangerment of data), I mean a structured and yet generative emotion that is shared across a group of actors at a particular time. It may be inchoate, but it is also recognizable: "a particular sense of life, a particular community of experience hardly needing expression," Williams wrote of structures of feeling. Such things are highly particular but elusive of expression (Payne and Barbera 2010, 670; Williams 2001, 64). During the time Williams developed the term from its first use in 1954 to its fullest exposition in the 1970s, he was careful to avoid the connotation of a "spirit of an age." Likewise, when speaking here of second-order endangerment – and, it should be said, also first-order endangerment – I do not mean to imply that this was an idealist notion or that it hovered over social facts and material realities, nor was it one that all anthropologists universally shared. To the contrary: it was, as we will see, a particular way of arranging objects and ordering systems. It was a conduit.

2

Initially, it may sound as if second-order endangerment simply echoes classic "salvage ethnography" as it was first developed in the early nineteenth century and expressed, for example, by John Wesley Powell, founder of the very Bureau of

American Ethnology Lévi-Strauss celebrated: "The field of research is speedily narrowing because of the rapid change in the Indian population now in progress; all habits, customs, and opinions are fading away; even languages are disappearing: and in a very few years it will be impossible to study our North American Indians in their primitive condition, except from recorded history. For this reason, ethnologic studies in America should be pushed with utmost vigor" (in Gruber 1970, 1295). Powell's timeline for fading away is different but the fact that there is a timeline, and a date of expiration on their objects of study, at first glance draws the two closely together. A melancholy sensibility pervades the statement, even as there is no direct call to action to prevent the disappearance of the Indian population from occurring. The phrase "fading away" emphasizes the fact that Powell and others saw it as a nearly natural and certainly unavoidable process: peoples and their ways were fading away like the colors of a cloth left out in the sun for a long time. (A variant of salvage can be observed in the overweening concern for museums that housed the imperiled materials of Victorian anthropology [Stocking 1991].)

Yet there was something significantly different in the Smithsonian 1966 alarum, for Lévi-Strauss there spoke as if he had boarded a spaceship approaching a distant planet: if he marked the passing away of native cultures rather unsentimentally, he mourned and celebrated all the more vividly *their documentation* in the form of the ghostly volumes, the Bureau of Ethnology itself with its standards of value and scholarship, the great projects of the end of the American nineteenth century and early twentieth. In short, it was the research data itself that entered the picture of *what exactly was endangered* in the middle of the twentieth century: not only are cultures disappearing unavoidably, but their records are too. The *emotion* is reserved for the latter. Thus, the endangerment sensibility (Vidal and Días this volume) that accompanies second-order endangerment is one of a diffuse vulnerability. The collection – for example, the Smithsonian volumes – while not actually at risk of disappearing from all libraries, was nonetheless vulnerable to degradation and the accidents of circulation (for example, an irresponsible person might buy the copies for the price of a pocket of small change). They were "sacred" and the data they held bore some of the qualities of a fetish. This is second-order endangerment. It is no longer the things and trappings that must be preserved against ruin, but the representations of the things and trappings – their data and the repositories in which they rest.

Still, there is a sense in which first-order endangerment also concerns data, and this should be mentioned nonetheless. Jacob Gruber, in a classic statement on the roots and dynamics of salvage anthropology, describes how, as a response to the threat of cultural devastation, "[d]uring the nineteenth century people began to sense the urgency of collection for the sake of preserving data whose extinction was feared" (Gruber 1970, 1290). Even at this moment of the emergence of the anthropological discipline, data could serve as a form of compensation. Gruber further stressed the effect of salvage sensibility on data itself, examining in particular how the shaping and storage of salvage data affected (and limited) the theoretical conclusions drawn therefrom. "The very operation of the collection itself infused the data with a sense of separateness, a notion of item discontinuity that encouraged the use of an acontextual comparative method and led only to the most limited (because they alone were observed) ideas of functional correlation" (1297). Such views bolstered the comparative method (see e.g., Strathern 2011). This limiting effect does not operate in regard to second-order endangerment, however. Instead, the gathering of data and the emergence of a self-conscious endangerment sensibility around data itself were explicitly designed to stimulate theoretical speculation – in the present or at an unnamed but anticipated future date – rather than narrow or rigidify it.

By the mid-twentieth century, things that were once simply endangered in terms of their existence were now endangered at another level, for their traces were endangered too, and more pressingly. Ethnological documents began to share, at this moment, among certain ethnographers and psychological experts, a twin fate with that which they collected (ways of life, ethnographic portraits, accounts of cultures). Both were somehow endangered – but endangered in different ways. Concern shifted from one to the other, loosely speaking from the threat of cultural apocalypse to the threat of informational apocalypse. One target of endangerment replaced another. If first-order endangerment refers to the imperilment of whole ways of life, languages, and natural species or sub-species, then second-order is concerned with the modes of memorializing and remembering these things. The accompanying elegiac emotion is at times even more piercing for the "ideas about the thing" rather than the "thing itself." It is sometimes more fragile too. The apparatuses and modes of keeping, it is discovered, are themselves vulnerable to the same forces that have doomed the object or person under study. Data stores meant to keep knowledge safe are themselves fragile things.

Among certain key subsets of American practitioners, many of them trained in the "culture and personality" school, a particularly urgent consciousness of "endangered data" flourished. This resulted in a flowering of preservationist projects. Before describing an example of the kind of project such scholars engaged in, we must briefly describe the culture and personality school, and ask what it was. The once-fashionable appellation "culture and personality school" may not ring bells of recognition among a general scholarly audience today, but during key interwar and postwar decades, the school formed the leading edge of innovation among two generations of American anthropologists, especially those who wished to collaborate across the behavioral sciences (Cohen-Cole 2014, 102). It was the local form functionalism took in the American context. The first generation of the culture-and-personality school, boasting such Boas-trained personnel as Margaret Mead, Gregory Bateson, Ruth Benedict, Otto Klineberg, Edward Sapir, and Ruth Landes, is better known individually than as a school. Yet they shared more than training. "In their influence and their public standing, the culture-and-personality scholars stood at the center of an American scholarly elite, but they were hardly representative of it [as their] networks included a surprising number of immigrants, Jews, women, and people known (to their friends then and to scholars now) as gay or lesbian," as historian Joanne Myerowitz characterizes the conflicted insider-outsider status they embodied and the left-leaning stance they tended to adopt (Myerowitz 2010). As a group, they pursued the question of what the psychoanalyst Abram Kardiner called the "basic personality structure," asking whether there was a personality type that accompanied each cultural formation (Stocking 1986; see also Manson 1988).

A discussion extracted from Mead and Bateson's groundbreaking visual-textual study *Balinese Character* illuminates this kind of finding, in which cultures operate systematically to shape the individual psyche and structure responses to environment:

> [I]n Bali the same attitude of mind, the same system of posture and gesture, seems able to operate with great contrasts in content with virtually no alteration in form. So also for climatic contrasts, and contrasts in wealth and poverty: the mountain people are dirtier, slower, and more suspicious than the plains people; the poor are more frightened than the rich, but the differences are in degree only; the same types of dirtiness, of suspicion, and anxiety are common at all levels.
>
> (Bateson and Mead 1942)

A pervasive form of Balinese character ran through all variety of Balinese people (with the exception of the North Bali, the Kesatryas, and the Vesias), and this structural type could be charted and studied. This was not only useful for understanding the Balinese, but also to pursue a vast team project in which most culture-and-personality workers were engaged, whether explicitly or implicitly: the painstaking empirical understanding of many such cultures around the world would contribute to the building of a new entity, "a culture richer and more rewarding than any that the world has ever seen" (Bateson and Mead 1942, xvi). The sum of cultural knowledge would be a guide for a future culture. Such a culture would also shape – the word "canalize" was sometimes used – the individuals who constituted it. Along the way, by drawing out such complex portraits of the interactions of culture and personality, they would build a "disciplined science of human relations" and so help to "plan an integrated world" (ibid.). Note that a sweeping "we-can-rebuild-culture" idealism often coupled with hard-headed empiricism during this initial period.

After World War II, a second-generation, tool-conscious, and forward-looking iteration of the culture-and-personality enterprise emerged (causing it to assume at times the moniker "movement" rather than the tamer-sounding "school"). This later generation, sometimes with the help of still-prominent prewar members such as Mead and Bateson, became significant for its focus on methodology of all kinds, which has been characterized elsewhere as a vogue for meta-methodological thinking (Erickson et al. 2013; Lemov 2011). In tune with this shared methodological turn, the second generation, including researchers such as Melford Spiro, George and Louise Spindler, Walter Goldschmidt, George deVos, Erica Bourgignon, Cora DuBois, Bert Kaplan, William Henry, and Jules Henry, pursued intensive data-gathering, sometimes promoting it as a worthy end in itself. What earlier researchers investigated was now taken for granted: "That human personality structure is a product of experience in a socialization process and that the resulting structure varies with the nature and conditions of such experience can scarcely be doubted" (Hallowell 1953, 608). Cultures might disappear but the mechanics of their propagation could be understood through firm research techniques and systematic data collecting. According to an emerging sensibility of second-order endangerment, cultures and people were not cared for so much as the *records* of what was felt to be disappearing, which inspired curatorial passion and, in due time, an endangerment-style concern.

As will become clear, I am not arguing that this urgent consciousness of "endangered data" was found in the work of all social scientists dealing with culture during the period loosely from 1941 to 1965, but that, among some of the most innovative and self-consciously interdisciplinary anthropologists pursuing the project of a unified social science, this displacement of the locus of the endangerment sensibility occurred. After all, it was easier or perhaps more satisfying to be an activist for data-preservation than to attempt to combat the forces of colonialism, decolonization, and even cultural change itself. This was particularly so when, at the heart of the scholastic project, lay an immense social engineering endeavor linked to a universalist project (Lévi-Strauss 1966; Meyerowitz 2010, 1084 but cf. Mandler 2013). In this way, even though it was far from everywhere shared, the second-order endangerment sensibility nonetheless played a notable role among a subset of postwar anthropologists and psychologists in the culture and personality school who styled themselves most fervently future-oriented.

Still, the two orders are related, and indeed often inseparable. In some cases, the difference is found in a tiny shift of emphasis. In other cases, it is more deliberate and obvious. One way of looking at it is as an index of the success of the science of anthropology. Sanguine about their professional role, practitioners urged themselves onward with visions of vulnerability not of the Trobrianders but of the Trobriand data. What is going is nearly gone (the thinking went), but the future of the profession and humanity itself must be provided for.

3

The twenty-five years that elapsed between Lévi-Strauss' purchase of his first Bureau volume in 1941 and his address to the Smithson commemoration were the pivotal timespan for a project one might call the "Database of Dreams" but that its builders somewhat less fancifully labeled the *Microcard Publications of Primary Records in Culture and Personality*, vols. 1 through 4. For the past decade or so, I have been researching the history of this archive, one of the most promising and yet strangely forgotten undertakings in American social science – a dizzyingly ambitious 1950s-era project to capture people's dreams in large amounts and store them in an experimental data bank. This storehouse would be made easily accessible, according to its planners' vision, by installing two miniaturized, micropublished copies in every research library in the world, where viewers could tap into their reserves of unique "subjective materials" by means of a special machine called the Readex. Although that scale of distribution never quite happened, the data remained available, if difficult to read – for the Readex had become obsolete decades before – at the Library of Congress and several university libraries across the United States (see Lemov forthcoming 2015). It was not only dreams: researchers also filed away the records of waking thoughts and an extensive variety of "subjective materials." Beginning with the first data bank volume, released in 1956, and pausing if not ending with the last installment, issued in 1963, the archive brought together the fruits of a generation of field workers and human-science laborers. In other words, it preserved the painstakingly gathered data of others – namely, the key figures of second-generation culture-and-personality studies, of which its roster of compiled data sets reads like a Who's Who. It was a double experiment (Kaplan 1958) in format and content.

Another way to look at this memory machine is as a monument to second-order endangerment. As Kaplan observed, social science "has been very wasteful of its empirical materials. Instead of squeezing the last bit of meaning and significance from our data, we are usually content to skim the surface for what is most easily available and then discard the remainder in favor of a new set of data which can be more easily mined" (Kaplan 1958, 53). Such data, although painstakingly gathered, found itself exiled in file cabinets, attic storage boxes, and, most regrettably, waste baskets. Kaplan designed his analog data clearinghouse to bring these exiled data sets together and preserve them for future use. His project shows how second-generation culture-and-personality scholars engaged with first-generation concerns, and, in engaging, altered them. Namely, they carried forward some of the more sweeping aims of their forerunners, and in particular the goals of developing social engineering capacities and of producing a unified science of human relations. At the heart of their project was the urge to store and protect the data-rich files and experimental archives that were the patrimony of anthropology.

Kaplan's databank of un-interpreted "raw" data arose out of these ambitions. Made up of sixty-some data sets collected painstakingly over previous decades, its constituent data extended back to the 1930s and included dozens of life histories and "expressive autobiographies," hundreds of dreams and ritual hallucinations, and many thousands of projective tests results from instruments such as the Rorschach, the Thematic Apperception Test (TAT), and the Draw-A-Person test. Psychologists and anthropologists touted projective tests as special apparatuses that acted as "X-rays of the self," offering direct access to the inner life. The data they produced, in the form of protocols registered on standardized forms, lists of responses, and quantified scoring charts, could be amassed, stored, and compared. For this reason, the Rorschach and TAT records of several dozen accused Nazis held at Nuremberg were considered a vital repository – a sort of gift to future researchers who would be able, it was hoped, to understand from this treasure trove of data the nature of political and personal pathology. Beginning during World War II and extending through the early 1960s, projective tests experienced a frenzy of development in the behavioral sciences, and this coincided with culture-and-personality scholars' second wave. It was not surprising, therefore, that its adherents adopted projective techniques with the eagerness up-to-the-minute fashions, whether intellectual or sartorial, often engender (Henry et al. 1955; overview in Spindler 1978, 7–38). Such tests could be combined and recombined. When administered in sets, they formed what was called a battery: that is, a relatively standardized group of three to eight tests that psychologically oriented scholars promoted as aids to fieldwork. Rigged to be portable, housed in wood boxes with handles and other special carrying cases, equipped with instructions for cross-cultural use, the tests traveled around the globe. With researchers trekking to faraway spots bearing batteries of tests and bent on administering them in unconventional situations from beachheads to straw huts, data stacked up.

All this scholars considered a new kind of data: particularly hard to get, especially ephemeral, and distinctly in need of preservation. Some of them took steps to do so. The prime movers of the culture-and-personality archive were psychologist Kaplan (described above) and a committee of ethnographers, child sociology experts, and micropublishing executives – including, most prominently, A. Irving Hallowell, John W. M. Whiting, Melford Spiro, Roger Barker, and Webb Thompson, Jr., in a group, the Committee on Primary Materials, sponsored primarily by the National Science Foundation and the National Research Council. They referred to their quarry as "subjective materials," in contrast to the kinds of cultural materials and less transient things anthropologists had traditionally collected. Their clearinghouse for all such data was designed to preserve it against the exigencies of academic retirement and the indignities of poor archival curation. The resultant mass grew to approximately 20,000 pages of now-forgotten data. The sum total rests as if in suspended animation in Washington, DC, in the Microform Reading Room of the Library of Congress where, beginning in 2007, I began spending hours trolling through hard-to-access documents – hard to access, in part because archivists often had trouble locating them in the basement, and also because, apparently, no one had asked to view them in recent memory. "Not much demand for these records!" a librarian there confirmed. By the end of the decade, microcard machines were commonly discarded and could occasionally be found on the trash heaps of Harvard University and other institutions.

Despite their latter-day fungibility, these data-storage cards and their reading machines were the one-time answer to the problem of the ephemerality of documents and the lack of space on library shelves, as well as the dangers wartime and political instability posed to collections of traditional books. Invented in 1945 by the librarian and jack-of-all-trades Fremont Rider, microcards were 3-x-5-inch opaque cards that neatly listed their data set on the top row, and serially below printed twenty to forty miniaturized pages on each card. Such pages could hold any kind of data, quantified or text-based. The low cost of rendering – half a cent per page – was at one time a rallying cry for social scientists whose data would otherwise not be publishable and might lie neglected in boxes after a single use. The microcard formed in the mid to late 1940s the leading edge of the micropublishing movement, which offered seemingly unlimited promise: the "application of the camera to the production of literature ranks next to that of the printing press," announced one visionary (Baker 2001, 73). The microcard, however, despite its early promise, rapidly ended up an outmoded format and would be succeeded within a decade or so by the ungainly but cheaper microfiche format (see Auerbach and Gitelman 2007; Gitelman 2014; Rider 1944; on the longer history of microphotography, see Luther 1950) (see Figure 3.1).

Figure 3.1 The Microcard Reader as illustrated in a brochure from the Microcard Corporation, West Salem, Wisconsin. These reading machines, a caption observes, "occupy little space, can be readily carried from one room to another, are very simple to operate, and provide a clear, sharp image even in a well-lighted room." An optional Fresnel screen provides "even greater screen brilliance" for desktop use while a 110-volt pocket reader "slightly larger than a package of king-size cigarettes" is available for intermittent use.

(Courtesy Bert Kaplan Papers, Santa Cruz, CA).

The resulting microcard archive holds endangered data about endangered cultures. Its contents are miniaturized copies of records collected over several decades in the middle of the twentieth century. These records amount to a refracted account of what philosopher Jonathan Lear calls the massive "cultural devastation" tribal and non-Western groups experienced in the face of modern forces and the inexorable need to adapt (Lear 2006). Just as the microcard represented the devastation of the physical book (a conclusion eagerly sought by some librarians and mourned by others), the contents of this archive testified to the apparent devastation of non-literate cultures. Much of the research conducted on acculturation – the long tail of the process by which tribal groups and dominant cultures accommodated – revealed that the bulk of the accommodation was carried out by the former. This data provided first-hand materials and relatively unmitigated voices testifying to the experiences people had undergone – "raw data," as Kaplan termed them, "in the form of life histories, projective test protocols, dreams, and interviews" (1958, 54).

During this chapter of the second-generation culture-and-personality school, there seemed to be no evident memorial urge or protective calling toward non-literate groups, even in the face of what was seen as an inexorable process of cultural extinction. As mentioned, only when it came to the documents themselves did a protective calling arise. In February 1956, the committee creating the "database of dreams" announced it was considering expanding its purview to a "broader function in collecting . . . data," foreseeing further vigorous growth and financial support under Kaplan as Executive Secretary (Finch 1956). In its meetings in the mid-1950s, the committee debated precisely *which* data were most in need of preservation and how this should be done. Yet everyone agreed the mission was important: to revolutionize the storage of social scientific data sets that were effectively like the California condor or the roseate spoonbill, facing extinction from forces of neglect or active harm (such as floods, fires, and poor storage conditions). Above all, Kaplan wished that the heroic-scale collective effort expended in amassing unique data around the world not be wasted. Toward this end, the committee decided to do two things: focus on preserving the most intimate data available (concerning waning types of subjectivity among non-literate groups) and experiment with the most spectacular storage system. Equally clearinghouse and cheerleader, the board tasked Kaplan, as its executive secretary, to "take the initiative in helping workers in certain areas, i.e. dreams, to organize themselves to get archives formed." With a new cash influx from the National Academy of Sciences in hand, he volunteered to spend the next academic year on a peripatetic mission delving into researchers' attitudes about making their data available. He traveled from coast to coast, interviewing leaders in social scientific data collection, conducting a sort of fieldwork tour in data-gathering itself.

Overall, the committee envisioned several "pilot projects" that would "dramatize" the possibilities of large-scale data-gathering: they did not feel they had to decide among all possible areas but to complete a pilot project of vital concern. This would dramatize the possibilities of their double experiment. After this, they hoped, the creation of social science databases would become standard and widespread procedure among all kinds of data from the records of the primates at the Yerkes Laboratories at Orange Park, Florida, to Kansan schoolchildren at play to Arawak and Pilaga people completing tests or telling stories to female drug addicts in New York City. This urge to preserve at-risk data was akin to Lévi-Strauss' nostalgia for the Smithsonian volumes. It was second-order endangerment, which can be defined as a set of ideas and practices that shifted the locus of concern from the rescue of primary objects, shared languages, and material collections to the preservation of data about these things. Much as the database of dreams was an archive of archives, the underlying assumptions on which it was based postulated the endangerment of endangerment documents. The archive itself became the cause of concern and the unsure fate of its data stimulated calls for preservation.

What sort of displacement took place between endangered people and endangered data? The first thing to note is a striking absence in the committee's discussions: when describing the disappearance (of ways of life), they did not express the once-typical late-nineteenth-century emotional tone of rue, regret, sorrow, or self-reproach, or any of the dominant-culture responses that often gather under the tag of the endangerment sensibility and are provoked by endangered cultures. Much as these researchers saw tribal groups as endangered, perhaps even past endangerment into extinction, much as they may have personally devoted themselves to solidarity with their subjects, it was nonetheless the case that the "disappearing" cultures they chronicled were – in being documented – presented as so anxiety-ridden and often pathological that a rapid assimilation to Western values could only be seen as expedient. By a kind of displacement, it was the data recording this loss, rather than the loss itself, that became valuable and necessary to protect. There was no "noble savage" in the *content* of documents, but to the *form* (that is, to documentation itself), a kind of scientific nobility adhered.

4

Here, I will examine a portion of the stored data held in the *Microcard Publications of Primary Records in Culture and Personality* 1956–1962. It concerns the Menominee American Indians, a group who resided in the woods and small towns of rural Wisconsin. Since the late nineteenth century, the tribe had found its music, cultural artifacts, rituals, and psychological states collected in an "explosive scramble," as one scholar termed it: "The impetus behind this explosive scramble was the commonly held non-Indian belief that Indian cultures were rapidly dying. Collectors rushed to get these items and data before they disappeared entirely or before someone else got them" (Beck 2010, 158). A 1910 official report reflected the dominant attitude toward disappearing material culture: "It was the endeavor to obtain as many as possible of ceremonial objects for the reason that the old ceremonial material is very much in demand, and when once it is picked up, it is gone forever since no more of these old objects be made" (as cited in Beck 2010, 146). By analogy, a later generation of anthropologists felt that personalities too could become "gone forever." By the mid-1950s, teams of scholars were arriving to gather up and document the Menominee "self," as well as that of neighboring tribes, as it underwent changes. Largely, the view was that the traditional self was vanishing as the result of the stressful onrush of the modern American labor and industrial system and the increasingly consumerist way of life that accompanied it. These forces influenced Menominee life more and more.

The husband-and-wife anthropology team George and Louise Spindler did fieldwork among Menominee in the late 1940s and early 1950s documenting an apparent gradual loss among the Menominee tribe of Menominee culture itself in the form of personality change. It was as if the culture were draining away, a loss tracked differentially in the heart and interior life of each Menominee individual. In a social scientific era of tool-centered experimentation and meta-methodological zeal, the Spindlers developed sophisticated innovations to track this loss. Postwar social scientists faced the "struggle to make sense of the dizzying expansion of their tool kits," notes historian Joel Isaac, and this justifies its appellation as a "Tool Age" (Isaac 2010, 135). In this spirit, the Spindlers studied – via the then- and still-popular paradigm called acculturation and via specific tools such as projective tests and methodological innovations to the fieldwork method – the process of loss experienced by a particular group of American Indians. The Spindlers demonstrated via five stages the procedure by which a particular culture seeped away. The seeping away registered in the behavior of five groups of Indians, each progressively less Indian and more assimilated.

A collection of five sets of the Spindlers' Menominee data subsequently appeared within the "database of dreams," and constitutes altogether, at approximately 3,000 pages, one of the most substantial contributions of any archival donor. Menominee culture, it seems on an initial reading, was radically endangered – for they were surely, according to the research and the testimony of individual Menominee themselves, ceasing to exist meaningfully in the terms they had in the past. However, according to the Spindlers, the moment of radical endangerment and its consequences of attempted preservation had already passed by the time the data were gathered. All that was now left was to observe the gradual alteration of ways of life and the gradual pathologization or assimilation of people whose personality structure could (or so ethnographers asserted) no longer respond to the exigencies for which it was formed. Alcoholism was rampant. Anomie reigned. Some, however, were able to forget Menominee ways and attain to middle-class lives.

In order to understand the complex relationship between *endangered cultures* (how they were studied) and *endangered data* (how it was collected and treated), one must look briefly at the acculturation framework. If there was a common thread running through the results of the projective test movement and the dream-taking movement of the twentieth century – much of it later collected in the database of dreams – it was the dynamics of acculturation. In scholarly terms, acculturation meant "sociocultural and psychological adaptation on the part of a contemporary native population to the conditions of life created by the impact of Western civilization" (G. Spindler 1957b, 1). Altogether, if in shreds and patches, the accumulated materials added up to a portrait of people experiencing change-under-stress due to contact with what George Spindler termed "whiteman instrumentalities."

Old-style "salvage" anthropologists from the Golden Age of anthropology (Stocking 1986) had a slightly different project. In the U.S., they followed the lead of Franz Boas, a visionary who in the years between World War I and II trained anthropology's greatest generation, including Alfred Kroeber, Margaret Mead, Edward Sapir, Ruth Benedict, Alexander Goldenweiser, Paul Radin, Zora Neale Hurston, Hortense Powdermaker, and many others. A salvage anthropologist such as the towering figure of Bronislaw Malinowski (Polish-trained, residing in Motu and the Trobriand islands, teaching in England and the U.S.) rued the "melting away" of his discipline's subject matter just at the moment, paradoxically, when anthropology had advanced sufficiently as a science to capture it. As Malinowski framed it at least in 1932, the tragedy often appeared to concern the scientist rather than the people: "Ethnology is in the sadly ludicrous, not to say tragic, position, that at the very moment when it begins to put its workshop in order, to forge its proper tools, to start ready for work on its appointed task, the material of its study *melts away with hopeless rapidity*" (Malinowski 1932, 1). Yet what has not melted away was this particular construction – salvage. As the following section details, it has only spread from professionals to the broader world's popular culture.

5

Salvage anthropology may seem to be irrelevant to second-order endangerment, which dwells on preservation technology rather than on the Malinowskian "materials" of study that regrettably melted away. Yet there is evidence that salvage anthropology did not entirely die. The revived intellectual fascination with the "Last Man" forms a sort of updated salvage anthropology that combines with second-order endangerment sensibilities. At its heart is a wish to be present, with the aid of cutting-edge technologies, to witness a culture's exact moment of dying out – and perhaps to reverse its fate.

An extreme version of this desire appeared in 2010 in Rondonia, Brazil, where, as headlines announced, "The Most Isolated Man in the World" was discovered residing. *"He's alone in the Amazon, but for how long?"* asked the online *Slate* article asserting his existence. Here was a naked man living all alone in a leaf-thatched hut who, "officials concluded," was "the last survivor of an uncontacted tribe" (Reel 2010). For fifteen years, government workers and ethnographers had attempted to engage him in peaceful conversation only to be repulsed by well-shot arrows. Although only the single surviving member of a putative tribe, this nameless man has been, since 2007, the focus of an experimental policy of the Brazilian government: within a 31-square-mile zone mapped around his domicile – forbidding trespassing and logging, so that officials dubbed it the "Policy of No Contact" – he is monitored via satellite and granted, via high-tech intervention, an engineered "solitude," ever watched by solicitous eyes implanted in cloud-based cameras. He is watched through advanced technology, which only reinforces his elusiveness.

This seemingly anomalous Last Man is part of a long tradition of high-tech documentation and surveillance from the intensive observations of fieldworkers to the bleeps on a satellite monitor. This entails post-Foucauldian surveillance through tracking and the creation of a "data double" at work (Haggerty and Ericson 2000, 611). An apparent cultural death can be literalized and embodied in the exact moment the final living speaker of a language breathes his last. As Stephanie Gänger (this volume) points out, the "last speaker" is a problematic notion. Examining the case of the Araucanian "living relic" Pascual Coña usefully historicizes this concept. European and Chilean imaginations alike in the nineteenth century fed on the idea of the last man as the occasion for a process of mourning and lamentation for something just about to disappear forever. The Last Man was a shared sensibility. Last traces, words, and deeds therefore required thoughtful preservation, and by the mid-1920s, this cyclical dynamic of mourning and documentation had become fodder for new generations of anthropological fieldworkers, who extended their efforts beyond funerary and bodily remains to more ephemeral materials such as voices, customs, memories, and mental artifacts (Gänger this volume; see also Stafford 1994; on Ishi as last man, see Kroeber 1973 and Starn 2005).

Paradoxes of the Last Man aside, the operative question has been, "Can that moment be extended?" Also: "Can technologies of documentation be used to reverse the extinction process and, if not, at least to preserve its traces?" If the loss of once-enduring ways of life was inexorable, how then should social scientists approach the phenomenon? The short answer was they should make the most of this moment for scientific purposes. Among classic "Golden Age" anthropologists, it was considered "quite a coup" to be present for the "dying gasp" of a language or culture, and their appointed task was to reconstruct for ever after that which had just disappeared (Fenton 1953, 169–170). So well worn did this way of speaking of "the last" of something become, that it turned into a near taboo: "It's a hackneyed and by now totally discredited trope of ethnographic filmmakers to film cultures on the wane, or in their death throes – it's called salvage ethnography," commented filmmaker and curator Ilisa Barbash recently in an interview (Barbash and Taylor 2010). Yet, she continued, "I don't mean to sound callous about the tragedy of these disappearing life ways, nor to denigrate the importance of documenting them." Note that for Barbash, the disappearance of "life ways" is both a real and tragic occurrence and also, when represented in ethnography, a "discredited trope."

Salvage anthropology became taboo for professionals after the discipline-altering work that began with Dell Hymes' and Talal Asad's contributions of the 1970s, consolidated influence during the 1980s with George Marcus and James Clifford's landmark *Writing Culture*, and turned by the 1990s into an anthropological encomium to train in rote disciplinary self-suspicion and a heightened awareness of the dangers of employing literary tropes such as salvage ethnography. Reflexivity reigned as method. New researchers favored focusing on the way cultures resisted rather than succumbed to colonial forces or economic imperialism. Traditional cultures did not wane so much as change and last men were more salient as literary devices than real things. Clifford's 1988 essay "The Pure Products Go Crazy" was a paean in praise of culture seen as a hybrid hodgepodge of impurities, accidents, and haphazardness (Asad 1973; Clifford 1988; Hymes 1972; Marcus and Clifford 1986; see also Rees et al. 2008). Perhaps the "pure" savage had never existed, or perhaps the question of his ontological "truth" should be bracketed, and was no longer the most urgent one to ask. Decades of critique made the point that anthropology in its pursuit of knowledge had neglected properly to account for the power relation that constituted and emerged out of its epistemological arrangements. As the result of this sea change, a transfer of concern from *disappearing cultures* to *disappearing documents* naturally occurred. Once culture and therefore salvage anthropology became unfruitful topics, it was easier to discuss, on the one hand, the circulation of power effects and, on the other, the extraction and preservation of data.

Still, awareness of the Last Man trope as a trope – with its ideological contradictions and epistemological limitations – has not done away with its enduring potency. As James Clifford shows in *Returns*, much as experts understood the eventual disappearance of "native" cultures and "native" selves to be inevitable in the post-World War II years, in fact, people were reinventing what it meant to be indigenous during that same time (Clifford 2013). They resisted being seen as last men. Meanwhile, a new generation of post-World War II anthropologists, psychologists, and sociologists, especially those who adopted the projective test as a privileged instrument and who targeted dreams and dream-like materials, also wanted to preserve the last gasp of cultures, and in particular, the waning moments of certain kinds of "selves."

For a long time in Euro-American literary and public circles, it was not possible to know the "self" of a purportedly disappearing American Indian, and not interesting. It was not important to see from the point of view of, say, Sitting Bull, what it was like to undergo the cultural apocalypse of the Sioux way of life. Nineteenth-century Americans were interested in Sitting Bull to the extent he embodied an ancient culture, not because he was experiencing the loss of that culture. Sitting Bull was politically and physically in danger, but he was not *endangered*. By the twentieth century this changed, and it has continued to change in the twenty-first century. People *did* want to know what such a loss felt like, and there was a fascination with the American Indian seen as if from the inside, from the psychological depths, with a focus on angst and anxiety – which partially explains the spread during the 1960s and 1970s of odd documents such as Chief Seattle's 1854 words, originally recorded second-hand, commenting on the end of things and the coming of technologically driven civilization:

> Our departed braves, fond mothers, glad, happy hearted maidens, and even the little children who lived here and rejoiced here for a brief season, will love these somber solitudes and at eventide they greet shadowy returning spirits. And when the last Red Man shall have perished, and the memory of my tribe shall have become a myth among the White Men, these shores will swarm with the invisible dead of my tribe, and when your children's children think themselves alone in the field, the store, the shop, upon the highway, or in the silence of the pathless woods, they will not be alone. In all the earth there is no place dedicated to solitude. At night when the streets of your cities and villages are silent and you think them deserted, they will throng with the returning hosts that once filled them and still love this beautiful land. The White Man will never be alone.[1]

This text and the subsequent less flowery but somehow more romantic versions penned in the civil-rights era circulated popularly and fed a nascent environmental movement. The Last Man had spoken, looking out over the ruins of his race and summoning up a future filled with its murmuring ghosts haunting Americans' fields, stores, shops, and highways. Only traces in the form of ghosts would remain, he is said to have said. There was a time when no dorm room would be complete without Chief Seattle's words – the later, likely less accurate version.

Around the mid-twentieth century, new tools developed to extract, as if from a living specimen or laboratory subject, the stream-of-consciousness experience of undergoing this change. Not just shards in the form of second-hand recollections were to be passed around. Researchers could now collect and perhaps preserve whole onrushing realities through new kinds of data. One methodological innovation was the life history, which emerged as a sociological genre in the Chicago school of the 1920s and 1930s. Within American anthropology, the "as-told-to" era of the life history dawned in 1942 with the publication of the autobiography of a Hopi Indian named Don Talayesva, whose life story, *Sun Chief*, appeared as the result of the efforts of Yale anthropologist Leo Simmons to collect Talayesva's oral narration and diary entries and then shape and edit the materials (Simmons 1942; Talayesva 2013).[2] In its wake followed a flood of contributions such as Vada Carlson's 1964 recording of Polinggaysi Qoyawayma's *No Turning Back*, Louis Udall with Helen Sekaqueptewa in 1969's *Me and Mine*, and Harold Courlander editing Albert Yava's 1978 *Big Falling Snow* (see Langness 1965). By the 1970s, the self-authored American Indian autobiography succeeded these assisted tales.

The Spindlers' data sets, in contrast to their published work, provided access to something closer to "raw data" – materials recorded in the field, random observations while driving in their Oldsmobile picking up friends and driving them to pow-wows. If nothing else, the microcard analog collection, with all its holdings along these lines, is testimony from Sioux Indians and South Seas islanders in their own contexts: *this is what it was like*. It holds traces of "native" lives, and also traces of a widespread social scientific obsession in the twentieth century with not fading away. In turn, these shored up fragments testify to the perishability of memory and the desire to cold-store modern life itself as it underwent vast changes.

6

The Menominee were ideal subjects in these terms. According to anthropologists, as a tribe, they were experiencing a fading away that, insofar as it left traces in living bodies and personalities, seemed visible and sufficiently slow to be properly studied. Before the Spindlers came Felix Keesing in 1930, a prominent expert in acculturation who found the tribe to be an ideal focus for study. Old names such as Tomah, Oshkosh, and Shawano to the elders (c. 1930) would

> call up intimate associations with the past. Mention a name to them, and they will give a flood of reminiscence concerning men of the past and events of the past, Indian and white, far exceeding in quantity and vividness that recorded in even the best documents. But mostly such people are very old, and every year takes its toll of their ranks. Facts that could have been gotten two years ago are now inaccessible, and those available today may be gone within two months. Yet so far little or no attempt has been made by any student of history to set them down systematically – this all the more strange seeing that these folk will be found not only willing but in some cases keen to give them to a sympathetic listener. Most are well used to drawing on their stock of knowledge for white students, but so far these letters have had almost exclusively ethnological interests. The young Indians of today 'take little stock of this old stuff,' and so *all this very accessible source material is passing unrecorded*.
>
> (Brown quoted in Keesing 1987, foreword)

Soon, scientists would step forward to take advantage of this "old stuff" and render it as data to be stored, circulated, and readied for future use.

The Menominee, according to Keesing, were distinctive for being extremely acculturated yet still identifiable as a group: after three centuries of "contact" with European and Euro-American forces, their history and ways of surviving could be documented (records were available). "Though their ways of living are vastly changed from pre-white times they still retain their identity as a group" (Keesing 1987, 1). Thus, they offered an "excellent opportunity for study of the processes of cultural contact and change" (ibid.). From aboriginal times through the early bartering, missionizing, and warfare eras, through the fur trade, lumbering, and finally agricultural and industrial development, their "fortunes can be followed." The Menominee began as more or less sedentary wild-rice-and-fish-eating villagers, then scattered in mobile bands for fur-trade hunting. Later, they settled on a reservation where they made many economic and social adjustments. A vigorous logging industry and summertime tourism only increased in economic importance. "Today," Keesing wrote in 1939, tribal members ranged from ultra-conservative to ultra-progressive . . . so that "the historical process of change itself appears to be in large degree spread out in living personalities."

In a kind of push-me-pull-you dynamic, acculturation in anthropology was a response to first-order endangerment that then led into second-order endangerment. Once the "thing itself" – the Menominee way of life as it once was in an indeterminate past – has been transformed out of existence, it in effect exists only in memory and in the data sets (and publications) developed to keep and preserve its attenuated traces and "last men." Yet these traces themselves often proved to be fragile and in need of protection. As Kaplan observed, these data sets, if not provided for, might linger in the corner of an office gathering dust for decades before suffering the final indignity of being thrown away.

Thus, there is a difference between acculturation and endangerment: one succeeds the other, and sometimes would seem to be successive theories or paradigms that exclude each other. But in fact, acculturation relies on an understanding of endangerment in order to be pursued and, as often happened, turns into practices that were built upon. The Spindler anthropological team agreed with their precursor, Keesing. Having worked with the Menominee, they, like him, decided the community offered an ideal site for a research experiment in which they would try to capture acculturation more precisely and more scientifically by innovating in method. For their purposes, they divided their Menominee subjects into five subgroups or "acculturative categories," estimating the groups' relative level of acculturation using twenty-four items from a "sociocultural field schedule" (G. Spindler 1957b, 2).

In gauging each person's level of acculturation, they also employed projective psychological tests, primarily the Rorschach and Thematic Apperception Test, to form accounts of the inner life and inner conflicts that accompanied the process of loss (of Menominee culture) and gain (of "white ways" and socioeconomic stratification). The first, least-acculturated group centered around the village Zoar, and were the most traditional of the Menominee, participating in pow-wow, the Medicine Lodge (*met-a-won*), and Dream Dance (*ni-mik-twan*). Group A often put on shows for tourists at nearby Kashena Falls. The middle three gradations included Group B, the Peyote Cult group, whose members devised what the Spindlers deemed their "special solution" to the strains of the adaptive process to white life via peyote-assisted worship; Group C, the "transitionals," experienced in both Menominee and Western-style religion, but truly affiliated with neither; and group D, the lower-status acculturated, who were thoroughly Western in their way of life, but lacked much influence or funds. They lived in sad shacks without modern plumbing. The most acculturated were Group E, the elite acculturated, who fully embraced Christianity and lived a life not much distinguishable from their middle-class white Wisconsin neighbors, including, Spindler mentioned, bridge clubs. Peyotists, as a group in the middle, were alienated from committed traditionalists (who were, really, revivalists, as the met-a-won and Dream Dance they practiced had been forgotten and subsequently borrowed from neighboring tribes or circulating movements) and not completely assimilated either.

According to George Spindler, they used the peyote religion as a protector to shield themselves from the cultural shocks of their seemingly inexorable transformation. (The Spindlers sat up all night for a peyote session, George eating several buttons, Louise abstaining. He was not sure he felt anything, but found the drumming sounded especially good. The microcard data collection includes a set of scores of peyote hallucinations gleaned and retold from such worship sessions.) Setting up a five-stop scale such as this implies that everyone will move inexorably along it. As critics have noted, the scale of acculturation from A to E implies inevitability, constituting what perhaps could be called a fallacy of onwardness and differential destabilization. That is, the diagramatic depiction suggests that cultural change and the extinguishing of the Menominee traditional "self" occurs in an orderly and step-wise fashion through accepted stages (Beck 2010, 178–180; Waldram 2004, 71; literature on acculturation following the Spindlers includes e.g. Trimble 2003). Likewise, the fact that the Spindlers labeled their construct the "acculturative continuum" and spoke of being "further along" in some cases (L. Spindler 1978, 185–187) contributes to the sense of ineluctability. In fact, it may be the case that things fall apart in different ways, not so neatly. Perhaps there are cycles or backsliding, or perhaps the assumption of a coherent traditional Menominee "self" is itself an artifact of projection and historical error. This was not the Spindlers' view, however. As George Spindler put it, comparing his work with the Menominee to the findings of Hallowell with the Ojibwe or Chippewa revealed that, despite differences in the "psychological data . . . the situation is essentially the same." In particular, "[t]he regressive breakdown in both cases consists particularly of decreases in emotional and intellectual control and in productivity. The personality structure shows signs of disintegration and loss. Its main outlines are discernible, but they are quite attenuated and highly generalized" (G. Spindler 1978, 934). The data the Spindlers collected for each of the five acculturative groups – expressive autobiographies, personal documents, and Rorschach test results – pointillistically and intimately painted a picture of progressive loss: the researcher could almost watch as a once-integrated cultural "self" shifted into abeyance.

Note the parallel rhetorical construction of MacGregor's *Warriors without Weapons* and the Spindlers' *Dreamers without Power*. (These were, incidentally, titles that made assertions to which Sioux and Menominee people took great exception.) At the back of acculturation research lies the persistent conviction that there is such a thing as "degrees of Indianism" (cf. Berkhofer 1978, 28–29). At the end of the continuum, the point at which researchers felt that Indianism was all but lost, the subject was seen as largely "without" an Indian self. Not all was deterioration, though. There was some persistence. Some native traits – whether in Ojibwe or Menominee, or, by implication, in any group undergoing acculturation – would last and, Spindler observed, "I believe that some recognizable features will continue to reappear in each new generation for a long time" – even if in the form of "attenuated and highly generalized features of the native personality structure" (G. Spindler 1978, 936). The process entailed "attenuated native features and regressive disintegration (particularly in emotional and intellectual controls, where the link with self image and role are most important)" (ibid.).

Consider Case #20 in the Spindlers' microcard data sample, an old man some portrayed as the *last Menominee* – the local Last Man, in effect – to have been raised in a traditional way and who, in the course of giving his life history to George Spindler around 1949, paused to collect his thoughts for some minutes. Then he began again. "We have come to the cross roads," he said.

> There came to us another white man . . . Skinner, that has written own records. What that white man predicted I still remember. He was talking to my old people. There were many of us there and we listened to him. [Here he referred to Alonson Buck Skinner, anthropologist and author of an ethnography of the Menomini, who visited in the first two decades of the twentieth century.] He told my father then that he was going to die sometime and he said, "This is your boy. When you die, then your boy is coming into the same place. If he has inherited the same wisdom he will be prominent as you are. But there are not many who will remember these things. Someday it will be gone.
>
> *That is now a reality*; there is nobody that I could say that represents the place of the old people.

Someday it will be gone . . . that is now a reality. A vision of obsolescence had come to pass, and he – his father's boy – was there fully to recognize it, to experience that which no longer was the case.

In a final twist, Case #20's own "expressive autobiography" testifying to this loss has sat, largely unremarked upon even in the Spindlers' own published writings, for the past fifty years in the data bank of dreams at the Library of Congress. Not only Case #20 and the Menominee but also *the account* of his travails was endangered, barely preserved, at risk of abandonment, as it turned out.

7

Even as the data bank's subjects – who were not only American Indians from north and south, east and west Americas, but also South Pacific islanders, Pakistani frontier dwellers from the Pathan region, Balinese villagers, and many others – experienced endangerment, so too would the project to memorialize their data. Their data too was in danger, as it emerged. The project of documentation held a double risk. As mentioned, the Menominee data, including the testimony of Case #20, ended up in the large compendium of social scientific data banks that the Kaplan committee envisioned and built starting in 1956.

The eminent psychologist Solomon Asch affirmed Kaplan's foresight in targeting data as a problem: "You are quite right to talk about the challenge of data" (Kaplan 1957, 3). The culture and personality data sets he collected were intended to be just the tip of an informational iceberg. Kaplan had in mind, eventually, targeting a full range of data about inner states of mind. A second pilot proposed to capture psychological records in microfilm. It would include Roger Barker's observational records of English and American classrooms, William S. Sosin's verbatim transcripts of interactions between married couples, Calvin Hall's dreams of American college students, and Howard Becker's life history of a woman drug addict: five hundred pages each. An additional whole data bank would hold psychoanalytic records. However, the funding tide was turning. Rejected in 1958, this never-realized fulsome database went on to have a checkered history, each document making its way or foundering under curious circumstances. Becker's heroin addict tale, for example, was eventually published under the title *The Fantastic Lodge: The Autobiography of a Girl Drug Addict*. Other rich sets of data languished in the basement of the Chicago Institute for Juvenile Research (Becker 2011). Perhaps each data set will turn out to have its own history of tragedy, misplacement, and possible resuscitation (for an example, see Shenk 2009).

Not for decades would social scientists understand the need to hold on to hard-won data, their own and others. Responding to Kaplan's pioneering vision for a massive data clearinghouse of rescued data sets, a senior psychologist in 1957 imagined him leading a roving experimental group of data experts who could lend a hand at the planning stage of any project by envisioning threats to data and possible future solutions:

> Your group could have a servicing agency and provide advice as to how to arrange raw data of the future so that they could be most useful. Your group should be present at the planning stages of research projects. You should experiment in this. Our own data: I'd be glad to have it done. . . . These things of ours should be arranged before they get too cold. Many of my interviews are on tape.
>
> (Kaplan 1957, 4)

Others, distressed at the "deterioration of valuable records on the Pacific Islands," as well as records drawn from other places, saw the urgency of massive micropublishing to form archive centers to preserve them. Another area where Kaplan's project was ahead of its time was in giving access to the "data itself" that lay behind the interpretations given in the studies where they ended up.

Finally, this second-order endangerment sensibility has only grown in our archive- and data-obsessed moment, when it is not only historians and anthropologists who involve themselves with storage formats and worry about disappearing troves of documents. Current popular obsession with "big data" argues for a qualitative shift and perhaps even a new paradigm for knowledge-making (Mayer-Schönberger and Cukier 2013); meanwhile, the Edward Snowden NSA document drop as well as the ensuing public debate over the necessity and consequences of near-ubiquitous, semi-clandestine data collection by governments bring these questions to the fore. The problem of "curating data" has only recently begun to inspire a sufficiently urgent response. In a 2011 *Science* magazine special issue on data, over eighty percent of scientists surveyed (representing the full range of sciences) said they lacked sufficient funding to preserve their lab or research group's data. "There are many tales of early archaeologists burning wood from the ruins to make coffee," remarked one environmental scientist responding to the survey on data preservation. "If we fail to curate the environmental archives we collect from nature at public expense, we essentially repeat those mistakes" (*Science* editors 2011, 693). From this point of view, it is not surprising that the project of saving the "raw data of the future" – precisely Kaplan's aim circa 1957 – failed to inspire funding agencies to continue their support. As often happens, being ahead of one's time meant running out of monetary support. The database of dreams project very rapidly became a sort of ruin.

As it turned out, Kaplan had chosen the particular style of microform doomed to fail most quickly. Declared officially dead by 1965, two years after the last Kaplan volume appeared, the microcard proved – to venture into anachronism – the Betamax of the data-storage world, losing out in the standardization battles of the 1950s and 1960s in library science. Meanwhile, Kaplan's committee met for the last time in 1959 in an undisclosed location chosen, as the minutes that day noted, "to save money," doubtless a step down from the La Salle Hotel, where they first met. Its final compilations appeared in 1962 and 1963. Within only a few years, digital storage capacities would exist to capture the scientific data on a scale Kaplan and others envisioned. During these retrospectively "interim" years, many social scientists shared a Waiting for Godot feeling: as another NRC group put it in 1961, "Until electronic data processing and data storage devices rescue scholars from the limitations of the traditional research tools," they must remain content with "abstracts, annotated bibliographies, and inventories" (NRC/NAS 1961). Electronic databases were just around the corner, and their would-be users languished in a historical pause. Some data compiled under the rubric of second-order endangerment was itself consigned to a state of suspended animation.

The Kaplan group's alternative of using available technology before dedicated machines actually existed to do what was needed by then held tens of thousands of otherwise inaccessible pages, and remained in a kind of limbo. To return to Lévi-Strauss' words, it constituted "a collection of oddments leftover from human endeavors," and even as it moved to the Library of Congress, technically available to all researchers, its technology platform slowly sank into oblivion. Today, that project appears prescient. It augured a new form of endangerment sensibility and a new object: endangered data.

Notes

1 Three versions of Seattle's speech exist; none is verbatim. Each becomes progressively more unlikely. The first, October 27, 1887, version was published in the *Seattle Star* and was reassembled from notes taken at the time of his 1854 speech in a Squamish dialect. It was translated into Chinook jargon and subsequently into English. The second and third are from the 1960s and later. http://www.synaptic.bc.ca/ejournal/wslibrry.htm (three versions are printed here).
2 Note that Simmons appeared as the primary author of the American edition until the 2013 edition of the work from Yale University Press, whereas the French and German editions credited Talayesva as author from the outset. The absence of authorial credit for a Hopi man's own autobiography was typical of this style of document.

Bibliography

Asad, T., 1973. *Anthropology and the Colonial Encounter*. Ithaca University Press, Ithaca.

Auerbach, J. and Gitelman, L., 2007. Microfilm, Containment, and the Cold War. *American Literary History*, 19, 745–768.

Baker, N., 2001. *Double fold: Libraries and the Assault on Paper*. Random House, New York.

Barbash, I. and Taylor, L., 2010. Interview. *Believer Magazine* (http://www.believermag.com/issues/201003/?read=interview_barbash_castaing-taylor).

Bateson, G. and Mead, M., 1942. *Balinese Character: A Photographic Analysis*. New York Academy of Sciences, New York.

Beck, D., 2010. Collecting Among the Menominee. *American Indian Quarterly*, 34, 157–193.

Becker, H., 2011. Pers. comm. [via email].

Berkhofer, R. F., 1978. The White Man's Indian: *Images of the American Indian from Columbus to the Present*. Alfred A. Knopf, New York.

Clifford, J., 1988. The Pure Products Go Crazy. *The Predicament of Culture*. Harvard University Press, Cambridge.

Clifford, J., 2013. *Returns: Becoming Indigenous in the Twenty-first Century*. Harvard UP, Cambridge.

Cohen-Cole, J., 2014. *The Open Mind*. University of Chicago, Chicago.

Daston, L. and Lunbeck, E., eds., 2011. *Histories of Scientific Observation*. University of Chicago Press, Chicago.

Erickson, P., Daston, L., Klein, J., Lemov, R., Gordin, M. and Sturm, T., 2013. *How Reason Almost Lost Its Mind: The Strange Career of Rationality in the Cold War*. University of Chicago Press, Chicago.

Fenton, W. N., 1953. Cultural Stability and Change in American Indian Societies. *Royal Anthropological Institute*, 83, 169–170.

Finch letter to President of NAS Detley Bronk, 1 February 1956, Records of Committee on Primary Records, National Research Council Archives, National Academy of Science, Washington, DC.

Gilbert, M. S., 2013. Foreword to the second edition. *Sun Chief: The Autobiography of An Indian*. Yale University Press, New Haven.

Gitelman, L., 2014. *Paper Knowledge: Toward a Media History of Documents*. Duke University Press, Durham.

Gruber, J., 1970. Ethnographic Salvage and the Shaping of Anthropology. *American Anthropologist*, 72, 1289–1299.

Guldi, J. and Armitage, D., 2014. *The History Manifesto*. Cambridge University Press, Cambridge.

Haggerty, K. and Ericson, R., 2000. The Surveillant Assemblage. *British Journal of Sociology*, 601, 605–622.

Hallowell, A. I., 1953. Culture, Personality and Society. In: Kroeber, A., ed., *Anthropology today*. University of Chicago Press, Chicago.

Henry, J., Nadel, S. F., Caudill, W., Honigmann, J., Spiro, M., Fiske, D., Spindler, G. and Hallowell, A. I., 1955. Projective Testing in Ethnography. *American Anthropologist*, 57, 245–270.

Hymes, D., ed., 1972. *Reinventing anthropology*. University of Michigan Press, Ann Arbor.

Isaac, J., 2010. Tool shock: Technique and epistemology in the postwar social sciences. *History of Political Economy*, 42, 133–164.

Kaplan, B., 1957. *Report of the Executive Secretary of the Committee on Primary Records*, 1 April. NRC/NAS Archives, Committee on Primary Records, Anthropology and Psychology Division, Washington, DC.

Kaplan, B., 1958. Dissemination of Primary Research Data in Psychology. *American Psychologist*, 13, 53–55.

Keesing, F., 1987 [1939]. *The Menomini Indians of Wisconsin: A Study of Three Centuries of Cultural Contact and Change*. University of Wisconsin Press, Madison.

Kroeber, T., 1973. *Ishi: The Last of his Tribe*. Bantam Starfire, New York.

LaFramboise, T., Coleman, H.L.K. and Gerton, J., 1993. Psychological Impact of Biculturalism: Evidence and Theory. *Psychological Bulletin*, 114, 395–412.

Langness, L., 1965. *The Life History in Anthropological Science*. Holt, Rinehart and Winston, New York.

Lear, J., 2006. *Radical hope: Ethics in the Face of Cultural Devastation*. Harvard University Press, Cambridge.

Lemov, R., 2011. X-Rays of Inner Worlds: The Mid-Twentieth-Century Projective Test Movement. In: Isaac, J., ed., *Journal of the History of the Behavioral Sciences*, 47, 251–278.

Lemov, R., forthcoming 2015. *Database of Dreams*. Yale University Press, New Haven.

Lévi-Strauss, C., 1966. Anthropology: Its Achievements and Future. *Current Anthropology*, 7, 124–127.

Lévi-Strauss, C., 1973 [1955]. *World on the Wane*. Atheneum, New York.

LeVine, R., 2010. The Six-Cultures Study: Prologue to a History of a Landmark Project. *Journal of Cross-Cultural Psychology*, 41 (4), 513–521.

Luther, F., 1950. The Earliest Experiments in Microphotography. *Isis*, 41, 277–280.

Malinowski, B., 1932. *Argonauts of the Western Pacific: An Account of Native Enterprise and Adventure in the Archipelagoes of Melanesian New Guinea*. Dutton, New York.

Mandler, P., 2013. *Return from the Natives: How Margaret Mead Won the Second World War and Lost the Cold War*. Yale University Press, New Haven.

Manson, W., 1988. *The Psychodynamics of Culture: Abram Kardiner and Neo-Freudian Anthropology*. Praeger, New York.

Marcus, G. and Clifford, J., 1986. *Writing Culture*. University of California Press, Berkeley.

Mayer-Schönberger, V. and Cukier, K., 2013. *Big Data: A Revolution That will Transform How We Live, Work, and Think*. Houghton Mifflin, New York.

Myerowitz, J., 2010. 'How Common Culture Shapes the Separate Lives': Sexuality, Race, and Mid-twentieth-century Social Constructionist Thought. *Journal of American History*, 1057–1084.

NRC/NAS Committee on Disaster Studies, 1961. *Field Studies of Disaster Behavior: An Inventory, Disaster Study*, 14 (http://ia600208.us.archive.org/9/items/fieldstudiesofdi00natirich/fieldstudiesofdi00natirich.pdf). Accessed 5 December 2014.

Payne, M. and Barbera, J.R., eds., 2010. *Dictionary of Culture and Cultural Theory*. Wiley Blackwell, New York.

Pomata, G., 2011. Observation Rising: Birth of an Epistemic Genre, 1500–1650. In: Daston, L. and Lunbeck, E., eds., *Histories of Scientific Observation*. University of Chicago Press, Chicago, 45–80.

Reel, M., 2010. The Most Isolated Man on the Planet. *Slate* (http://www.slate.com/articles/news_and_politics/dispatches/2010/08/the_most_isolated_man_on_the_planet.html). Accessed 5 December 2014.

Rees, T. with Rabinow, P., Faubion, J. and Marcus, G., 2008. *Designs for an Anthropology of the Contemporary*. Duke University Press, Durham.

Rider, F., 1944. *The Scholar and Research University*. Hadham Press, New York.

Science, eds., 2011. Online Survey on Data Use. *Science*, 331, 693.

Shenk, J., 2009. What Makes Us Happy? *The Atlantic* (http://www.theatlantic.com/magazine/archive/2009/06/what-makes-us-happy/307439/). Accessed 5 December 2014.

Simmons, L., 1942. *Sun chief: The Autobiography of a Hopi Indian*. Yale University Press, New Haven.

Spindler, G., 1957a. Autobiographic Interviews of 8 Menomini Indian Males. *Microcard Publications of Primary Records in Culture and Personality*, 2, 12.

Spindler, G., 1957b. Rorschachs of 68 Menomini Men at Five Levels of Acculturation. *Microcard Publications of Primary Records in Culture and Personality*, 2, 11.

Spindler, G., ed., 1978. *The Making of Psychological Anthropology*. UC Press, Berkeley.

Spindler, L., 1978. Psychology of Culture Change and Urbanization. In: G. Sprindler, ed., *The Making of Psychological Anthropology*. UC Press, Berkeley, 174–200.

Stafford, F., 1994. *The Last of the Race: The Growth of a Myth From Milton to Darwin*. Oxford University Press, Oxford.

Starn, O., 2005. *Ishi's Brain: In Search of America's Last 'Wild' Indian*. Norton, New York.

Stocking, G. W. Jr., ed., 1986. *Malinowski, Rivers, Benedict and Others: Essays on Culture and Personality*. University of Wisconsin Press, Madison.

Stocking, G. W. Jr., 1991. *Victorian Anthropology*. The Free Press, New York.

Strathern, M., 2011. Binary License. *Common Knowledge*, 17 (1), 87–103.

Talayesva, D., 2013. *Sun chief: The Autobiography of A Hopi Indian*. Yale University Press, New Haven.

Trimble, J., 2003. Introduction: Social Change and Acculturation. In: Chun, K., Organista, P. and Marin, G., eds., *Acculturation: Advances in Theory, Measurement, and Applied Research*. American Psychological Association, Washington, DC, 3–13.

Waldram, J., 2004. *Revenge of the Windigo: The Construction of the Mind and Mental Health of North American Aboriginal Peoples*. University of Toronto Press, Toronto.

Whiting, J., 1994. *Culture and Development: The Selected Papers of John Whiting*. Cambridge University Press, Cambridge.

Wilcken, P., 2011. *Claude Levi-Strauss: The Poet in the Laboratory*. London, Bloomsbury.

Williams, R., 2001. *The Long Revolution*. Broadview, New York.

Part II

Situated politics

The fact that endangerment-related issues in biological and cultural areas are eminently political at all levels (local, national and international) requires no demonstration. The scientization of politics often called for in the name of rational decision-making does not depoliticize those issues nor prevents the correlative politicization of science. Without exception, they are entangled in clashes and negotiations, which in turn result in distributions of power and resources. The state always participates and intervenes – as one interested party among others, as mediator, legislator, and policy-maker. Decision-making with regard to biocultural endangerment has become integral to global governance. But the adjective "global" is not an explanation, nor does it imply that something is everywhere the same, or that the different forms of the processes or phenomena to which it is applied derive from a single system and express the emergence of a homogeneous "global era." Rather, as Anna Lowenhaupt Tsing demonstrates in *Friction: An Ethnography of Global Connection*, the global emerges from "makeshift links across distance and difference." It evolves from and through situated contexts and interactions among individuals, communities and institutions.

For example, the UNESCO biosphere reserves (Stefan Bargheer's chapter) were envisioned in the late 1960s as field laboratories for research on sustainable development – for conducting experiments on ecosystems, not for protecting endangered species. But rather than being founded from scratch for their designated purpose, they were for the most part established on the basis of already existing national parks. Their role changed in the 1990s, when the notion of biodiversity substituted that of sustainable development as leading concept of nature conservation. The difficulties of combining the two goals of sustainability and biodiversity conservation were embodied in local practices that concurred to shape the global discourse about them, and to define scientific and political priorities at the national and international level.

The local and the global mutually determine each other in the appearance of endangerment sensibilities. The modern concern about deforestation provides a major example. Early on, both negatively and positively, the forest came to define Brazil as an autonomous political entity. The Brazilian Atlantic Forest had been devastated since colonial times, and almost 90 percent of it was destroyed by the late 20th century. Yet its fate did not attract as much scrutiny and concern as that of the Amazon forest, which began to be intensively exploited only in the 1970s. The attention showered on the latter was brought about by far-reaching interactions between locally situated political activities, and discourses and practices around the world; together they took part in transforming threatening jungles into endangered rainforests (José Augusto Pádua's chapter).

While the Brazilian Atlantic Forest was destroyed "with broadax and firebrand" (as reads the title of Warren Dean's 1995 book on the topic), the Chilean used the weapons of war to advance the country's frontiers. Until Araucanía was entirely conquered in 1882, its long-resisting inhabitants were considered as savages whose demise was necessary for building the Chilean nation. After their defeat, the conquerors themselves lamented not only the impending disappearance of the "noble Araucanian race," but also the future impossibility of studying it directly (Stefanie Gänger's chapter). Yet the Mapuche, as they prefer to call themselves, have become the largest native people of Chile, and have been in conflict with successive Chilean governments throughout the 20th century. They survived and grew not in spite of, but *within* the situated circumstances of their struggles, which both contributed to motivate global indigenous rights movements and were in return empowered by their values and idioms.

4 Conserving the future

UNESCO biosphere reserves as laboratories for sustainable development

Stefan Bargheer

The present chapter investigates the conceptions of nature enacted by the Man and the Biosphere (MAB) program set up by the United Nations Educational, Scientific and Cultural Organization (UNESCO) in the late 1960s. At center stage of this interdisciplinary research program stood the peculiar relationship between man and the natural world that would become known as sustainable development. Following the UNESCO Conference on the Rational Use and Conservation of the Resources of the Biosphere in Paris in 1968, a global network of biosphere reserves was initiated to serve as outdoor laboratories for conducting research on the various projects administered by the program. Biosphere reserves were intended to represent the different ecosystems of the world and to allow for experiments on their controlled transformation over time.

As such, the insights gained at these reserves were intended as models for the management of ecosystems outside the reserves. With this focus on controlled experimental change and intervention by people, biosphere reserves were envisioned as alternative models to national parks, which at the time had a focus on keeping natural processes stable and free from human influence. The role of biosphere reserves within the MAB program changed with the altered framework for conservation following the United Nations Conference on Environment and Development at Rio de Janeiro in 1992. In the wake of this conference, the then still novel notion of biodiversity was added to the concept of sustainable development and the meaning and implications of the latter transformed in the light of the former. The chapter traces the transformation of the status assigned to biosphere reserves and shows that far from being stable, the conservation goals of the UNESCO Man and the Biosphere program are as unstable as the ecosystems it aims to protect.

Today, it is the notion of biological diversity that provides the justification for the management of biosphere reserves. As the following sections will demonstrate, the rationale for protection at the reserves was altered over time to match changing policy priorities within the United Nations framework, while the majority of the protected sites stayed the same throughout this process. The chapter thus shows that in the case of the UNESCO Man and the Biosphere program, a sensibility for endangerment stood at the end rather than at the beginning of a historical development that resulted in the establishment of protected areas. This chapter accordingly looks at the notion of endangerment in reverse: that is, from the point of view of its practical consequences in the form of efforts at conservation.

Nature conservation is more than one thing. The various schemes for conservation at the local, national, and international level in existence today do not germinate from one shared essence or embody one timeless common denominator, but instead try to achieve different goals and follow divergent trajectories over time. It is through these various forms of conservation and their trajectories that we can gain a more fine-grained understanding of what endangerment is about. Put differently, for every particular way to conserve something there is a particular way to perceive it as endangered. The very entity that is perceived to be endangered has an impact on the way it is to be protected or conserved. The significance of this point becomes apparent when applied to entities such as ecosystems shaped by human intervention: that is, to an entity that is not stable but instead is the product of past modifications that can continue in the future. In the case of the MAB program, it was the potential for change in the form of a future use of nature by man that was to be conserved. It was a yet unrealized future potential, not a current state of affairs or an idealized past that was perceived as potentially endangered. Conservation aimed at the future.

This chapter looks at the establishment and transformation of the MAB program and the place of endangerment within this process. In the following section, the initial rationale of the MAB program as providing field laboratories for sustainable development is described. Biosphere reserves were intended to serve as sites for large-scale experiments in the functioning of ecosystems. The next section traces how these reserves were selected in practice and what kind of spaces were most likely to be recognized as biosphere reserves. The subsequent section shows how a change in policy priorities at the United Nations transformed the reserves in places for the enactment of the Convention on Biological Diversity. The actual places designated as biosphere reserves remained largely the same. The last section concludes by addressing the question of the role played by geographical space in the formation of a concern for endangerment in the context of the UNESCO MAB program.

The rational use of the biosphere

The notion of *biosphere* entered the vocabulary of nature conservation for the first time at the intergovernmental conference of experts on the scientific basis for rational use and conservation of the resources of the biosphere in September 1968 in Paris. The conference was organized by the United Nations Educational, Scientific and Cultural Organization (UNESCO) and attended by 238 delegates for 63 member states and 88 representatives from six United Nations Organizations and other groups. The notion of biosphere that stood at the center of the event was originally coined in the late 19th century and popularized by the Russian geochemist Vladimir Vernadsky in his 1926 book titled *The Biosphere*. The term refers to the entire space on Earth in as far as it is relevant to living beings:

the lower part of the atmosphere and the outermost layer of the solid planet, including the seas and all abiotic conditions of life (like climate, water, soil, etc.). The most important function of the biosphere addressed at the conference was the continuous reproduction of living substance through accumulating and conserving the sun's energy (UNESCO 1970).

The Soviet soil scientist Viktor Kovda, a former director of the department of natural sciences at UNESCO, presented the opening paper on "contemporary scientific concepts relating to the biosphere," in which he argued that mankind as a whole can make a more efficient use of the resources on Earth by basing the production of food and other raw materials on the principles of scientific planning and management of the biosphere. "In modern industrialized society, based on scientific planning and expedient use of the laws of nature and means of science, technology and industry, the biosphere can be manipulated as a man-controlled system, which will provide the most favourable conditions for the welfare of mankind" (UNESCO 1970, 18). The ecosystems within the biosphere were perceived as factories that transformed the sun's energy into living matter that can be used as food for mankind. The goal of the conference was to improve the scientific basis for the conservation and development of the resources of the biosphere: that is, for the long-term, planned increase of its productivity.

In such a perspective, the delegates at the conference made a plea for systematic research on the biosphere and called for the creation of a global network of research sites. In this way, the conference facilitated a novel way to think about endangerment and protection. Protected sites as representative samples of nature were envisioned as field laboratories. The UNESCO biosphere conference introduced a rationale for conservation at the international level that would be put into action by the organization's Man and the Biosphere (MAB) program, established in 1971 as an outcome of the conference. Biosphere reserves were established within the MAB program five years later to enable the kind of ecological research projects Kovda and other delegates at the conference had envisioned.

At the present, biosphere reserves are part of one of UNESCO's most successful and best-known programs. There are currently 564 such reserves in 109 countries worldwide. With their declared focus on scientific research about and for the management of ecosystems for humanity's use, biosphere reserves were initially conceived as alternatives to the national park. One of the greatest influences on how to think about nature conservation was exerted by Yellowstone National Park, established in 1872 in Wyoming in the United States. National parks can be described as being akin to large outdoor natural history museums – they focus on the preservation of nature as it is for posterity and intend to reduce the impact of humans to a minimum. It was this focus on wilderness preservation that became synonymous with the term "national park." Time and again, nature conservationists in many countries around the world copied Yellowstone as a model. Settler societies and colonies provided (and continue to provide) the ideal context for this model. At times, this vision could have utopian dimensions not necessarily because of a lack of local support for conservation, but for the more practical lack of space – the circumference of the fabled park exceeded the proportions of some smaller political units at other places in the world, most notably in Europe. Yellowstone was, however, not only a model for nature conservation at a large scale, but also for the expulsion of native populations from the land they inhabited. The park stands out both for the exclusion of any use of the land as well as for the accompanying failure to recognize that the nature in the park was to a certain extent the product of human intervention (Spence 1999). Dan Brockington (2002) has coined the term "fortress conservation" to describe the process of dispossessing native people from the land they traditionally inhabited in order to create wilderness areas. Over time, the fortress style conservation initially practiced at Yellowstone not only raised ethical questions concerning human rights, but also called into doubt the very notion of nature untouched by human beings that was constitutive for setting up the park in the first place (Haines 1996; Krech 1999; Nash 1967; Schullery 2004).

Not every scheme for the conservation of nature, however, aims at the protection of wilderness as places unaltered by human intervention. There are other programs that quite explicitly try to reconcile human use and conservation. The first global inventory of parks was published in 1961 by an organization now called the World Commission on Protected Areas (WCPA), and a classificatory system of management categories for these parks was created by the International Union for Conservation of Nature (IUCN) in 1978. Initially, all protected areas were assumed to fit into one of ten categories. These categories were not discrete: i.e. in many instances, they were different only by name. UNESCO biosphere reserves were the best example of this weakness. Listed under the headline of international programs, they sometimes overlapped with the management goals of national programs listed under different categories. The management categories were in effect more a list of names than a classificatory system.

Revised guidelines for protected management categories were published by IUCN in 1994 in order to remedy this situation. The revised list was reduced to six categories arranged following a logic of gradation of human intervention. Category Ia designates strict nature reserves; Ib, wilderness areas; II, national parks; III, natural monuments or features; IV, habitat/species management areas; V, protected landscape/seascape; and VI, protected areas with sustainable use of natural resources. Protected areas are thus classified in a way that takes into account that people can be present: that is, the preservation of nature untouched by human beings is only one among many forms to manage land. Within these revised management categories, biosphere reserves are no longer a separate category. Since every biosphere reserve is supposed to consist of three zones, as will be described in more detail below, each of these zones falls within a different category. It is in fact this assemblage of different management goals within one reserve that makes for the peculiarity of biosphere reserves (Bridgewater et al. 1996; Phillips 2004).

When the UNESCO global network of biosphere reserves was established within the MAB program, it was the most clear-cut example for the kind of controlled use of the environment in which human beings are not excluded, but instead are a central part of the process that is to be protected. The protection or conservation of endangered species, on the other hand, was initially mentioned only at the margins. The section on the protection of rare and endangered species for the final report was provided by the American conservationist Stanley Cain, who also served as Rapporteur-General on the Steering Committee of the 1968 biosphere conference. The section was by far the shortest of all the contributions and in large parts unconnected to the main conference theme. This lack of salience at the conference was not a reflection of an overall marginality of the topic within the international community of nature conservation. To the contrary, the decade had just seen the institutionalization of what is now the Red List of Threatened Species at the IUCN. In fact, the originator of the Red List, the British naturalist Peter M. Scott, was among those who provided comments and additions to the final report of the biosphere conference (Fitter and Fitter 1987).

Concern for endangerment, however, was not what the Man and the Biosphere program was initially about. The program was envisioned as a tool for conducting experiments on ecosystems, not for protecting endangered species. A prominent example for the kind of experiments envisioned by the program was the ecosystem study carried out since 1963 at the Hubbard Brook Experimental Forest in the white mountains of New Hampshire. It is one of the best known ecosystem studies in the U.S., in which all the vegetation on one of the experimental sites was cut, and the vegetation regrowth was inhibited for three summers by periodic application of herbicides. This experimental manipulation was designed to determine the effects of fundamental biological and chemical relationships within the forest ecosystem. Not only was the resulting information of scientific interest, but it could potentially be used for rational planning of resource management. This study represents the type of ecological research that was envisioned by the MAB program, and the Hubbard Brook Experimental Forest became subsequently a UNESCO biosphere reserve (Risser 1979).

Field laboratories for sustainable development

The Man and the Biosphere (MAB) program grew out of the experience of the International Biological Program (IBP) conducted between 1964 and 1974. The IBP was an international research program that had a date of termination and the MAB program was intended to prolong this research on ecosystems into the future. It focused on solving specific environmental problems and, for that purpose, on transferring scientific knowledge at the international level. The MAB program hoped to be able to build upon the substantial advances in knowledge in ecology that came from IBP research. The director of the MAB program became Francesco di Castri, who had been one of the principle investigators of an IBP research project (Golley 1993).

The current state of the art of knowledge in the field of ecology was central to the MAB program. From the 1930s to the 1960s, the science of ecology provided a scientific justification for nature conservation. One of the most influential textbooks in the field dealing with the "Fundamentals of Ecology," first published in 1953 by the American biologist Eugene P. Odum, outlined a formula that became known as the diversity-stability hypothesis. It states that an ecosystem is the more stable the more species it comprises.

> It may be safe to state, at our present level of knowledge, that violent oscillation of density is a definite population characteristic of certain populations living in a simple ecosystem where there are but few food species or predators, parasites, and disease-producing organisms capable of interacting with the population in question.
>
> (Odum 1953, 141)

In short, diversity fosters stability. It can be considered as an extended version of the ancient idea of the balance of nature, this time supplemented by the novel notion of ecosystems (Bocking 1997; Egerton 1973; Goodman 1975; Worster 1994).

At the time, ecosystems were conceived as closed systems, something akin to a large-scale organism. Left to its own, the system was assumed to be stable: that is, in a state of equilibrium. Instability and change did not derive from internal processes, but only from interventions from the outside. Human use of the environment was one of these interventions and the aim of ecological research within the MAB program was to find out what kind of use and how much of it the various ecosystems of the world were able to sustain without a disruption of their long-term equilibrium. The rationale for research on ecosystems presented at the 1968 biosphere conference described above reflected these assumptions in the most explicit manner.

The establishment of the MAB program followed the biosphere conference. As a result of the resolutions adopted there, the UNESCO General Conference initiated deliberations on an international research program on Man and the Biosphere. In 1969, discussions about the design of the MAB program started and scientists from all over the world made project proposals. In October 1970, the 16th UNESCO General Conference decided to establish the MAB program. The logo of the program became the ancient Egyptian sign of ankh, which stands for eternal life. The member states soon set up MAB National Committees and scientists began discussing the details of the program, especially in terms of concrete interdisciplinary strategies and projects for a modern environmental policy, one that combines protection and yield of the biosphere. MAB was designed as an intergovernmental program from the very beginning, and decisions about its strategic agenda were taken (and still are) by the International Coordinating Council (ICC), consisting of representatives of the member states. During the first meeting of the ICC in 1971, initially, 13 major research projects had been identified on topics such as tropical rain forests, deserts, mountain regions, coastal areas, and islands. The list was later enlarged to 14 projects (UNESCO 2000; 2006).

The assumptions about ecosystems dominant at the time – first of all, the diversity-stability hypothesis – formed the scientific basis of the program. The report of the first ICC session in 1971 made the point.

> The basic design of the ecosystem is that of a machine capable of intercepting radiant energy from the sun, converting it into chemical energy through

photosynthesis and distributing this chemical energy in such a manner as to ensure the maintenance of its functional structure. . . . Control mechanisms, often closely related to species diversity, enable ecosystems to maintain, or reestablish if exposed to disturbance, their functional structure.

(UNESCO 1972, 10)

The MAB Council defined the overall objective of the program as being to provide a scientific basis for the rational use and conservation of the resources of the biosphere, as well as for the improvement of the global relationship between humans and the environment. Its aim was to predict the consequences of today's actions for the world of tomorrow and thereby to increase humanity's ability to manage natural resources efficiently.

Research was carried out with an eye on human use of the land, thus the name Man and the Biosphere. It was intended to highlight that people are a part of the ecosystems under investigation. Critics today point out that it would have been preferable to refer to Man *in* the Biosphere for that matter, or still better, *Humans* in the Biosphere, given that men and women were addressed alike. Yet such relabeling would not change the fact that the MAB program was concerned with very different matters at the time of its inception than it is at the present. Initially, the program did not plan to establish any reserves or protected sites whatsoever. The designation of reserves followed several years after the establish-ment of the research program. This is not to say that reserves were not previously considered an option. The first reference to biosphere reserves was made in 1970 in the initial plan for the MAB program, but they were intended exclusively for research and monitoring purposes, not for conservation.

The idea of conservation was introduced at the first meeting of the MAB International Coordinating Council in 1971 as part of project No. 8, which was aimed at the conservation of natural areas and of the genetic material they con-tain. Within this project, reserves were proposed as "basic logistic resources for research where experiments can be repeated in the same place over periods of time, as areas of education and training, and as essential components for the study of many projects under the Programme" (UNESCO 1972, 27). The Coun-cil recommended that each participating country designate biosphere reserves containing representative areas of each of the major or otherwise relevant ecosys-tems within the nation's boundaries. These representative samples of natural sys-tems in the major ecological regions of the world were intended to serve as a basis for worldwide networks of biological reserves and other protected areas. They were moreover assumed to facilitate research into the functioning of the undis-turbed biosphere, and thereby provide a baseline against which the stability and performance of modified and managed systems could be checked and compared.

At the second ICC session in 1973, multiple functions of reserves were deter-mined and the United Nations Environment Program (UNEP), together with a special working group of UNESCO, compiled guidelines for the establishment of biosphere reserves in the following year. The first reserves thus represented a research approach rather than a category of protected areas. The MAB project 8 was set up to insure that there would be regions available for implementing the research results. The idea inspired dozens of proposals for biosphere reserves presented to UNESCO and the MAB program. Sites were supposed to form a representative network of the world's ecosystems. A 1975 publication commissioned and published by UNESCO – Miklos Udvardy's *A Classification of the Biogeographical Provinces of the World* – provided the scientific basis for the selection.

Biosphere reserves were envisioned as experimental sites – they can be described as outdoor or field laboratories (on the topic of field laboratories more generally, see Kohler 2002). Laboratories are set up to carry out controlled experiments. In such an experiment, at least two settings are compared with each other. In one setting, one single aspect is altered by an intervention and the effect compared to the setting in which no change was introduced. The aim of such methodologically controlled experiments is causal explanation. The simple observation of a case over time alone is not sufficient if causal evidence is desired – one additionally needs a comparative control case. The MAB program aimed for such controlled experiments, and when reserves were established as part of the program in the early 1970s, it was with that purpose in mind. All biosphere reserves taken together were intended to constitute a global network that represented all types of ecosystems in the world. Each reserve was to encompass three areas or zones: a core zone, a buffer zone, and a transition (or regeneration) zone. The zoning served the practical purpose of conducting controlled outdoor experiments. The core zone constituted the control area, while interventions were carried out in the transition zone. As its name indicates, the buffer zone had the function of preventing a spillover or contamination of the core zone by the experimental intervention in the transition zone.

The arrangement can thus be described as an outdoor laboratory not merely in a metaphorical sense but in the full sense of the word – it was designed to function along the same methodological lines as those experimental arrangements commonly referred to as laboratories. Yet not every use of technological tools for scientific ends can be described as a laboratory setting. Many biological research projects conducted in the field continued to follow the long-established goal of describing and classifying species and estimating their abundance, not the goal of explaining the causal relationships between species and populations. The former set of objectives was an entirely descriptive enterprise that consisted mainly of large-scale data gathering. At the present, this kind of field research is commonly referred to as monitoring. It resembles the descriptive research traditionally carried out in a museum, not the kind of experimental research done in a laboratory. To qualify as a laboratory in a literal rather than a metaphorical sense, a research project conducted in the field requires an arrangement of sites that allows for methodologically controlled comparisons of cause and effect.

Controlled biological experiments were carried out outdoors well before the MAB program came into being, not the least by the International Biological Program of which it was a continuation. However, MAB represented the first time that the idea of organizing such experiments was translated into a scheme for selecting and administering research sites on a global scale. Examples for research sites that matched the envisioned goals come from many places in the world. The La Amistad biosphere reserve in Costa Rica, for instance, where research is conducted along those lines, consists of 15 different units, including two national parks, two biological reserves, a forest reserve, a wildlife reserve, a protected watershed, seven indigenous reserves, and a botanical garden.

At a 1981 UNESCO workshop on ecology in practice, Francesco di Castri referred to the MAB projects as enormous outdoor or field laboratories in which different approaches to the management of natural resources in a rational and sustained manner could be tried out. As in an indoor laboratory, experiments conducted in outdoor laboratories could potentially be unsuccessful.

> [The] MAB philosophy implies a willingness to accept the possibility of failure. . . . The whole idea of field laboratory includes acceptance of the notions of numerous (often unsuccessful) experiments, the approach of trial and error, the need to innovate and to understand how things work through evaluation of the past.
>
> (di Castri and Hadley 1984, 6)

With such an emphasis on experiments allowing for success as well as for failure, biosphere reserves would not just constitute a UNESCO label for nature reserves following the museum logic, but instead be at the very forefront of scientific research on the relation of human civilization to the ecosystems in which it exists.

Envisioned in this way, biosphere reserves served as a conceptual alternative to national parks. While the national park continued to build on the idea that the impact of people on the land should be reduced to a minimum, the biosphere reserve had its point of departure in the opposite idea that the administered land is the product of human use and should be managed for it. Biosphere reserves contained the more conventional nature reserves as part of their core area, but they were much more in addition to that. These areas were considered in their relation to the surrounding land used by humans and this relationship was to become the object of scientific research. Biosphere reserves as alternatives to national parks were envisioned as models for the future. A commentator in the 1982 IUCN Bulletin developed truly visionary enthusiasm in describing the workings of this novel kind of reserve.

> A biosphere reserve is not just a pretty place, it's an idea and an approach to management. In an ideal world all protected areas would be managed in a 'biosphere reserve manner,' with a zoning system which includes strictly protected core areas and buffer zones, institutionalized relationships with the surrounding land and people, management-related research and training programmes, and links with national and international monitoring programmes. In this sense, all of the world's protected areas may one day be 'biosphere reserves' as well, or at least managed in a 'biosphere reserve manner.'
>
> (McNeely 1982, 59)

This inclusion of people and their practices of using the land that made the biosphere reserve so different from the national park was part of the overall goal to facilitate ecological research.

Translating this vision into practice was not always as easy as fostering agreement with its principles. During its first decade of existence, the program relied heavily on already established sites. Often, biosphere reserves did not meet the ideal standards. In some cases, they were identical with already existing national parks and consisted only of a core zone. This was not due to a lack of interest in the program. At the end of 1975, no less than 38 countries had proposed to UNESCO a total of approximately 155 areas to be designated as biosphere reserves (Gilbert 1979). Yet when the first biosphere reserves were established in 1976, the main criterion that was applied in their selection turned out to be their conservation role. Most of them originated from already protected areas, among them several national parks. In fact, no less than 28 out of the 57 reserves designated in the first year were areas in the United States that had the status of national parks or wilderness areas. Given that biosphere reserves were intended to contain wilderness areas as their core zone, this selection made sense, yet not all of the newly minted reserves comprised the additional buffer and transition zones. In some cases, already established national parks thus simply received the additional UNESCO biosphere label without undergoing any changes in their boundaries or management policies.

The issue was addressed at a conference on the selection, management, and utilization of biosphere reserves organized in 1976 by the United States and the Soviet Union. The American plant ecologist Jerry Franklin contributed a report on the use of biosphere reserves and the conceptual basis for their selection in his country. He talked about the unrealistic goals of biosphere reserves as outlined in UNESCO documents: "This idealized Biosphere Reserve would be a rare situation in any country. We quickly find that most potential areas lack one or more essential feature, whether it be protective status, level of existing research program, or overall size" (Franklin and Krugman 1979, 4). Two categories of land in the United States were likely candidates for biosphere reserves: national parks and wilderness areas on the one hand, and reservations used for field-oriented research on the other. Given that the former kind of reserves frequently showed an incompatibility between the conservation and experimental research objectives of the MAB program, an extension of the zoning concept, called reserve cluster, was proposed. The United States MAB committee for project 8 decided that the best solution to their dilemma was to match selected large, natural reserves with nearby reserve-rich, experimentally oriented reservations. A pair of areas, separated in space but designated as one reserve, would fulfill the functions of a biosphere reserve in each biotic province.

Yellowstone was one of the national parks in the United States designated as a biosphere reserve. It was among the first 57 reserves to attain this status in 1976. It is furthermore one of the places in which the entire reserve is largely identical with a core zone: that is, it does not live up to the initial intention of the MAB program to establish three zones with different functions. In the case of Yellowstone, the national park is the biosphere reserve. In 1987, following the designation of the park as a biosphere reserve, the U.S. Forest Service and the National Park Service tried to use this status and made an attempt to implement an 18 million acre buffer zone through a planning process commonly called the "Vision" exercise. A first draft of the plan was produced in 1990 and a final one in 1991. This document was met with a huge outcry by the public and the states of Idaho, Montana, and Wyoming, and the plan to create a buffer zone in compliance with the UNESCO guidelines was ultimately not realized (Lichtman and Clark 1994; on changes in management practices and conservation goals at the park, see also Jax 2001). Yellowstone is thus one of the places that in practice matches the idea of the biosphere reserve rather incompletely.

Whether fully successful in practice or not, one of the central functions of biosphere reserves was confirmed by the rise of the concept of sustainable development. It describes a form of resource use that aims to meet human needs while preserving the environment so that these needs can be met not only in the present, but also in the future. The concept of sustainable development was for the first time placed center stage in international environmental policy through the World Conservation Strategy (WCS) published in 1980 by the International Union for Conservation of Nature. The document was commissioned and partially financed by UNEP and sent to various UN organizations, UNESCO among them, for approval before publication (McCormick 1986; 1989). The strategy stressed the assumption that economic development and environmental management are not incompatible. Development was defined as the modification of the biosphere and the application of human, financial, living, and not-living resources to satisfy human needs and improve the quality of human life. "For development to be sustainable," the document argued, "it must take account of social and ecological factors, as well as economic ones; of the living and non-living resource base; and of the long term as well as the short term advantages and disadvantages of alternative actions" (IUCN 1980, 1).

The MAB program picked up on this strategy. In 1983, UNESCO, UNEP, and IUCN jointly organized the First World Congress of Biosphere Reserves in Minsk, Belarus. The most tangible outcomes were the creation of an International Advisory Board of experts for the biosphere reserves and the development of an "Action Plan for biosphere reserves." The Action Plan formulated at the Minsk Conference and presented at the eighth ICC session in 1984 stressed the importance of biosphere reserves for the implementation of the WCS. "Successful biosphere reserves constitute models of the harmonious marriage of conservation and development. They provide visible examples of the application of the World Conservation Strategy – sustainable development in action" (UNESCO 1985, 4). Yet in the case of many reserves in the global network, this was more of a possibility than a reality. The Action Plan pointed at many still existing deficiencies of the MAB program and outlined possible activities for the 1985–1989 period to overcome them.

Practical tools for biodiversity conservation

The Action Plan formulated at Minsk was not the last word on the matter. Following the conference on Environment and Development at Rio de Janeiro in 1992, the notion of biodiversity took over the status as leading concept of nature conservation previously assigned to the notion of sustainable development (McConnell 1996; Takacs 1996). The Convention on Biological Diversity (CBD) produced at Rio became the central background for the work of the MAB program. In addition to instantiating additional fields of research and action, the CBD also redefined already established projects within the program. At the twelfth session of the ICC in Paris in January 1993, just about half a year after the CBD had been formed, the 14 MAB project areas were reduced to 5 priorities:

1   conserving biodiversity and ecological processes;
2   exploring approaches to land use planning and sustainable management of resources in regional landscapes;
3   formulating and communicating policy information on sustainable resource management and promoting environmentally sound behavior;
4   building human and institutional capacities for land use planning and sustainable resource management; and
5   contributing to the Global Terrestrial Observing System.

Biodiversity conservation now stood at the top of the list and the decisions of 1993 mainly referred to biosphere reserves for implementation and research. These priorities are in slightly modified form valid until today. These changes do not merely represent a reduction of the number of goals of the program, i.e. from 14 to 5, but an almost complete change of direction or focus. The previous list of MAB projects described substantive areas of ecosystem research, such as, for instance, research on desertification, not general policy orientations described by the five priorities above. The change was in effect an abandonment of the independent research agenda of the MAB program. Instead of initiating and coordinating ecosystem research, the program now served the function of implementing the CBD. Annex 3 of the Final Report of the session advanced a framework for the future development of the MAB program in the 1996–2001 period in which the CBD played a central role.

> The Convention on Biological Diversity was signed in Rio, without most people knowing what is meant by biological diversity. There are many uncertainties surrounding biodiversity . . . Key scientific questions in need of answers include: Do we need all those species for the functioning of the biosphere? What is the role of biodiversity in ecosystem functioning, and therefore sustainability?
>
> (UNESCO 1993, 41)

The authors of the report seemed not to remember that the answer to this kind of question – as, for instance, in form of the diversity-stability hypothesis – was among the central reasons for setting up biosphere reserves in the first place. Yet the diversity-stability hypothesis had lost academic credibility almost at the same time as the MAB program came into being. The first critical voices that pointed out a lack of empirical evidence could already be heard throughout the 1960s. The theoretical plausibility of the hypothesis was further challenged in the early 1970s by the work of the Australian born physicist Robert May, who played a key role in the development of theoretical ecology throughout the decade. May used linear stability analysis on models constructed from a statistical universe and found that diversity tends to destabilize community dynamics (May 1973; for the rise and fall of the diversity-stability hypothesis more generally, see Worster 1994). The once dominant theory of ecology that had been so prominent at the 1968 biosphere conference and had justified the creation of reserves was gone – it was now the concept of biodiversity discussed at the 1992 Rio conference that took over this central role.

An almost complete revision of the MAB program followed in the next years at a Second World Congress of the MAB program at Seville, Spain, in 1995. The purpose of the Seville meeting was to review progress and to determine the role biosphere reserves could play in the light of the Rio conference, in particular in implementing the CBD. The "Seville Strategy" defined the role of biosphere reserves for the 21st century and the congress additionally adopted a "Statutory Framework of the World Network of biosphere reserves." Since then, there are three minimum criteria for areas joining the MAB program. Each biosphere reserve must

1   identify three zones (core area, buffer zone, transition zone);
2   provide three functions (protection, research, development); and
3   conduct regular, every 10 years, evaluations to review the state of the biosphere reserve.

Thus, since 1993 and 1995, the MAB program increasingly focused on the world network of biosphere reserves, not any longer on the various ecological research projects that made up the program throughout the first years. Now the Seville Strategy replaced the Action Plan that grew out of the meeting at Minsk more than a decade earlier.

> While much of this action Plan remains valid today, the context in which biosphere reserves operate has changed considerably as was shown by the UNCED process and, in particular, the Convention on Biological Diversity. . . . The major objectives of the Convention are: conservation of biological diversity; sustainable use of its components; and fair and equitable sharing of benefits arising from the utilization of genetic resources. Biosphere reserves promote this integrated approach and are thus well placed to contribute to the implementation of the Convention.
>
> (UNESCO 1996a, 3)

The Seville Statutory Framework went on to point out the intention to combine sustainable development and biodiversity conservation. Yet a great amount of such attempts to combine the old and new conceptual focus of the MAB program do not venture beyond putting the word "and" between the two concepts. It is a true Sunday speech declaration. How the two policy goals work together in practice is not spelled out. The possibility that sustainability and biodiversity might be goals that are not only not identical but might even conflict with each other is not even mentioned. The Seville framework does not venture beyond the simple affirmation of the fact that the UN family wishes to have the best out of these two worlds. How to reconcile these goals in the case of possible conflict is not discussed. It is likewise not explored why one should assume that the already selected biosphere reserves should be places particularly suited for protecting biodiversity.

The assumed fit between the goals of the CBD and the reality of the MAB biosphere reserves is – given the absence of conclusive data – by far more easily discernible in UNESCO reports on the aims of the program than in reference to real life results and achievements. A coordinated effort at monitoring the biological diversity of the entire global network of biosphere reserves currently does not exist. This should not be taken as a lack of good will to implement the Seville Strategy and Statutory Framework. A complete match would have been something of a surprise for the simple reason that what is now called biodiversity was not among the criteria for selecting biosphere reserves. The CBD defines biodiversity in article 2 as "the variability among living organisms from all sources including, inter alia, terrestrial, marine and other aquatic ecosystems and the ecological complexes of which they are part; this includes diversity within species, between species and of ecosystems" (http://www.cbd.int/convention/articles/?a=cbd-02). The CBD thus explicitly addresses biodiversity at three levels (genetic, species, and ecosystem), while the MAB program operates at the level of ecosystems: that is, it is prepared to protect one but not all three of the levels the convention deals with. The match with the goal of the convention in regard to ecosystems does not automatically extend to species diversity and genetic diversity.

There can of course be an overlap between all three levels of diversity at a certain place, but this is not a given: some ecosystems, for instance, have a rather low species diversity. Similarly, certain populations of rare species are marked by a very low genetic diversity. In short, it is not necessarily possible to address all the dimensions of the CBD at one and the same place and at one and the same time. Even more than this, the assumption that the places currently on the MAB list are the ideal sites to attain the goals formulated by the CBD is something one might hope for, but it is by no means certain. Biosphere reserves had been selected with very different ends in view. The selection of biosphere reserves was initially conducted with an eye to the diversity of ecosystems and facilitated by Udvardy's work on the biogeographical provinces of the world (1975). There is currently no comparable map or dataset available that would allow for the assessment of the overlap between ecosystem, species, and genetic diversity on a global scale, albeit a monitoring scheme for biosphere reserves was established in more recent years. Beginning with a meeting in Strasbourg, France, in 1991, there has been a concerted effort among many nations to coordinate and centralize the provision of data about biosphere reserves.

The effort was termed the Biosphere Reserve Integrated Monitoring (BRIM) program. Its goal is to harmonize monitoring programs among biosphere reserves and to make it possible that data and information can be shared and used at scales greater than that of the reserves themselves. The first product of BRIM was a directory called ACCESS that provided summary information on 175 EURO-MAB biosphere reserves (Secretariat 1993). This model was extended by the UNESCO MAB office to the rest of the world's biosphere reserves. Special emphasis was placed on looking at the quantity and quality of 72 categories of data (UNESCO 2004). Yet systematic data of this kind is not available for the selection of new biosphere reserves and the results of monitoring already existing reserves are likewise not used to make decisions about the further existence as reserve.

Even for representing the world's ecosystems, the MAB network of biosphere reserves was from the very beginning rather deficient. The assessment of the current situation of the MAB program in the 1984 Action Plan was particularly outspoken when it came to the many gaps in the network.

> Many important representative types of ecosystem are still to be included, especially of coastal and aquatic ecosystems. Only a few biosphere reserves established so far cover the full range of purposes for which biosphere reserves were intended. Few reserves have been established, which include centres of high biological diversity and endemism.
>
> (UNESCO 1985, 9)

The Action Plan called for surveys on the already existing reserves and for the provision of data for the selection of future reserves. The situation had not changed much by the time the new Strategy and Statutory Framework was established. An evaluation of the implementation of the 1984 Action Plan presented at the Seville meeting by Tomas Azcarate Bang pointed out that one should recall that the biosphere reserve concept, while very attractive in theory, is, in fact, very difficult to put into practice in the real world. Thomas Lovejoy noted that "many existing biosphere reserves had remained in name only, or had remained as 'intellectual deserts' without research. He also felt that some biosphere reserves could probably not conserve their biodiversity because of the inappropriate size and shape of the core area" (UNESCO 1996b, 10). Yet even today, the monitoring data necessary to make informed judgments on the overlap between genetic, species, and ecosystem diversity is not even in existence.

The CBD is currently the major rationale for the MAB program. Some commentators even go so far as to argue that it was only following the convention that the biosphere reserve concept truly came to life. The important status assigned to biosphere reserves was emphasized at a 1998 workshop on the Role of UNESCO MAB Biosphere Reserves in Implementation of the Convention on Biological

Diversity. In the proceedings, Jane Robertson Vernhes, a program specialist at the UNESCO Division of Ecological Sciences, made the point that

> although the biosphere reserve concept was conceived in the early 1970s, it can be seen today that it appears to have anticipated the Convention on Biological Diversity and is tailor-made to contribute to its implementation. The three main objectives of the Convention are embodied in the multipurpose functions of biosphere reserves and their translation into different zones on the ground. . . . Thus biosphere reserves offer a concrete, geographical space for putting the Convention on Biological Diversity into practice.
>
> (UNESCO 1999, 8)

The statement ignores that the MAB program underwent a substantial change since it was established. It is currently consistent with the goals of the CBD because it was remodeled after the fact to match the convention. The mentioned tools for implementation can only be seen as "tailor-made" because they had been stripped of their initial function to serve as field laboratories. The importance of zoning was reaffirmed in the Seville documents, but little to nothing was left of the initial justification of this zoning for the use of reserves as field laboratories. Instead, the different zones are increasingly used as independent areas for combining separate conservation aims, ranging from wilderness conservation to sustainable development.

Patchworking in the biosphere

Biodiversity is UNESCO's current core concept for dealing with the conservation of nature. It is used to integrate the various programs of the UN organization that have developed since its inception. The reserves in the Man and the Biosphere program are key sites where this concept is put into action. It is one of the advantages of biosphere reserves that they allow to combine in practice what might not match in theory. The reserves are something akin to a garbage can in which all goals the organization is committed to are put together. The zoning of the reserves into core, buffer, and transition area is central in this respect. When the model of the biosphere reserves was initially created in the early 1970s, the zones were to operate as an experimental setup in the form of a field laboratory. While there are still biosphere reserves that are run as laboratories, most of them no longer follow this model. Following the developments of the 1990s, the purpose of the zones was redefined. Their function now is to allow for the implementation of different conservation goals.

Throughout this process, old institutional forms were invested with a new meaning. The change reflected new policy priorities at the UN level. Endangerment emerged as the top priority with the establishment of the Convention on Biological Diversity. It is now the monitoring and conservation of species diversity, rather than experimentation on the functioning of ecosystems, that defines the rationale of biosphere reserves. It is because of this change in rationale that the MAB is a success. The number of reserves is continuously growing.

Yet the continuity and success of the MAB program cannot be explained by constantly changing policy priorities. Space – that is, the materiality of biosphere reserves – is an important factor in this development in at least three regards. First, the availability of already protected spaces had a decisive impact on the selection of places as biosphere reserves. Their limited number explains why, despite its novel conservation goals, the MAB program ended up with largely the same old places, i.e. national parks, on their list of protected areas.

Second, the logic of space contributed to the endurance of the program over time: that is, the spaces persisted even though the rationale for their protection disappeared. At the beginning of the program, it was the diversity-stability hypothesis that provided this rationale; once it was gone, the notion of biodiversity took its place. This change was not brought about by the MAB program, but reflects changes in the field of ecology and nature conservation more broadly. Yet the difference between the two justifications for conservation became particularly visible at the UNESCO biosphere reserves. Despite those differences, the transformation was not accompanied by a replacement of the places designated as reserves. The reasons for their suitability were simply reinterpreted in the light of changing policy priorities.

Last but not least, space mattered in that the existence of three separate zones within each biosphere reserve allowed for the realization of different conservation goals within a single reserve, e.g. it is possible to combine wilderness areas without human intervention with areas allowing for controlled population management within the same reserve. Yet far from being integrated, these different goals simply run parallel to each other. This parallelism has in practice the same effect as establishing separate reserves. The division of each reserve into three zones thus helps to produce the impression that some kind of integration of the various conservation goals enacted in the different zones has been attained. Spatial proximity is used to make up for a lack of conceptual integration. It is a form of peaceful co-existence of incommensurable conservation schemes that is made possible by biosphere reserves. Biosphere reserves came into being to enable field research – they stay in existence because they allow for conceptual patch working. In this way, the MAB program has provided a flexible tool for keeping up with UNESCO's quickly changing projects and goals.

This story demonstrates how biosphere reserves were shaped by space as a material factor as much as by an evolving concern for preserving nature. The two cannot be studied in isolation from each other; the relationship between them is not one-directional, but rather a continuous interaction that unfolds over time. Thus, in the case of UNESCO biosphere reserves, a sensibility for endangerment stood at the end rather than at the beginning of a historical development.

Bibliography

Bocking, S., 1997. *Ecologists and Environmental Politics: A History of Contemporary Ecology.* Yale University Press, New Haven.

Bridgewater, P., Phillips, A., Green, M. and Amos, B., eds., 1996. *Biosphere Reserves and the IUCN System of Protected Area Management Categories.* Australian Nature Conservation

Agency, World Conservation UNION, UNESCO Man and the Biosphere Programme, Canberra.

Brockington, D. D., 2002. *Fortress Conservation: The Preservation of the Mkomazi Game Reserve*. Indiana University Press, Bloomington.

Di Castri, F., Baker, F.W.G. and Hadley, M., 1984. *Ecology in Practice*. Tycooly International Pub./Unesco, Dublin/Paris.

Egerton, F. N., 1973. Changing Concepts of the Balance of Nature. *The Quarterly Review of Biology*, 48, 322–350.

Fitter, R.S.R. and Fitter, M., 1987. *The Road to Extinction. Problems of Categorizing the Status of Taxa Threatened with Extinction*. International Union for the Conservation of Nature and Natural Resources, Gland, Switzerland.

Franklin, J. F. and Krugman, S. L., eds., 1979. *Selection, Management, and Utilization of Biosphere Reserves: Proceedings of the United States-Union of Soviet Socialist Republics Symposium on Biosphere Reserves, Moscow, USSR, May 1976*. Pacific Northwest Forest and Range Experiment Station, U.S. Dept. of Agriculture, Forest Service, Portland, OR.

Gilbert, V.C., 1979. Biosphere Reserves in the United States and their Relationship to the International Program Man and the Biosphere. In: Franklin, J. F. and Krugman, S. L., eds., *Selection, Management, and Utilization of Biosphere Reserves: Proceedings of the United States-Union of Soviet Socialist Republics Symposium on Biosphere Reserves, Moscow, USSR, May 1976*. Pacific Northwest Forest and Range Experiment Station, U.S. Dept. of Agriculture, Forest Service, Portland, OR.

Golley, F. B., 1993. *A History of the Ecosystem Concept in Ecology: More than the Sum of the Parts*. Yale University Press, New Haven.

Goodman, D., 1975. The Theory of Diversity-Stability Relationships in Ecology. *The Quarterly Review of Biology*, 50, 237–266, SS.

Haines, A. L., 1996. *The Yellowstone Story: A History of our First National Park*. Yellowstone Association for Natural Science University Press of Colorado Yellowstone National Park, Wyoming, CO.

IUCN, 1980. *World Conservation Strategy: Living Resource Conservation for Sustainable Development*. IUCN, Gland, Switzerland.

Jax, K., 2001. Naturbild, Oekologietheorie und Naturschutz: zur Geschichte des Ökosystemmanagements im Yellowstone-Nationalpark. In: Höxtermann, E. and Kaasch, J., eds., *Berichte zur Geschichte und Theorie der Oekologie*. Verlag für Wissenschaft und Bildung, Berlin.

Kohler, R. E., 2002. *Landscapes and Labscapes: Exploring the Lab-Field Border in Biology*. University of Chicago Press, Chicago.

Krech, S., 1999. *The Ecological Indian: Myth and History*. W. W. Norton & Company, New York.

Lichtman, P. and Clark, T. W., 1994. Rethinking the "Vision" Exercise in the Greater Yellowstone Ecosystem. *Society and Natural Resources*, 7, 459–478.

May, R. M., 1973. *Stability and Complexity in Model Ecosystems*. Princeton University Press, Princeton.

McConnell, F., 1996. *The Biodiversity Convention – A Negotiating History: A Personal Account of Negotiating the United Nations Convention on Biological Diversity, and After*. Kluwer Law International, London/Boston.

McCormick, J., 1986. The Origins of the World Conservation Strategy. *Environmental Review*, 10, 177–187.

McCormick, J., 1989. *Reclaiming Paradise: The Global Environmental Movement*. Indiana University Press, Bloomington.

McNeely, J. A., 1982. Why Biosphere Reserves? An Introductory Note. *IUCN Bulletin*, 13 (7–9), 59.

Nash, R., 1967. *Wilderness and the American mind*. Yale University Press, New Haven.

Odum, E. P., 1953. *Fundamentals of Ecology*. Saunders, Philadelphia.

Phillips, A., 2004. The History of the International System of Protected Area Management Categories. *Parks*, 14, 4–22.

Risser, P. G., 1979. Characteristics of Research Programs at Established U.S. Biosphere Reserves. In: Franklin, J. F. and Krugman, S. L., eds., *Selection, Management, and Utilization of Biosphere Reserves: Proceedings of the United States-Union of Soviet Socialist Republics Symposium on Biosphere Reserves, Moscow, USSR, May 1976*. Pacific Northwest Forest and Range Experiment Station, U.S. Dept. of Agriculture, Forest Service, Portland, OR.

Schullery, P., 2004. *Searching for Yellowstone: Ecology and Wonder in the Last Wilderness*. Montana Historical Society Press, Helena, MT.

Spence, M. D., 1999. *Dispossessing the Wilderness: Indian Removal and the Making of the National Parks*. Oxford University Press, New York.

Takacs, D., 1996. *The Idea of Biodiversity: Philosophies of Paradise*. Johns Hopkins University Press, Baltimore.

Udvardy, M.D.F., 1975. *A Classification of the Biogeographical Provinces of the World*. International Union for Conservation of Nature and Natural Resources, Morges, Switzerland.

UNESCO, 1970. *Use and Conservation of the Biosphere: Proceedings*. Unesco, Paris.

UNESCO, 1972. *International Co-ordinating Council of the Programme on Man and the Biosphere (MAB). First Session. Paris, 9–19 November, 1971. Final Report (SC/MD/26)*. UNESCO, Paris.

UNESCO, 1985. *International Co-ordinating Council of the Programme on Man and the Biosphere (MAB). Eighth Session. Paris, 3–8 December, 1984. Action Plan for Biosphere Reserves (SC-85/WS/26)*. UNESCO, Paris.

UNESCO, 1993. *International Co-ordinating Council of the Programme on Man and the Biosphere (MAB). Twelfth Session. Paris, 25–29 January, 1993. Final Report (SC/MD/102)*. UNESCO, Paris.

UNESCO, 1996a. *Biosphere Reserves: The Seville Strategy and the Statutory Framework of the World Network*. UNESCO, Paris.

UNESCO, 1996b. *International Co-ordinating Council of the Programme on Man and the Biosphere (MAB). International Conference on Biosphere Reserves. Seville (Spain), 20–25 March, 1995. Final Report*. UNESCO, Paris.

UNESCO, 1999. *Role of UNESCO MAB Biosphere Reserves in Implementation of the Convention on Biological Diversity. International Workshop, 1–2 May 1998, Bratislava, Slovakia*. Slovak National Committee for the UNESCO, Bratislava.

UNESCO, 2000. *Biosphere Reserves: Special Places for People and Nature*. UNESCO, Paris.

UNESCO, 2004. *Progress in the Implementation of BRIM and Suggestions on the Future Phase of its Work Plan*. UNESCO, Paris.

UNESCO, 2006. *Sixty years of science at UNESCO 1945–2005*. UNESCO, Paris.

U.S. MAB Secretariat, 1993. *Access: A Directory of Contacts, Environmental Data Bases, and Scientific Infrastructure on 175 Biosphere Reserves in 32 Countries*. U.S. MAB Secretariat, Washington, DC.

Vladimir Vernadsky [1926] 1998: The biosphere, translated by David B. Langmuir; revised and annotated by Mark A.S. McMenamin. New York: Copernicus.

Worster, D., 1994. *Nature's economy: a history of ecological ideas*. Cambridge University Press, Cambridge/New York.

5 Indigenous evanescence and salvage in the conquest of Araucanía, 1850–1930*

Stefanie Gänger

Some four decades after the conquest of the Araucanian territories by the Chilean state, in 1927, Pascual Coña dictated his life story to a Capuchin missionary. The friar wrote in his preface to the account that Pascual Coña was "a legitimate indigene of the ancient Araucanian race," and that he died in old age, on the same day he dictated the last lines of his life story (Moesbach 2006 [1927]), 22). Pascual Coña gave his own reasons for bearing testimony as follows.

> I am already old; I believe I am more than 80 years old. During this long life I have learned the manners of the people of the old days, I treasure the different moments of their lives; they had good habits and bad ones. I shall speak of all of this now: I shall narrate the whereabouts of my own existence, and also the manner of living of the ancestors. In our days life has changed, the new generation has become Chilean [*se ha chilenizado mucho*]; little by little they have forgotten the . . . nature of our race; a few more years and they might even forget their mother tongue. And when that time comes [*entonces*], they can at least read this book.
>
> (Pascual Coña, in Moesbach 2006 [1927], 25)

Pascual Coña represents an iconic form of the Indian, one that underlay the imagination of Chileans and Europeans alike in the nineteenth century: the old, the dying Indian, aware that the end of his time had come and that only his memories, treasured by the conquerors, were to survive him and his kind.

Endangerment sensibility has recurred like a wave, from the early 1800s into our present, in Europe, its colonial possessions and beyond. Like other "sensibilities" that traveled along the veins of global colonial and post-colonial connections and through time, endangerment is mutable in its expressions; its contents – prescribed actions, objects and the validation of endangerment – have shifted in innumerable ways, and yet change happened continuously within the same general outline: a concern with the extinction or evanescence of particular natural, material or human objects associated with primeval authenticity and an urge to preserve what could still be saved, or to memorialize what was already lost. Indigenous extinction or evanescence discourses – the assumptions, literary narratives and scientific theories that arose to explain the decimations of indigenous groups in imperial expansion – have long constituted a standard theme in the literature on "postcoloniality," on the cultural impact of British or French imperial expansion and New England-style settler colonialism in the nineteenth century (Brantlinger 2003). Endangerment thinking, however, while deeply linked to, and ultimately a consequence of, perceptions of the "Indians'" evanescence, has rarely been historicized and has usually been taken to be self-evident, an expression of values, such as cultural diversity, supposed to be universally accepted.

This chapter studies the gradual appearance of endangerment discourse, of the modern sensibility that had materialized so clearly by the 1920s in the words of ethnographic observers like Moesbach and even of their objects, men like Pascual Coña. It traces how the mourning and lamentation of what was not yet lost but would unfailingly be lost soon took hold in Chilean discourses about Araucanians – who today prefer the self-referential name Mapuche (Boccara 1999) – during the occupation, or as it came to be called, the "Pacification" of the formerly independent territories of Araucanía in southern-central Chile. Following discourses about Araucanians' impending end through Chilean military intelligence reports, missionaries' writings and naturalist treatises, this chapter traces the roots, the conditions and the exact purpose of the idea that, as Araucanians were nearing extinction, their words and deeds, their language and thought urgently demanded treasuring and preservation.

Evanescence

Before the mid-nineteenth century, Araucanian Indians were not considered endangered, nor was there any idea that they were evanescent or vanishing. After a general Araucanian uprising in 1598, the Bio-Bio River had marked the boundaries of Spanish colonial rule. From the early seventeenth century, the Spanish stationed a small standing army in the south to patrol the frontier, and after Chile's independence from Spain in 1810, the Chilean government continued this colonial policy for several decades (Collier and Sater 2004). By the early nineteenth century, marriage alliances within Araucanian groups, with Pampas and Tehuelche bands, and with Chilean and European traders, outlaws and frontier settlers, as well as intermarriage with Chilean and Argentine "captives," had created a multifarious network of ethnic relations in Araucania. In an active cross-frontier trade, the inhabitants of Araucania supplied Chileans with ponchos and cattle in exchange for hardware, wine and European manufactured goods (Gascón 2007; Hidalgo Lehuedé 1984; Jones 1999). Around the mid-century, Araucanian power had reached its zenith, extending from the Pacific almost to the Atlantic, with chiefdomships in Patagonia and the Pampas (Bechis 2002). Even though the myth of the noble savage – romantic narratives about Araucanians' indomitableness inspired in *La Araucana*, a sixteenth-century epic – arose side by side with that of the untamed and ferocious savage, "co-existing with it in the relation of light to darkness" (Pocock 2008, 160), fear of Araucanians' untamed barbarism prevailed in Chilean discourses about the "free" territories into the 1850s.

From the early 1800s on, in various parts of the world, a vast array of writings was devoted to the inevitable disappearance of "primitive races" caused by the encounter with "civilization," to the firm conviction that wherever "primitive" and "modern" societies met, "primitive" societies would not resist for long and would give way to modernity (Brantlinger 2003; McGregor 1997). In Chile, discourses about Araucanians' nearing extinction began to surface significantly in politicians' speeches, naturalists' treatises and military reports from the moment the Chilean state contemplated the annexation of independent Araucania in the 1860s. By the mid-nineteenth century, new markets had opened for Chilean wheat as a result of the California gold boom and demand for the rich farming land in southern Chile pushed settlers into the vast and fertile territories of Araucania (Jones 1999, 176). Communities of predominantly German-speaking settlers were established around the southern exclaves of Valdivia and Osorno, both colonial foundations, and, following in 1852, next to Lake Llanquihue (Heberlein 2008). At the same time, Araucanians' independence and hostility came to be perceived increasingly as a breach of Chilean national sovereignty and safety (Bengoa 2009; Casanueva 2002; Pinto Rodríguez 1996). In 1861, Rudolf Philippi, the curator of Chile's National Museum from 1853 to 1897, himself an immigrant from Kassel, authored an article in a German periodical about the impending "extinction of the Araucanians." One noted it everywhere, Philippi wrote, the "odd fact that the number of Indians has declined steadily." Even if prosperous and well treated, they were being decimated by a combination of proneness to disease and a superstitious resistance to "adequate" medical treatment. Soon, Philippi concluded, the time would come when they would dissolve into the "white population," especially if the number of immigrants into Araucania increased (Philippi 1861). The idea of evanescence was performative in the sense that it acted on the world as well as described it (Brantlinger 2003, 4). Rudolf Philippi had long been involved in campaigns promoting German immigration to the region around Lake Llanquihue, assisting his brother Bernardo, a colonial agent of the Chilean government in Germany (Heberlein 2008; Schell 2013); his article on the Araucanians' extinction addressed a German audience of potential settlers. Given the decline of the Araucanians, taking their land seemed only natural. Philippi did not mourn their impending demise; on the contrary, Araucanians' end was desirable, for it would remove a threat to settlement and progress in the area.

In an 1868 parliamentary debate, congressman Benjamin Vicuña Mackenna, who was to become Santiago's mayor in 1872, argued in favor of military operations in Araucania. Vicuña Mackenna held that a victory against the Indian – vicious and barbarian, "the enemy of civilization" – was now possible: the population of the Republic had grown at a "gigantic" pace, while Araucania had long been "shrinking, in its territory and in its population" (Vicuña Mackenna 1868). Historians studying evanescence have long argued that the belief that primitive races were doomed – that nature had ordained their vanishing in the face of civilization – was a mantra for advocates of British imperial expansion, American manifest destiny and British settlers living on colonial frontiers in Australia, New

Zealand and North America: it justified, and it even worked toward, extinction (Brantlinger, 1995; Brantlinger 2003, 9; Maybury-Lewis et al. 2009). Chileans had pursued frontier colonization in explicit adherence to the New England-style model of settler-colonial progress and westward expansion, premised upon the displacement and even the annihilation of an economically dispensable population (Vicuña Mackenna 1868). Discourses like Philippi's or Vicuña Mackenna's about evanescence likewise mirrored British imperial and New England literary narratives about "the last of the Mohicans" or scientific discourses about the "extinction of the Tasmanians," Charles Darwin's main example in his considerations of racial extinction (Stafford 1994, 239). In Chile, as in other parts of the world, the disappearance of native populations gave rise to mixed feeling; it was, however, principally considered desirable, since it was expected to further military victory, national unity and settlement.

Even though there was considerable controversy over the means by which the incorporation of Araucania should be effected, the advocates of violent measures to displace the local population would eventually prevail (Pinto Rodríguez 1996, 84–95). Through the gradual encroachment of settler colonies, the undermining of internal alliances, and, from 1868, increasingly systematic military assaults, the Chilean state gradually advanced into Araucania. In 1868, the Chilean army secured the line marked by the Malleco River and, after years of Araucanian assaults and Chilean punitive raids, a more southerly line was established along the Traiguén River in 1878 (Collier and Sater 2004). Whereas some Araucanian communities responded with violence to the Chilean penetration into their territory, others opted to seek redress working within the governmental system, and others migrated east over the cordillera to join the growing encampments in the *pampas* (Jones 1999, 176). In 1883, after the War of the Pacific (1879–1883) with Peru and Bolivia over nitrate beds in the Atacama Desert, the Chilean army resumed the campaigns and finalized the conquest of independent Araucania.

Endangerment

Discourses about Araucanians' death or assimilation, though present throughout the progressive invasion of their territories, underwent profound transformations once Chile's politicians, military men and missionaries rested assured of their success: once the campaigns were over and the victory nearly won, the celebration of the end of "savagery" as a necessity for social progress, was gradually but steadily, giving way to the mourning of that demise. Around 1882, discourses lamenting the Indians' impending death began to surface: those who observed Araucanians' imminent downfall began to call for the protection and preservation of their remnants. Sensibilities, after all, are not beyond time or place; they may wither, travel, and make their appearance in a new environment, abruptly and even, as in the Araucanian territories, somewhat unexpectedly.

In his 1882 instructions to hydrographical expeditions, the Chilean military official and director of the Hydrographical Office (1874–1891) Francisco Vidal Gormaz ordered his men in "the Araucanian territories" to study not only the area's routes, topography, resources, and rivers, but also "Araucanian ethnography and anthropology," to excavate "in any place that may enclose objects of interest to our museums," to collect "measurements of the body in general, principally of the cranium and the limbs and the facial angle." His men were to observe and make notes regarding the inhabitants' "physical characteristics, physiological and pathological," their weapons, domestic utensils and clothes, and they were to document their religion and traditions; they also had to point out to the accompanying photographer the "best objects of study . . ., the best types of the indigenous race, their houses, and . . . ruins," and record their language, compiling "a small vocabulary of the usual words." The soldiers and marines were to "write down any observation," and the acquisitions made were to "be catalogued and preserved as state property." Vidal Gormaz also interfered with the renaming of places, prohibiting the widespread practice of European settler toponymy. "You will give preference to the indigenous geographical names," he ordered his men; "no one has a right to substitute those names for other capricious or more modern ones." If the authorities had imposed new names, these had to be adopted; still, his subordinates were to "preserve the old ones, recording to which of the [newly given names] they corresponded." In the last paragraph of his instructions, Vidal Gormaz laid out reasons for collecting material evidence of Araucanian cultures:

> . . . the planned occupation of Araucania will inevitably produce the assimilation of the indigenous element in its independent and savage form. Within no time the noble Araucanian race will have disappeared, and it will no longer be possible to study it except through the ancient chroniclers and modern explorers. What is still preserved of it will in the future gain significance in a way we can only have a vague presentiment of at the moment. . . .
>
> (Vidal Gormaz, 1882, 702)

Much like Philippi or Vicuña Mackenna years earlier, Vidal Gormaz felt Araucanians were nearing extinction and disappearance, if not into physical death, then at the very least, into cultural or linguistic "assimilation". The conclusion he drew from this apprehension, however, differed markedly from that of his predecessors: "noble" rather than "barbarian", Araucanians' vanishing had already acquired a soupçon of nostalgia – a bitter aftertaste. And it was in this altered atmosphere, owing to this novel sentiment, that Vidal Gormaz and the others who, like him, were present to witness Araucanians' demise, began to feel the moral obligation to treasure and preserve their relics and memory, by way of words, images and things.

Between 1884 and 1929, when Araucanians were relocated and placed in reservations, the need for the protection and preservation of their remnants came to be felt more acutely, by Chile's handful of professional scientists like Philippi but also by the clergymen, military officials and colonists directly or indirectly involved in the occupation and settlement of the Araucanian territories. In his "History of Araucanian Civilization," published close to 1900, Tomás Guevara, a teacher in the occupied territories and an amateur ethnographer, asserted that studying the "physical characteristics" of the remaining "authentic" indigenes was a pressing matter, for "the time of their complete disappearance was near" (Guevara 1898–1899, 306). Removal and displacement, but also the mere fact of the Indians' sustained contact with "modern life" as the priest-ethnographer Martin Gusinde (1916, 40–1) phrased it, were affecting Araucanians' cultural practices and, as a consequence, diminishing the material record available for scholarship. It was, therefore, high time to rescue and guard the material remains that could still be found in Araucania. It was high time to rescue and guard the material remains that could still be found in Araucania. As Guevara put it in 1900:

> The race is about to become extinct and with them the objects they have used in the different periods of their existence. How many instruments, furniture and pots could one find in their houses, in their possessions and cemeteries! How easily could one form a complete picture of their means of existence and activity, their habits and religious thought!
>
> (Guevara 1900, 138)

The concern with the preservation of remnants that is traceable in the writings of Vidal Gormaz or Guevara evolved in dialogue with an interconnected trans-Atlantic scientific community. It took up, in particular, the discourses and practices of European and North American "salvage anthropology" – a disciplinary orientation that embodied a moral and scientific concern with the "savage" vanishing before the spread of civilization (Gruber 1970; Sahlins 2000). A Chilean branch of "salvage anthropology" took shape primarily around the study of Araucanians' remnants. Chile's National Museum had received some artifacts of Araucanian provenance as donations from the mid-nineteenth century, often from Philippi's contacts in the German settler communities, but it was only after the end of the Araucanian campaigns that museum directors, collectors and amateur ethnographers like Guevara began to gather artifacts of Araucanian provenance on a larger scale, and to generate scholarly narratives and create institutional structures around them. The increased urge to collect and study what was about to disappear prompted an unprecedented expansion of Araucanian collections, both at the hands of private individuals as well as in Chile's public museums: the prehistory section of the National Historical Museum and the anthropology and ethnology section at the Museum of Natural History were those that received the largest share of Araucanian material culture in the decades around 1900 (Gänger 2014, Ch. 3). The end of the conquest also spurred the publication of some of the first significant anthropological studies about Araucanians in Chile – Diego Barros Arana's 1884 *General History of Chile* or José Toribio Medina's 1882 *Chilean Aborigines* – and it led to the foundation of a series of learned anthropological and archaeological societies devoted to the collecting and study of Araucanian material culture and oral traditions, such as Santiago's Society of Folklore (*Sociedad de Folklore*, 1909) (Gänger 2014, Ch. 3;

Orellana Rodríguez 1994; 1996). As endangerment became the prevailing sensibility in relation to Araucanians, scholars in Chile came to gather artifacts, to generate narratives, and to create institutional structures around them.

Whereas observations of Araucanians' customary ways of doing things had been collected for decades together with their material culture and physical remains, it was only in the second and third decade of the twentieth century that scholars in Chile began to systematically preserve also Araucanians' own voices, their memories and minds. A series of "amateur ethnographers" – school teachers like Guevara, but also missionaries or settlers – set out to fix in writing testimonies delivered by Araucanians: they interviewed "old Indians" and questioned the "ancient families of the south" about burial practices, beliefs and meanings (Gotschlich 1913; Guevara 1898–1902, 279). In 1913, Tomás Guevara edited a collection titled *The Last Araucanian Families and Customs*. Its objective, Guevara explained, was to describe the customs that had persisted "in the race in the last period of its existence, and those that disappeared upon contact with progress and the necessities of a new life" (Guevara 1913, 5).

Pascual Coña's 1926 narrative in a way both epitomized a pinnacle and, at the same time, anticipated the end of this testimonial "vogue". By the time the old man dictated his life story, it had become believable to the Chilean public imagination that the kind had been further reduced to a single representative – that all hope of regeneration had vanished, as only one very old man, on the verge of death, burdened with his memories, spoke for his dying race (Stafford 1994, 1). Pascual Coña's memories not only foresaw, they also sealed the end of a time: his memories would supersede not only Araucanians' ways and lives, but even the last of their kind.

Contextualizing endangerment

The appearance of an endangerment sensibility in Chilean discourses is somewhat unexpected. The "New England-type" of settlement colonialism of North America, Australia or French Algeria, where European settlement followed colonial rule, intent on making the conquered territory their permanent home and on expelling the former population for that purpose, is often prevalent in anthropological literature (Elkins and Pedersen 2005; Wolfe 1999). As a consequence, evanescence and "salvage" have become commonplace in discourses about colonialism. In Latin America, however, Chile was, together with neighboring Argentina, exceptional both in effectively displacing many of the inhabitants of Araucania and in producing ideas about evanescence and endangerment. Indeed, outside the British New England colonies, Canada, Australia or French Algeria economic dependence on indigenous or imported labor force entailed very different ideas about the future of indigenous populations. In colonial Spanish America, and other exploitation colonies such as British India, French Indochina, German Togo or Japanese Taiwan, autocratic government entailed rather a paternalistic care for the native population. Most Spanish American successor states perpetuated the colonial discourse of tutelage over a population that was very much alive, levying tribute on the colonized rather than displacing them. In neighboring Peru, for instance, as in the other Andean countries, social reality would have rendered any lamentation of indigenous disappearance both counterproductive and absurd. Indigeneity as a category and Indian labor were foundational and necessary elements of the Peruvian economy and population: up to the guano boom, Peru depended economically on Indian tax revenue. As a consequence of that necessity, the division of Peruvian society along the Indian/non-Indian fault line – a classification carried over from colonial fiscal practice – continued to ground relations between state and society. The guano boom allowed for the abolition of tribute in 1854, but caste divisions transcended this moment. Indian tribute was re-introduced once more under the government of Mariano Ignacio Prado (1876–1879) under a new guise, the "personal contribution." This decision was justified in racial terms, premised upon the idea that the Republic needed to coerce the congenitally lazy and frugal Indians into the labor market by imposing a monetary head-tax (Larson 2004, 146–155; Walker 1999, 11). Because the outright elimination or displacement of the Indian population was not an option that Peruvian politicians and intellectuals could realistically consider, Indians' perceived cultural stagnation was thus interpreted differently. While in Chile, the Indians' backwardness implied their impending death in the face of civilization, in Peru, their abundance and their inalienability suggested the necessity of Creole governance as a means of taking them out of their perceived stagnation (Majluf 1996, 43). In Chile, in contrast to Peru, indigeneity was – and it could be – seen to exist only on the fringes of a homogeneously white Chilean nation; it was and would long be considered quite dispensable (Applebaum 2003). The endangerment sensibility thus reflects a particular socioeconomic reality as well as an ideological conception of society: it is the product of a gritty, tangible materiality; it is not entirely imaginary, and yet, there is something deeply imagined to it (Huggan 2005, 5).

Guevara's anxiety rested, like Vidal Gormaz's twenty years earlier and like Moesbach's twenty years later, on a cognitive level; Araucanians' material culture, bodies and ideas were items of information necessary to the understanding (Delord 2006, 659) – knowledge to become useful for a purpose as yet unknown, or as Vidal Gormaz would have said, "in a way we can only have a vague presentiment of at the moment." Vidal Gormaz or, indeed, Guevara did not endeavor to preserve and to treasure the endangered themselves, protecting them from violence or physical death, but to fix their words and ideas in writing, their bodies in measures, and their things in collections – in short, to preserve not Araucanians' selves from disappearance, but their relics and memory, in anticipation of their inevitable disappearance. The anxiety at the heart of Chilean endangerment discourse around 1900 was thus ultimately about change, about transformations, in those roughly 78,000 Araucanians who had survived the occupation and who had remained in Araucania (Course 2011, 12). It embodied uneasiness with the vanishing of the "indigenous element," as Vidal Gormaz wrote, with the disappearance of "customs," as Guevara argued, or as Pascual Coña said, with the sense that they were forgetting "the nature of their race," and that they would soon forget even their language, "their mother tongue." It was, ultimately, a concern with extinction: not the biological "ending of a reproductive lineage" (Delord 2006) but the disappearance of a culture, a "race" in the period's voice, through cultural assimilation.

Again, Pascual Coña's observations were not imaginary, but reflected a reality of change and disruption. Pascual Coña probably had a very particular layer of the "new generation" in mind in his memoirs. Men like himself, the bilingual and Christian sons of the Araucanian leadership, were swift to constitute an important and visible presence in the Chilean public and political sphere (Pavez 2003, 28). Familiar with the cultural codes of Chilean society, they organized themselves only a generation after the conquest in a Society of Mutual Protection (*Sociedad de Protección Mutua*) founded in 1916, and politically, in the Araucanian Federation and in an Araucanian Congress (Crow 2010; Foerster and Montecino 1988; Ménard and Pavez 2005). They spoke Spanish as an active choice because not to do so would have prevented them from pursuing their socio-economic needs and political concerns. Like the abandonment of certain cultural practices, language disappearance in relation with "indigeneity" is usually lamented and equated with cultural impoverishment; it is, still today, hardly ever seen as a rational response to changed communicative and socio-economic needs (Mufwene 2005).

Cultures, however, perpetually renew themselves and this renewal is not coterminous with disappearance (Sahlins 2000). Endangerment and its many myths are – like the idea of a culture disappearing, resisting or going extinct – bound up with the rise of the belief in the early 1800s that, whatever may have been its original unity, humanity was fragmented into a series of discrete and stable groups (Gruber 1970, 1294). The "last-of-the-race" myth emerged as a recognizable figure in late seventeenth-century Britain, when time came to be perceived as a line and change as irreversible, and when last things ceased to be associated with a collective and apocalyptic ending of mankind, when a notion of evolution was first suggested and, within it, the possibility that new "species" could appear and others vanish forever (Stafford 1994). To Pascual Coña's contemporaries, the abandoning of customs and language appeared particularly deplorable because it stood in contradiction with the very conception of indigeneity they entertained. It seemed to happen within a peculiar cultural and epistemological framework that imagined a peculiar meaning and the impossibility of change in societies described as indigenous – constitutively defined through their authenticity, understood as the antonym of progress and change. Araucanians could only cease to be indigenous and disappear culturally into modernity – or die; reform or evolution was impossible. From the early-nineteenth century, indigeneity has thus been an archetype of the quintessentially endangered; evanescence and the need for protection have been one of its constitutive and defining qualities precisely because change and progress were not.

There is thus a socioeconomic reality to the possibility of endangerment thinking, and a specific epistemological moment, but there is also temporal contingence. Even though the discourse of endangerment had been traveling along trans-Atlantic intellectual networks for decades, it was unheard of and unconceivable in Chile prior to the campaigns in 1882. The esteem that is necessary to the possibility of endangerment thinking, and the reverence that is conducive to it, is usually possible only after the danger or inconvenience the endangered might once have represented had been overcome: gypsies (Bogdal 2002), Cherokees or Highland Scots were celebrated only after they ceased to be perceived as a threat. Like Araucanians, these other "barbarians" had long harbored a latent potential for being romanticized, complementary to and, ultimately, inherent in hostility and derision (Pocock 2008, 160). As in early-nineteenth century North America, guilt and a concern with the ethics of colonialism moved to the fore after Araucanians became harmless (Ellingson 2001; Osterhammel 1998, 269). Endangerment discourses that involve opponents – say, colonial masters and anthropologists, or the managers of the hydro-electric power plants that effect ecological change and environmental activists seeking to prevent it – obscure the temporal quality inherent in the ambiguity of preserving what is not yet lost, but must disappear if progress is to be achieved.

Missionaries like Moesbach, marine officers like Vidal Gormaz or settlers like Guevara, who turned amateur ethnographers at the end of the military campaigns, personify and lend temporality to the tension inherent in the very concept of endangerment and in the societies that produce it: the "personal union" of the perpetrator and the protector reveals that doing and undoing, destruction and protection, progress and preservation are potentially a sequence, a progression, rather than the two sides of an antagonism. Endangerment is a selfish lamentation – albeit often a proleptic one – of the irretrievable loss of something or someone precious, valued by societies who feel involved in bringing about this loss. The new reverence and nostalgia for the past came in the late 1700s, when humans realized not only the possibility of change in a supposedly static Creation, but also that these changes could be anthropogenic (Stafford 1994, 117). The very repression of "indigenous" societies, as historians have argued for the Mohicans, the Sioux or the Apache, roused "a sense of loss that demanded the preservation of what was disappearing" (Stafford 1994, 242). In the posthumous elegies of defeated men, as Fiona Stafford observes, writers found expiation (Stafford 1994, 243). Endangerment sensibilities succeed the verification of the accomplished damage. The lamentation of change is not alien to its causation but contained in it; only, it manifests itself once change can no longer be undone.

Coda

Endangerment-thinking in nineteenth-century Chile was first and foremost an intellectual and an affective concern. It was not a yearning to return to the past or to preserve Araucanians alive, but was a cognitive anxiety over a race that was about to be and must be lost. Throughout the nineteenth and twentieth centuries, endangerment thinking has been defined by reluctance toward change – disappearance, extinction or assimilation – in certain human, natural or material objects. But while fatalism and a deeply finite quality underlay nineteenth and early-twentieth century endangerment sensibility, from the second half of the twentieth century, it came to be driven by a belief in the possibility and the desirability of preservation. Intervention, the urge to prevent change and to reverse decline and loss, neither obliterated nor entirely superseded memorialization and salvage. Once more, temporal, material and historical contingence alternately prepared or barred the ground for such discourses. Araucanians' end was not as near as Pascual Coña's words had suggested it in 1927; he was neither the last, nor even among the last, of his kind. The Mapuche are today the most numerous among Chilean "indigenous peoples," with, according to the 2002 national census, over 604,349 individuals, corresponding to approximately 4% of the Chilean population. These figures have been criticized for their restrictive standards; in the 1992 census, the number of men and women self-identifying as "Mapuche" amounted to 928,060 individuals, then about 10% of the Chilean population (Di Giminiani 2011, 62).

Araucanians have not only not disappeared, they have maintained conflictive relationships with the Chilean government throughout the twentieth century. Mapuche groups repeatedly experienced state repression and marginalization (Foerster and Montecino 1988; Mallon 2005). Open conflicts over land rights have further aggravated relations between Mapuche groups and the successive Chilean governments. For a long time after Pascual Coña's "dying gasp," it may have appeared unwise to Chilean authorities to encourage living representatives of the Mapuche to reverse decline and loss; the best strategy for the Chilean state resided in treasuring the memory of the dead and their remnants. Indeed, long into the second half of the twentieth century in Chile and several other parts of Latin America, endangerment sensibilities with regard to native communities rarely moved beyond the fatalism of indigenous extinction or the wishful thinking of cultural assimilation and salvage. It required more than a global call to prevent change and to reverse decline and loss in "indigenous groups" for governments to take action in that respect. In Chile, it was only after relations between the state and Mapuche groups were amended with the end of Augusto Pinochet's dictatorship in 1989 that the Mapuche were granted – hesitantly, and still today but partially – legislation that recognized a traditional use of land or coastal areas. Only today, in rural community schools in the south of Chile, there are intercultural, bilingual education programs, agendas for the recovery of Mapuche "ancestral knowledge" and teaching programs of the Mapudungun language for all those who have indeed forgotten what had once been their ancestors' "nature" and "their mother tongue."

Note

* I would like to thank Nélia Dias, Fernando Vidal, and Josh Berson for their comments on various drafts of this chapter. This article draws on, recapitulates and develops parts of the third and fourth chapter of my doctoral dissertation, which has in the meantime been published as a book. For the dissertation, see Gänger 2011. For the book, see Gänger 2014.

Bibliography

Applebaum, Nancy P., A. S. Macpherson, and K. A. Rosemblatt, eds. Race and Nation in Modern Latin America. London: Chapel Hill, 2003.

Bechis, M., 2002. The Last Step in the Process of "Araucanization" of the Pampa, 1810–1880: Attempts of Ethnic Ideologization and "Nationalism" among the Mapuche and Araucanized Pampean Aborigines. In: Briones, C. and Lanata, J. L., eds., *Archaeological and Anthropological Perspectives on the Native Peoples of Pampa, Patagonia, and Tierra del Fuego to the Nineteenth Century*. Bergin & Garvey, London, 121–132.

Bengoa, J., 2009. Chile Mestizo; Chile Indígena. In: Maybury-Lewis, D., MacDonald, T. and Maybury-Lewis, B., eds., *Manifest Destinies and Indigenous Peoples*. Harvard University David Rockefeller Center for Latin American Studies, Cambridge, MA/London, 119–144.

Boccara, G., 1999. Etnogénesis Mapuche: Resistencia y Restructuración entre los Indígenas del Centro-sur de Chile (siglos XVI-XVIII). *The Hispanic American Historical Review*, 79 (3), 425–461.

Bogdal, K.-M., 2002. "Menschen sind sie, aber nicht Menschen wie wir." Europa erfindet die Zigeuner. In: Gutjahr, O., ed., *Fremde*. Königshausen & Neumann, Würzburg, 159–184.

Brantlinger, P., 1995. "Dying Races": Rationalizing Genocide in the Nineteenth Century. In: Nederveen, J. and Parkeh, B., eds., *The Decolonization of Imagination. Culture, Knowledge and Power*. Zed Books Ltd, London/New Jersey, 43–56.

Brantlinger, P., 2003. *Dark Vanishings: Discourse on the Extinction of Primitive Races, 1800–1930*. Cornell University Press, Ithaca/London.

Casanueva, F., 2002. Indios malos en tierras buenas: visión y concepción del mapuche según las elites chilenas. In: Boccara, G., ed., *Colonización, resistencia y mestizaje en las Américas (siglos XVI-XX)*. Ediciones Abya-Yala, Quito, 291–327.

Collier, S. and Sater, W., 2004. *A History of Chile, 1808–1994*. Cambridge University Press, Cambridge.

Coña, P., 2006 [1927]. *Lonco Pascual Coña ñi tuculpazugun / Testimonio de un cacique mapuche. Texto dictado al Padre Wilhelm Ernesto de Moesbach*. Pehuén Editores, Santiago de Chile.

Course, M., 2011. *Becoming Mapuche: Person and Ritual in Indigenous Chile*. University of Illinois Press, Urbana.

Crow, J., 2010. Negotiating Inclusion in the Nation: Mapuche Intellectuals and the Chilean State. *Latin American and Caribbean Ethnic Studies*, 5, 131–152.

Delord, J., 2006. The Nature of Extinction. *Studies in History and Philosophy of Biological and Biomedical Sciences*, 38, 656–667.

Di Giminiani, P., 2011. *Ancient lands, contemporary disputes: land restoration and belonging among the Mapuche people of Chile*. Unpublished PhD thesis, University College London.

Elkins, C. and Pedersen, S., eds., 2005. *Settler Colonialism in the Twentieth Century: Projects, Practices, Legacies*. Routledge, New York/Oxon.

Ellingson, T., 2001. *The Myth of the Noble Savage*. University of California Press, Berkeley/Los Angeles/London.

Foerster, R. and Montecino, S., 1988. *Organizaciones, líderes y contiendas mapuches (1900–1970)*. Centro Estudios de la Mujer, Santiago de Chile.

Gänger, S., 2011. *The Collecting and Study of pre-Columbian Antiquities in Peru and Chile, c. 1830s-1910s*. Unpublished PhD thesis, University of Cambridge.

Gänger, S., 2014. *Relics of the Past. The Collecting and Study of Pre-Columbian Antiquities in Peru and Chile, 1837–1911*. Oxford University Press, Oxford.

Gascón, M., 2007. *Naturaleza e imperio. Araucanía, Patagonia, Pampas: 1598–1740*. Editorial Dunken, Buenos Aires.

Gotschlich, B., 1913. Llanquihue i Valdivia. *Boletín del Museo Nacional de Chile*, 6, 1.

Gruber, J.W., 1970. Ethnographic Salvage and the Shaping of Anthropology. *American Anthropologist*, 72, 1289–1299.

Guevara, T., 1898–99. Historia de la Civilización de Araucanía. *Anales de la Universidad de Chile*, 616–653, 279–317.

Guevara, T., 1898–1902. *Historia de la Civilización de la Araucanía*. Imprenta de Cervantes, Santiago de Chile.

Guevara, T., 1900. Museos etnológicos americanos. In: Porter, C., ed., *Cuarto Congreso Científico (Primero Pan-Americano)*. Imprenta, Litografía i Encuadernación Barcelona, Santiago de Chile.

Guevara, T., 1913. *Las últimas familias i costumbres araucanas*. Imprenta, Litografía i Encuadernación Barcelona, Santiago de Chile.

Heberlein, R.I., 2008. *Writing a National Colony: the Hostility of Inscription in the German Settlement of Lake Llanquihue*. Cambria Press, Amherst.

Hidalgo Lehuedé, J., 1984. The Indians of southern South America in the middle of the sixteenth century. In: Bethell, L., ed., *Colonial Latin America. Cambridge Histories Online. The Cambridge History of Latin America*, I. Cambridge University Press, Cambridge.

Huggan, G., 2005. Introduction. In: Huggan, G. and Klasen, S., eds., *Perspectives on Endangerment*. Georg Olms Verlag, Hildesheim/Zürich/New York.

Jones, K., 1999. Warfare, Reorganization, and Readaptation at the Margins of Spanish Rule: The Southern Margin (1573–1882). In: Salomon, F. and Schwartz, S., eds., *South America. Cambridge Histories Online. The Cambridge History of the Native Peoples of the Americas*, III 2. Cambridge University Press, Cambridge.

Larson, B., 2004. *Trials of Nation Making: Liberalism, Race, and Ethnicity in the Andes, 1810–1910*. Cambridge University Press, Cambridge.

Majluf, N., 1996. *The Creation of the Image of the Indian in 19th-Century Peru: The Paintings of Francisco Laso (1823–1869)*. Unpublished PhD thesis, University of Texas.

Mallon, F., 2005. *Courage Tastes of Blood. The Mapuche Community of Nicolás Ailío and the Chilean State, 1906–2001*. Duke University Press, Durham.

Maybury-Lewis, D., Macdonald, T. and Maybury-Lewis, B., eds., 2009. *Manifest Destinies and Indigenous Peoples*. Harvard University Press, Cambridge/London.

McGregor, R., 1997. *Imagined Destinies: Aboriginal Australians and the Doomed Race Theory, 1880–1939*. Melbourne University Press, Melbourne.

Ménard, A. and Pavez, J., 2005. El Congreso Araucano. Ley, raza y escritura en la política mapuche. *Política*, 44, 211–232.

Moesbach, P.W., 2006 [1927]. Prefacio. In: Moesbach, P.W., ed., *Lonco Pascual Coña ñi tuculpazugun / Testimonio de un cacique mapuche*. Pehuén Editores, Santiago de Chile, 22–24.

Mufwene, S.S., 2005. Globalization and the Myth of Killer Languages: What's Really Going on? In: Huggan, G. and Klasen, S., eds., *Perspectives on Endangerment*. Georg Olms Verlag, Hildesheim/Zürich/New York.

Orellana Rodríguez, M., 1994. *Prehistoria y Etnología de Chile*. Bravo y Allende Editores, Santiago de Chile.

Orellana Rodríguez, M., 1996. *Historia de la arqueología en Chile (1842-1990)*. Bravo y Allende Editores, Santiago de Chile.

Osterhammel, J., 1998. *Die Entzauberung Asiens. Europa und die asiatischen Reiche im 18. Jahrhundert*. C.H. Beck, München.

Pavez, J., 2003. Mapuche ñi nütram chilkatun / Escribir la historia Mapuche: Estudio posliminar de trokinche müfu ñi piel. Historias de familias. Siglo XIX. *Revista de Historia Indígena*, 7–53.

Philippi, R. A., 1861. Das Aussterben der Araucanier in Chile. *Mittheilungen aus Justus Perthes' Geographischer Anstalt über wichtige neue Erforschungen auf dem Gesamtgebiete der Geographie von Dr. A. Petermann*, 155.

Pinto Rodríguez, J., 1996. Del antiindigenismo al proindigenismo en Chile en el siglo XIX. In: Pinto Rodríguez, J., ed., *Del discurso colonial al proindigenismo. Ensayos de historia latinoamericana.* Ediciones Universidad de la Frontera, Santiago de Chile, 83–116.

Pocock, J.G.A., 2008. *Barbarism and Religion.* Cambridge University Press, Cambridge/New York.

Sahlins, M., 2000. "Sentimental Pessimism" and Ethnographic Experience or, why Culture is not a Disappearing "Object". In: Daston, L., ed., *Biographies of Scientific Objects.* The University of Chicago Press, Chicago/London.

Schell, P. A., 2013. *The Sociable Sciences. Darwin and his Contemporaries in Chile.* Palgrave Macmillan, New York.

Stafford, F. J., 1994. *The Last of the Race. The Growth of a Myth from Milton to Darwin.* Clarendon Press, Oxford.

Vicuña Mackenna, B., 1868. La conquista de Arauco. Discurso pronunciado en la cámara de diputados, en su sesión de 10 de Agosto 1866. In: Universidad de Chile, ed., *Discursos parlamentarios.* Imprenta del Ferrocarril, Santiago de Chile.

Vidal Gormaz, F., 1882. Oficina hidrográfica de Chile. Instrucciones impartidas por la oficina hidrográfica. *Anales de la Universidad de Chile*, 61, 697–702.

Walker, C. F., 1999. *Smoldering Ashes. Cuzco and the Creation of Republican Peru, 1780–1840.* Duke University Press, Durham/London.

Wolfe, P., 1999. *Settler Colonialism and the Transformation of Anthropology. The Politics and Poetics of an Ethnographic Event.* Cassell, London/New York.

6 Tropical forests in Brazilian political culture

From economic hindrance to endangered treasure

José Augusto Pádua

Endangered forests?

The perception that there is a major ecological crisis in the Amazon rainforests is probably one of the most widely held in the globalized world. The crisis is frequently cited as a threat to the global community, which is rapidly losing one of its most important biological treasures, with grave consequences for the planetary environment. Given the severity of this crisis, it is not uncommon to hear arguments in the United States and Europe advocating that countries in the region, particularly Brazil, which is home to approximately 60% of this rainforest region, give up some of their sovereignty and allow international monitoring to ensure the preservation of a resource considered to be the birthright of all humanity.

A comparative examination of the statistical record, however, offers some surprises, and brings to light certain incongruities that give nuance to widely held stereotypes. Today, 56% of Brazilian territory is covered with forests, and Brazil's Amazon still retains nearly 80% of the forest cover that existed at the beginning of European colonialism. The forest base of the Amazon region exceeds the high levels found in Japan and Sweden (69%). In fact, Japan's status would allow for interesting reflections on the cultural construction of global environmental perceptions, given that the country is rarely associated with images of forest landscapes. The percentage of overall forest base in Brazil, and particularly in the Amazon, is also higher than that of countries such as Indonesia (52%), Russia (49%), the United States (33%), France (29%), China (22%) and the United Kingdom (12%). In addition, the forest base in Brazil is composed mainly of natural forests, not planted ones (Veríssimo and Nussbaum 2012, 6).

Why are the countries cited above not considered examples of forest crises far worse than that which affects the Amazon? Why is this region constantly being put forth as the heart of the planet's forest problem? It would obviously be naïve to deal with the problem in purely quantitative terms, and to argue that there is no such thing as an ecological crisis in the Amazon, or that it is irrelevant. The knowledge provided by quantitative data can be misleading, especially when it is detached from thorough geographical and historical scrutiny. The 56% level of the Brazilian forest base can give the impression of a territory broadly covered with forests. But the concentration of forest masses in the country's northern region, the Amazon, biases the national figure. In fact, many regions of Brazil suffer the environmental consequences of a low presence of forests. On the other hand, the 80% figure for the Amazon region seems to be high. But its actual significance changes when we consider that around 19% of that huge forest of around 4 million square kilometers was destroyed in the last 40 years. The rate of the Amazon deforestation in the last decades is much more worrying than its total amount.

The essential point, however, is that perceptions of this environmental crisis, as other manifestations of an endangerment sensibility in different latitudes, go beyond and deeper than mere figures. They must also be analyzed qualitatively, in light of specific historical and cultural contexts, and as constructed through the interplay of objective and subjective factors. This chapter focuses less on the objective processes of deforestation than on conceptual shifts regarding the value and status of tropical forests in Brazilian political culture. It considers how these shifts, which have taken place on a global scale, have manifested themselves in a developing country whose territory includes large tracts of tropical forest.

To begin with, we must take into account the specificities of the notion of danger of disappearance as it applies to living ecological units such as forests. Except in cases of extreme ecological change, forests – unlike documents or art works, whose physical destruction may be definitive – always offer a certain level of resilience through their ability to restore themselves, albeit with differing degrees of coverage, biological strength and variety. It is rare that the total mass of a big forest is actually lost. Yet it is precisely this specter of total loss that has infused collective beliefs about the Amazon region since the 1970s. One example of such perception was the title of a book by Robert Goodland and Howard Irvin that had a major impact upon its publication in 1975: *Amazon Jungle: Green Hell to Red Desert?*

As we shall see, the idea of forests being transformed into deserts was already present in Brazil's political thought in the early 19th century. But, ironically, the Amazon region in 1975 still possessed 98% of its original forest cover! In reality, the idea of total loss masks the complexities of forest endangerment. Forests contain particular elements, such as plants and animals, which are not homogeneously distributed and could be driven to extinction through the disappearance of relatively small percentages of forest mass. The tropical world is particularly rich in endemic species located within limited geographic areas. On the other hand, forests of different sizes can continue to exist even with the general impoverishment of their biodiversity. From an external standpoint, the landscape might appear stable. However, its contents change significantly, and undergo processes of genetic erosion. Moreover, total loss is not necessary for the production of systemic effects. The loss of a significant percentage of forest vegetation, even if it is far from complete, could have considerable impact on climate and/or hydrological systems. It all depends on the specificities of each ecological region.

When discussing the making of the image of the Amazon forest as an endangered reality, it is significant that there is a far more serious case of tropical deforestation in Brazil's environmental history: the cumulative destruction of the

Atlantic Forest, which once covered the Atlantic coast, from the northeast to the south of modern-day Brazil. This loss began in the 16th century, and was particularly acute during the first eight decades of the 20th century (Dean 1995). Only 12.5% of the forest cover that existed at the beginning of European colonialism – around 1.38 million square kilometers – still exists. Yet even this figure is controversial, reflecting the quandaries of discussing forest endangerment. If we take into account only the fragmented remnants bigger than 100 hectares, no more than 8.5% of the original forest cover is left. This difference of over 4% is due to the inclusion of fragments smaller than 100 hectares. Both figures are offered in the yearly assessments of a major satellite-monitoring project (Fundação SOS Mata Atlântica 2013). According to some ecologists, these fragments are not statistically relevant, since it is unlikely that they can contribute to the survival of the Atlantic Forest as a whole or to the richness of its biodiversity. Others, however, believe that any fragment, regardless of its size, is relevant to the continuity of the forest system.

The debate about these fragments, furthermore, reveals an additional aspect of the social and cultural construction of the endangerment sensibility – an example of what Joan Martinez-Alier (2003, Ch. 4) calls conflicts of "ecological distribution" and "ecological valuation." A hypothetical forest fragment of 50 hectares may be considered by a local community as essential for its livelihood and well-being, yet at the same time be rejected by scientists as being too small for the production of ecological value or for a conservation effort. This scientific assessment, which is itself likely to be contested by other representatives of the scientific establishment, could inform a legal report on a business project's right to destroy this fragment. Who then defines the right of a forest to exist as a whole or in fragments? Which social valuation, considering the different perceptions produced by different social actors, must be supported by a country's legal system as the most acceptable one? Which geographical index must be used by a public policy of quantification and conservation of the remaining pieces of a forest?

To return to the point about the social construction of the endangerment sensibility, the massive case of deforestation in the Atlantic Forest has remained largely invisible to the international community. Of course, historical timing has played an important role in this. The destruction of the Atlantic Forest started to take place before words like "rainforest" and "biodiversity" became popular in contemporary culture. It is interesting to note the diffusion of the word "rainforest" in our common parlance by the late 20th century, in place of words such as "wilderness" and "jungle," which evoked more somber and threatening images of tropical forests.

According to Candace Slater, the term "rainforest" first appeared in English around 1898, as a translation of the German word *Regenwald*. However, by the end of the 20th century, it had gained new meanings with the public: "the rainforest that emerged in the 1960s and 1970s had acquired an aura of science" (Slater 2002, 138). These forests henceforth appeared as "delicately poised natural laboratories that invite use even as they help to regulate and ensure the planet's environmental balance" (ibid.). Furthermore, the term became even more familiar to the public at large in the early 21st century, through a "growing com-modification that has helped to change the spelling of rain forest (a two-word noun) to rainforest (a single-word adjective) and has placed rainforests on a wide array of shampoo bottles and cereal boxes" (ibid., 147). Of particular relevance to this chapter is Kelly Enright's observation that the diffusion of the rainforest ideal sparked a transformation of the image of tropical forests from "a place that endangered human lives" to "an endangered place invested with the power to save human lives" (Enright 2012, 7).

Most of the destruction of the Atlantic Forest happened prior to this con-ceptual revolution, during the period from the 1930s to the 1960s, or while the dissemination of the "rainforest" idea into the global and Brazilian public con-sciousness was still underway, particularly in the 1970s and 1980s. It is important to emphasize that this conceptual change also took place in countries with large expanses of tropical forest – Brazil being one of them. In other words, it should not be seen as an externally driven dissemination aimed at the tropical world, but as a cultural shift that was also being forged within the universe of tropical societies – at least some of them – through a complex interaction of exogenous and endogenous forces.

In any case, the historical context for the growth of the idea of the Amazon rainforest as deeply endangered is rather different from that which an analysis of the Atlantic Forest reveals. First, it is significant to note that the Amazon rain-forest already offered an almost archetypical image of a "jungle," if for no other reason that it was around three times the size of the Atlantic Forest (or five times if we consider the whole South American Amazon). It is interesting to observe, for example, how the Amazon forest, much more than the Atlantic one, was the exotic scenario for internationally popular writers of adventures since the 19th century, like Arthur Conan Doyle and Jules Verne (Gondim 2007). In addition, the timing of its most intense destruction, beginning in the 1970s, coincided with the international emergence of current Western environmentalism, when the rainforest begins to be seen as an "endangered place" as opposed to a "place that endangers." The intensity and speed of deforestation in the Amazon was par-ticularly aggressive, constituting what was referred to as the "decades of destruc-tion" (Cowell 1990). Over a relatively short period of time – from 1970 to 1995, approximately – 12% of that enormous forest was lost (Pádua 1997). Without question, the speed of such destruction and its reporting by modern mass media was an objective factor in the creation of an image of the Amazon rainforest as an "endangered" reality, interacting with the subjective factors cited above.

The different historical moments of the massive deforestation in the Amazon and the Atlantic forests must also be considered in relation to the local popula-tions of these macro regions, most especially of the inhabitants of the forests as such. These populations are frequently ignored in narratives about the destruc-tion and conservation of the world's forests. As Candice Slater (1995, 114) observed, in the case of the Amazon, the view of the forest as "a kind of Eden" fosters "a skewed and largely static approach toward a multilayered and decidedly fluid reality." She continues: "The problem is not just that this vision is often false and exaggerated but that it obscures the people and places that actually exist there" (ibid.).

At other times, however, the inhabitants of the forest, the "indigenous peoples" (in Portuguese, *indios* – a very misleading concept inherited from the early modern world), are included in the picture, but as generic, ahistorical and static entities, as if they were a part of nature and the tragic and passive victims of the destruction processes. Sometimes, they are considered as one aspect of the negative image that contributes to the justification of the conversion of the dark forests into "civilized" landscapes. At other times, these communities have been represented as part of a positive image that helps portray the forests as charismatic landscapes that merit conservation.

The words "multilayered" and "fluid" are useful for characterizing the recent historical emergence of what have been called the "forest peoples" in Brazil. The concept of "forest peoples" embraces indigenous peoples, rubber tappers, Brazil-nut gatherers, fishermen and other social actors living in the forest and from its resources. The historical moment of the destruction of the Amazon forests offered much broader political, cultural, institutional and communication tools for the presence of these actors in the decision making processes about the fate of their environment (Hecht and Cockburn 2010).

The struggles of the Amazon forest peoples' communities against deforestation successfully forged political alliances with other expressions of the endangerment sensibility toward deforestation, more concerned with the loss of biological richness or ecological services. These struggles also converged with the reaction of the scientific community – along with a considerable portion of public opinion, both in Brazil and abroad – against the continued destruction of woodlands in the region. Of course, similar alliances and social connections were not available during the older and completely different sociological scenarios at hand when the Atlantic Forest was undergoing deforestation.

The junction of all these social processes and cultural changes in the last decades had major repercussions in what I will refer to, in rather broad terms, as Brazilian political culture.[1] This, of course, is not a homogeneous ideological reality that has a permanent and complete hegemony. A political culture is an historical product that can change according to the movements of the social life along a period of time. Antagonistic views, counter-discourses and social resistances can challenge it. But, at the same time, it does have a certain conservative drive and cannot be considered as an ever-changing reality. Thus, if the hypothesis that will be discussed in this chapter – namely, that the dominance of a "conversion imperative" gave way to that of a "conservation imperative" in relation to the tropical forests in contemporary Brazilian political culture – were to be confirmed by the historical outcomes of the next years, then we would have witnessed a major historical transformation.

The possibility of such a change in Brazilian political culture can be linked to concrete political decisions that led to a significant reduction in forest destruction, despite the fact that these decisions have not come immediately and that the process as a whole has experienced ups and downs. By the end of the 20th century, both in the case of the Amazon and the Atlantic Forest, a certain political consensus against deforestation began to emerge across party lines. The decision to preserve the remnants of the Atlantic Forest from clearcutting, even in zones outside of protected areas, was taken by a 1993 federal decree that was turned into law by the Brazilian Congress in 2006. Relatively speaking, the current levels of deforestation are rather small – approximately 239 square kilometers per year – alongside forest growth through natural and induced regeneration in different regions (Fundação SOS Mata Atlântica 2013). In the case of the Amazon rainforest, even though the current level of deforestation is much higher, there was a reduction of around 80% in the annual deforestation from 2004 to 2013 from 27,772 to 5,891 square kilometers (INPE/PRODES 2014).

Thus, the historical and geographical contexts of the two processes of deforestation reduction are rather different. In the case of the Atlantic Forest, the legal and political effort is focused on saving the last remnants and supporting regeneration. In the case of the Amazon forest, the significant cultural and political shift cited above became intertwined with the early stages of the deforestation process, allowing for concrete political and legal measures to be taken while 80% of the original forest cover still remained.

On the whole, both processes of deforestation reduction must be taken into account when discussing the impact of ideas and cultural perceptions on the course of environmental history. And we must be careful when faced with idealistic assessments of the increase in forest conservation as a victory of "pure ecological reasoning" – something that would be very difficult to define – over political and economic interests. While there is some truth to such an interpretation, the dominant political mode through which change has taken place in this domain in Brazil, particularly in the early 21st century, can also be viewed as the development of a new form of understanding economic and political interests in relation to the rainforests, rather than as a victory over these interests.

In any case, it is important to discuss the modes through which a conceptual change in relation to the importance of forest conservation is taking place in the Brazilian political culture. A broad historical perspective must be adopted in order to show just how culturally transformative this change is. The construction of Brazilian society, together with the development of its territory, has some essential characteristics that are diametrically opposed to the logic of forests as "endangered places" to be preserved. The majority view with respect to forests, which has its foundation in the centuries of Portuguese colonialism in America, was based on an opposite vision that I call the "conversion imperative." This vision did not recognize the value of the tropical forests as landscapes, even if some individual species of the local flora and fauna were strongly valued for economic or cultural reasons. Moreover, forests were not perceived as endangered, but rather as abundant and excessive, as well as a hindrance to the creation of a Western-like civilization. The conversion of wild forests to useful and civilized landscapes was one of the main aims of the dominant social actors. The destruction of the Atlantic Forest started to take place in the context of such a perspective. The same can be said, in a different historical moment, of the opening of the "decades of destruction" in the Amazon region.

In what follows, I will examine two significant aspects of this history: 1) the model that led to the construction of the Brazilian territory by way of a "conversion imperative," and 2) the evolution of the debate and the main arguments that were introduced and took hold in Brazilian political thought, in dialogue with global discourses, in favor of recognizing the value of forests and the possibility that they could become an "endangered" resource. These arguments, taken together, seem to be giving rise to a new political consensus in favor of a "conservation imperative," particularly with respect to the future of the Amazon rainforest.

The conversion imperative

When in the 16th century Europeans began to explore the Atlantic coast of what would become Brazil, they found a striking continuum of different types of tropical forests – those that are now unified under the category of Atlantic Forest. The forest was ubiquitous. The technology available at the time, obviously, did not allow for a precise measurement of the size of the continent's green mass, which stretched 100 to 500 kilometers inland. But from the outset, the landscape illustrated by colonial documents was grandiose and unlimited. The first description of the Portuguese encounter with this new land, written by Pero Vaz de Caminha in 1500, explained that "look as we would, we could see nothing but land and woods" and added that these woods were "so abundant and large and so vast, with foliage of such quality that they cannot be measured" (Caminha 1963 [1500], 59, 67).

The notion that a piece of the natural world may be "endangered" involves perceptions of finitude, scarcity and the risk of a partial or total loss: these were all absent from objective and subjective assessments of the forests during the initial construction of Brazil. Another central component of the endangerment sensibility, the idea of value motivating the protection and preservation of an entity, appeared only selectively. During the ongoing process of colonization, ecclesiastical and lay writers hailed the natural resources of the Brazilian tropics. However, the tendency that can be observed in the first centuries of the colonization of both the Atlantic Forest and the Amazon forest regions placed much more emphasis on the remarkable elements of the local flora and fauna than on the surrounding forest that served as their habitat. Parrots, monkeys, cashews and passion fruits – or, particularly in the case of the Amazon region, turtles, manatees, sarsaparillas and copal trees – received much more attention than the forest as such. Even in the cases where positive observations were made about the wider landscape, the presence of fresh air and pure waters received primacy over the woodlands. Indeed, these last elements had a more direct correspondence with signals of health and perfection present in biblical lore and in the medical literature of Mediterranean antiquity, both of which greatly influenced the European post-Renaissance writers in the colonial tropics (Assunção 2001; Holanda 1968).

Interest in specific elements capable of generating wealth or awakening intellectual curiosity largely obscured a vision of the forest landscape as a whole. In
order to understand this phenomenon in its historical context, we must first
examine its subjective and cultural aspects. In Western thought and its expression
in the colonial world, the attribution of value to the forests occurred relatively
late, basically starting from the 18th century. As Keith Thomas shrewdly noted
with respect to English cultural history in the same period – and this could apply
to a large portion of Europe – value was assigned to green landscapes and wildlife
in the context of the growth of urban and industrial spaces, as well as capitalist
agricultural activities. In other words, the landscape of modernity served as the
locus from which value was assigned to the landscape of wilderness. Prior to this,
from the 16th to the 18th centuries, concern about caring for and preserving the
natural world was uncommon. As the author points out,

> at the start of the modern period, man's ascendancy over the natural world
> was the unquestioned object of human endeavor. By 1800 it was still the
> aim of most people . . . But by this time the objective was no longer unques-
> tioned. Doubts and hesitations had arisen about man's place in nature and
> his relationship to other species.
>
> (Thomas 1983, 289)

It could be said that the emergence of new sensibilities with respect to nature,
including the attribution of value to forests, appeared within two major cultural
currents. The first of these was Enlightenment science, particularly by way of
the concept of a "system of nature." The image of a dynamic interdependence
between a variety of elements in the natural world – of an "economy of nature" –
helped increase the value of forests as instruments contributing to the quality of
the climate, soil fertility and a regular supply of water (Worster 1994). One particularly relevant theoretical formulation was the so-called "desiccation theory."
This theory, which was possibly the first modern scientific conception of the
risk of anthropically induced climate change, equated the destruction of native
vegetation with a decline in humidity, rainfall and water sources (Grove 1995,
153–165). The perception arose that desertification by deforestation posed a risk
to the security of states and their territories.

The second cultural current was Romanticism, which attributed to large forest
landscapes' intrinsic aesthetic and spiritual value. Natural wilderness, embodied
in great forests, mountain ranges and deserts, was seen in a positive light through
the lens of the "sublime." Aesthetic reflections and literary descriptions by a variety of Romantic writers and thinkers went beyond the post-Renaissance tendency to reduce the positive aspects of the rural world to a landscape of gardens
and plowed fields (Gusdorf 1985; Nash 1982; Thomas 1983).

The basic model of occupation of the Brazilian territory was defined well
before the emergence of these new sensibilities. Its essential continuity into the
19th and most of the 20th centuries also sidestepped any specific ethos attributing tangible value to the forests. Obviously, we are not speaking of a monolithic

cultural tradition. Already in the colonial period and, more intensely, in the post-colonial world, there were some voices in favor of the forests. But they were a minority, and the dominant cultural framework, embedded in the quotidian pragmatism of social survival and economic productivity, was much closer to the image disseminated in 1711 by a Jesuit author, André João Antonil, in his book *Cultura e Opulência do Brasil por suas Drogas e Minas* (The Culture and Opulence of Brazil through its Drugs and Mines), the first extensive description of the Brazilian colonial economy. Antonil saw the forest as an obstacle to agriculture. Thus, as he wrote, "once the choice was made regarding the best land for sugarcane, it was burned and cleaned and cleared of anything that can be a hindrance" (Antonil 1976 [1711], 102). The tropical forest, which was later re-signified as a representation of ecological beauty and wealth, was an obstacle to be removed by iron and fire.

It is important to place this process within the broader framework of environmental history (Pádua 2010). The Portuguese colonizers, accustomed to the ecological limits of their homeland, saw the Brazilian forests as a vast green ocean, ever open to an ongoing occupation. Portuguese colonial diplomacy was capable of negotiating, particularly with Spain, formal possession of an enormous territory. Its overall occupation by neo-European dominance, however, was small. There were scattered points of greater demographic concentration, where the natural resources were exploited through plantations, ranching, mining etc. When the country gained independence in 1822, its population was around 4 million people in the areas effectively controlled by European descendants. But the formal territory, accepted by diplomatic treaties and represented in maps, was almost the size of modern-day Brazil. By 1900, the population counted in the Brazilian national census had reached the 17 million range (while at the same time, in comparison, the United States was already close to 76 million people). Most of the country, including the areas surrounding those strongly dominated by European descendants, was made up of uncolonized territories scarcely inhabited by neo-Europeans, with a strong presence of wild nature and varying levels of indigenous populations. These areas were called *sertões* in Luso-Brazilian culture.

Therefore, from the point of view of the dominant colonial and post-colonial elites, the equation between populations and territory was important for the establishment of an occupation model that could be considered destructive with respect to the large forests. Production methods were generally careless and extensive, based on a parasitic relationship with the natural world. The ubiquity of slash and burn clearings was the most obvious symbol of this mentality. Instead of mulching the soil in order to preserve its fertility, the choice was made to progressively slash and burn new forest areas, since the richness of the resulting ashes would guarantee a few years of good harvests, after which the soil would decline in productivity, overtaken by weeds and ants. The forest, therefore, was not just a hindrance, but also a source of biomass to be combusted. The burnings were at the heart of the dominant agricultural method.

The ease with which land was granted to or conquered by the landlords was another aspect of this model. The colonial sprawl took place through informal occupation or the requisition of land donated by different levels of authority within the colonial state. After 1850, a national law decreed that properties must be bought in the market. But informal occupation remained widespread and the abundance of supply stimulated low prices. The relative ease with which land was obtained encouraged careless use and the subsequent near abandonment of the degraded areas. As a consequence, the push continued in the direction of unexplored and unexploited forests. As wood was used up near the productive centers, the frontier kept moving toward regions where it remained abundant. The Jesuit Antonil, cited above, recognized that the rudimentary furnaces of the sugarcane plantations were "mouths that truly devoured the forests." But he offset this fact with the image of an inexhaustible natural world: "only Brazil, with its immense forests, could satisfy, as it has satisfied for so many years, and shall satisfy in the years to come, so many furnaces" (Antonil 1976 [1711], 115).

The model was thus based on the assumption that the native forests would have to be destroyed and converted into economically and socially useful landscapes, even though in practice they were generally transformed into degraded areas. This process was well captured in 1799 by mineralogist José Vieira Couto. Armed with the new tools of Enlightenment science, capable of assessing the value of these forests, Couto stated, "It seems that it is time to pay attention to these precious forests, these pleasant woods that the Brazilian farmer is threatening with total burning and destruction, with a broadax in one hand and a firebrand in the other" (Couto 1848, 319). The main cause of destruction was the fact that "the farmer gazes around and sees two or more leagues of forests, as if he were looking at nothing, and even before he finishes reducing them to ashes he is again gazing far away to take destruction elsewhere. He has no affection nor love for the land that he cultivates, because he knows well that it will not last long enough for his children" (Couto 1848, 319). The combination of the devaluation of the native forest landscape (into "nothing") and its abundance and relatively free occupation was practically the antithesis of the idea that they might be considered "endangered" and ought to be preserved.

Moreover, it is the view of the most renowned book of history from the period following Brazil's independence, Francisco Varnhagen's 1857 *General History of Brazil*, that such occupation should not be condemned because it represented the very essence of the country's heroic formation. According to the author, the English colonizers in North America encountered conditions similar to those of Europe, such as a temperate climate and an absence of poisonous animals. By contrast, in Brazil, the Portuguese and their descendants met with an environment marked by a blazing sun and forests covered with thorns, vines and snakes. The vigor and strength of human will needed to tame that wild territory was a badge of honor for the construction of the Brazilian nation (Varnhagen 1975 [1857], 15–16).

In any event, the model of territorial occupation based on the "conversion imperative" and inherited from the colonial past remained in effect, at least with regard to its primary aspects, throughout the 19th and 20th centuries. Moreover, it was implicitly clear that it would contribute to the defeat and displacement of the "rude and primitive" inhabitants of these forests, who were also perceived as a hindrance to progress and civilization. A deeper analysis, which is beyond the scope of this chapter, would consider a variety of nuances, qualifications, exceptions and regional peculiarities. But some of the main features of this occupation pattern, which are directly related to our focus here – such as the sense of an open frontier, the low value attributed to native vegetation as a whole and the contempt for forest peoples – long remained part of the ethos that dominated the territorial construction of Brazil.

Even though regional economic activities were significantly based on deforestation and the use of wood, the overall forest stock remained enormous at least until the mid-20th century. In 1954, for example, it was in the range of 76% of the territory (Veríssimo and Nussbaum 2012, 6). The size of the economy and the overall population, as the sum of its regional manifestations, was not enough to fill the vast expanses of national space. Brazilian society was concentrated along the coast, the domain of the Atlantic Forest, with very low levels of occupation in the Amazon region and the great savannah of central Brazil known as *Cerrado*. Brazil reached the mid-20th century with an essentially rural economy and population, with low levels of literacy and a high concentration of wealth among a small elite. The urban-industrial spaces, in turn, were geographically concentrated and limited in size.

The second half of the 20th century, however, was a major turning point, with the country registering some of the highest rates of economic growth on the planet, albeit interspersed with periods of crisis, stagnation and inflation. The total population grew from 51 million in 1950 to 190 million in 2010. And the proportion of those living in cities went from 36% in 1950 to 84% in 2010. This expansion was also fueled by the growth and remodeling of urban landscapes and the construction of major infrastructure, particularly hydroelectric plants and highways (Dean 1995; Hecht and Cockburn 2010; Pádua 2012).

It was in this context that large new frontiers were opened to agriculture and livestock in regions covered with tropical forests and other native ecosystems, be it in the Atlantic Forest, the Cerrado or the Amazon forest. Considering the aggressiveness with which this development was pursued, particularly during the rule of a military dictatorship between 1964 and 1984, we could expect a powerful renewal and unparalleled expansion of the traditionally dominant "conversion imperative." Indeed, the initial moves of the process led to countless clashes between the new agents of capitalist territorial occupation and local populations. However, the historical moment was quite different from the past and a wide coalition of resistance against deforestation, particularly related to the fate of the Amazon, was able to produce a new political framework for the discussion about the meaning and the future of the Brazilian territory.

The difficulties of access to the Amazon region in the pre-industrial world, especially because of the size of its rivers, the abundance of wetlands in the areas of easier access, the resistance of indigenous communities and the occurrence of diseases, hampered more intense settlement. The Portuguese were able to adopt a "low intensity" pattern of colonial domination in order to guarantee a relatively loose political and military command over some key areas of that huge region. The native communities remained living in an autonomous way in the vast areas outside direct colonial occupation, or had to deal with the settlers through a variety of options going from negotiation to open war. The relatively small number of settlements dominated by Euro-descendants were created mainly along the Amazon River, based on the selective extraction and cultivation of certain marketable resources from the forest, like spices and cocoa (Pádua 2011). After Brazilian political independence, during the "rubber boom" from 1850 to 1915, hundreds of thousands of migrants went to the Amazon to work as rubber tappers. Since the rubber economy did not require felling the trees, they had to live in a permanent forest landscape, and had to maintain the forest habitats where these precious trees were able to survive. The "boom" led to the remarkable growth of a few cities, such as Belém and Manaus, but the whole process stopped with the growing world market hegemony of rubber plantations in South East Asia after 1910 (Pádua 1997).

At the beginning of the 1970s, the Amazon forest was opened for massive economic occupation backed by the geopolitical ideology promoted by the military regime. The main goal was not to obtain short-range economic profits, but to guarantee the effective control of these vast "empty" spaces by the Brazilian state, thereby protecting their resources from the greed of the great world powers. The construction of roads and infrastructure and the allotment of subsidies for economic activity were carried through without a real sense of the ecological or social damage that might be done (Hecht and Cockburn 2010; Pádua 1997).

It must be remembered that the indigenous populations in Brazil who were best able to survive in an autonomous way were precisely those who lived in the most distant backlands from the Atlantic coast, or those who had taken refuge there. The Amazon and Center-West hinterlands are even today the locus of Amerindian cultures and populations in Brazil (around 890,000 people in 2010). However, it is important to not consider only the indigenous communities living in a more or less isolated way. Fishermen, rubber tappers and collectors of Brazil nuts and other local resources were able to create many communities of forest peoples along the rivers. Together, they made up an extensive forest population that resulted mainly from the cultural and physical mixing of Indians and descendants of the poor migrants from Northeast Brazil who came to the Amazon during the rubber boom.

In any case, the resilience of these communities, along with other cultural and political changes of the late 20th century, formed an essential component of the historical process that has been halting the simple reproduction in the Amazon of the "conversion imperative." Like the Indonesian rainforests studied by Anna Tsing, the Amazon forest became a space of "friction," of "awkward, unequal, unstable, and creative qualities of interconnection across difference" (Tsing 2005, 4). In many cases, this friction was tragic, as in the new epidemiological shocks suffered by the Indigenous societies after the 1970s or in the violent private appropriation of territories that destroyed the livelihood of entire communities after the 1980s. But it also generated unexpected outcomes, like the international scandal produced by the assassination of the union leader and environmentalist Chico Mendes in 1988.

The next section deals with this new historical context, in which important cultural-political shifts are contributing to a new tendency, maybe even a new consensus within Brazilian political culture. This favors the thesis that tropical forests, particularly the Amazon rainforest, are totally or partially endangered and ought to be saved. This is what I shall call here the "conservation imperative." Such a historical change is still very far from having implemented the best standards in relation to the conservation of forests and of the human and cultural rights of its inhabitants. But concrete improvements, above all in comparison with the historical past, are remarkable. To understand them, we must revisit some central features of Brazilian political and cultural debates since the 19th century. Like the "conversion imperative," the "conservation imperative" has quite a long history.

A conservation imperative?

Forests did not come to be valued in Brazilian cultural debates for the first time at the end of the 20th century. At least since the end of the 18th century, some voices, such as that of José Vieira Couto cited above, criticized the destructive model of land occupation and stated, "It seems that it is time to pay attention to these precious forests" (1848, 319).

As the nation state of Brazil was being built during the 19th century, a paradoxical situation arose with respect to the image of the forests. The economic relation with the forests was very destructive and based on the practical consideration of them as a hindrance to civilization. At the same time, the cultural value of forests increased, as they became rhetorical tools and national symbols. Since the country's independence, the vision of a vast territory and its natural wealth has been widely used as a symbolic resource for affirming national prestige. In fact, during the diplomatic negotiations surrounding the recognition of Brazilian independence by Portugal and England in 1824, one of the main arguments featured in the letters of instruction from the Brazilian Foreign Ministry relied upon the territory itself: "an empire so spacious and large, provided by nature . . . with a variety and wealth of natural products, must be a separate and independent power" (Valle 2005,156).

The national authorities sponsored a version of the Romantic culture that constantly took up the subject of the magnitude of the Brazilian forests and the value of the indigenous peoples as their primitive inhabitants. Even in dialogue with their European counterparts, Brazilian Romantic intellectuals clearly understood that the new country lacked the strong institutions valued in the Western countries. The presence of vibrant tropical nature and indigenous peoples, however, gave rise to an attractive exoticism that could be used as a cultural and political resource. As several authors have pointed out, notwithstanding its artistic qualities, such Romanticism was conceptually superficial, and subject to the political order of the monarchy (Costa Lima 1989). After centuries of decimation, it was particularly ironic to see largely imaginary indigenous heroes associated with the national identity. The same could be said for the native vegetation, which had also been devastated.

Images of forests, tropical plants and indigenous peoples occupied important places in countless cultural manifestations of the time. They appeared in architecture, literature, the visual arts and the official iconography of power. Museums sought to emphasize the diversity of the flora and fauna, promoting the image of a spectacular, picturesque and scientifically valuable natural world. Brazilian stands at international exhibitions focused on the wealth of the forests, together with indigenous crafts, thus encouraging through exoticism both the cultural admiration and economic interest of foreigners (Schwarcz 2004).

It was in the 19th century that certain traditions and images emerged and became prominent, that have come to define Brazil as an autonomous political entity. These images conveyed significant ambiguities. The positive evaluation of forests, for example, proudly praised by the botanist and emperor's personal physician, Francisco Freire Alemão, as the "royal seal of the Brazilian soil, admired by foreigners" (Alemão 1961 [1851]), also gave rise to controversy. The celebration of tropical nature and peoples contrasted with many aspects of European culture that remained a model for the Brazilian elite.

The geographic determinism of a mid-19th century author like Henry Buckle was a source of worrisome perspectives. Buckle saw an inverse correlation between the richness of nature and the advance of civilization: "Amid this pomp and splendor of Nature, no place is left for Man. He is reduced to insignificance by the majesty with which he is surrounded . . . The whole of Brazil, notwithstanding its immense apparent advantages, has always remained entirely uncivilized; its inhabitants wandering savages, incompetent to resist those obstacles which the very bounty of Nature had put in their way." The "physical impediments" were so significant that "during more than three hundred years the resources of European knowledge have been vainly employed in endeavoring to get rid of them" (Buckle 1930 [1857], 54–55).

Thus, two contradictory features coexisted in the making of Brazilian political culture in the 19th century: on the one hand, an idealized image of the forests and indigenous peoples in the world of culture and arts and, on the other, an implicit consensus that forests and indigenous populations were practical hindrances to national progress. The "conversion imperative" ruled the world of economic production and political decisions.

Within this same historical framework, however, a minority of thinkers began to adopt positions that differed greatly from both the rhetorical praise of the tropical forests and the defense of their aggressive conversion. Such voices had been taking shape since the end of the colonial period, in the context of the Luso-Brazilian Enlightenment.

The first consistent criticism against the devastation of the forests came from rationalist and scientifically trained authors. In an essentially political and anthropocentric reading – which with the proper caveats can be compared to the current concept of sustainable development – the environment was praised for its economic potential. Its reckless destruction was seen as a sign of ignorance and lack of concern for the future. True progress should involve conservation and the intelligent use of natural resources, based on scientific knowledge. The destruction of these resources was identified with archaic and rudimentary elements inherited from the colonial past, on a par with slavery and the big properties system. It was seen not as the price of progress, but as that of backwardness (Pádua 2002). Romantic intellectuals in Brazil, on the other hand, who spoke so much of nature and the forests, barely mentioned their destruction. Occasional lamentations were heard, but nothing in the sense of a radical defense of native landscapes and traditional peoples against the advancing modern world, as had been put forth by certain European and North American Romantic intellectuals.

The ideas of one of the main advocates of the rationalistic conservation viewpoint, José Bonifácio de Andrada e Silva, provide the most consistent synthesis of the criticism of deforestation at the time. It is interesting to note that this statesman/philosopher, similar in style to the founding fathers of the United States, was one of the most important figures in the construction of an independent Brazil, particularly for having led the ministry that advocated the split from Portugal in 1822 (Pádua 2004). In an important text in which he proposed to the National Assembly a plan to gradually do away with slavery, he took an almost apocalyptic stance with respect to the consequences of deforestation:

> Nature has done everything to help us, but we have done nothing to help Nature . . . Our precious forests are disappearing, victimized by fire and by the destructive axe of ignorance and selfishness. Our mountains and hillsides are daily balding, and in time there will be a shortage of the fertile rains that appease the vegetation and feed our watersheds and rivers, without which our beautiful land of Brazil will be reduced, in less than two centuries, to the condition of the empty plains and the arid deserts of Libya. The day (a terrible and fatal day) will then come when scorned nature will have completed its revenge against so many mistakes and crimes committed against it.
>
> (Silva 1973 [1825], 38)

Texts such as these must be included in the genealogy of the vision of forests as "endangered" in the modern world. There was no precise quantification of what was being lost and what might take place with its eventual extinction. But there was a very clear view that "our precious forests are disappearing," and also a new positive focus: the recognition of an asset to be preserved. The most significant point, in the context of this chapter, is the presence of two central utilitarian arguments that even today serve as the basis for global anxiety about deforestation: 1) the forest is a storage of very useful species, and 2) it possesses systemic value for the health of the environment, including climate equilibrium. The understanding of these key problems has of course been expanded and reformulated, according to cultural and political changes during the 20th century and up to the present day.

It is not possible here to offer a detailed discussion of the continuity of concerns about the risk of deforestation in Brazil up to the end of the 20th century. These concerns, though intellectually very fecund, have had almost no concrete impact, and have generated almost no effective political action. The first concrete, even if unsuccessful, proposals for forest preservation through the creation of protected areas, however, betrayed a certain resignation about inevitable loss. Upon proposing in 1876 the creation of national parks in Brazil, based on the model of Wyoming's Yellowstone National Park, the engineer and abolitionist leader André Rebouças preferred to emphasize regional progress, since (he claimed) these parks would attract rich European tourists and would present Brazil to potential immigrants as an attractive place, filled with natural wonders. Only at the end of his proposal did Rebouças mention the advantages of keeping intact, "free from iron and fire," some areas of particular beauty, so that "hundreds of years from now" their descendants would be able to see "specimens of Brazil, as God created it," featuring "a flora unrivaled anywhere in the world" (Rebouças 1876, 70, 73).

The same resignation persisted for most of the 20th century. The few voices critical of extreme deforestation were unable to change the "conversion imperative" and to stimulate truly effective conservation measures. Brazil's first national park, Itatiaia, in the state of Rio de Janeiro, was created only in 1937, followed by a handful of others up until the 1970s. In fact, until the end of the 20th century, there were almost no social and political forces that could transform intellectual criticism into effective conservation initiatives. The state was weak and had a limited tax base. Governments were highly dependent on the support of regional elites, whose power was based upon devastating productive practices. During the first decades of the Brazilian Republic, which was founded in 1889, the institutional framework actually worsened because the greater decentralization strengthened the control of local oligarchs over the economy, allowing for an even more careless use of the forests. The phenomenon was aggravated by railway construction into relatively unoccupied portions of the Brazilian territory.

Although some social thinkers, such as Euclides da Cunha and Gilberto Freyre, continued to denounce what the latter referred to in 1936 as "a state of war between man and forest" (Freyre 2003, 81), the political debate surrounding deforestation did not take off until the 1970s. The dominance of an ideology of progress and the desire for growth obscured any real concern about the danger of losing the forests. What did continue, paradoxically, was the superficial, rhetorical and generic sense of forests as an emblem of Brazil's greatness. This image, which had been conceived in the 19th century, persisted in both learned and popular cultural manifestations. But true criticism was almost totally confined to the academic realm, where natural scientists lamented the loss of rich natural treasures and encouraged the occasional establishment of small nature protection associations and the promotion of legislative measures that were limited in scope and implementation.

In 1930, there was a political revolution aimed at overcoming Brazil's economic stagnation and promoting progress. In the following decades, authoritarian governments (interspersed with occasional moments of democracy) adopted institutional modernization measures and promoted industrial growth. Some measures, largely institutional, were adopted in relation to the management of the territory, such as the creation in 1934 of laws aimed at controlling the exploration and use of natural resources: the Water Code, the Hunting and Fishing Code and the Forest Code. The latter represented a step forward at the theoretical level, insofar as it placed limits on property rights in the name of conservation. Some forest areas were considered to be protective, given that they sheltered watersheds and prevented erosion. Similarly, limits were placed on the quantity of forests on each property (a maximum of 80%) that could be converted to other uses (Dean 1995, 276). However, as far as Brazil's forests were concerned, these measures hardly represented a real change. In addition to the ideology of growth, the traditional power of large landowners hindered more consistent state actions to preserve native flora.

By mid-century, during the democratically elected administration of Juscelino Kubitschek (1956–1960), the motto "advancing 50 years in five" summed up the "developmentalist" approach. This included the construction of the country's new capital, Brasília – significantly located in the middle of the Center-West Cerrado – and a major boost for industry (particularly the automobile industry). In this context, forest conservation policies became even more difficult to promote. In the 1960s, despite an awakening of the ecological debate at the international level, the establishment of a military dictatorship restricted the possibility of critical debate and promoted economic growth at all costs. A clear indicator of this was the fact that forest management was carried out by the military governments through an agency, created in 1967, called the Brazilian Institute for Forest Development. Its objectives included "carrying out the necessary measures for the rational use, protection and conservation of natural resources." Yet, as the name itself indicates, the development of a forest economy took absolute precedence over any conservationist concern. Warren Dean's history of the destruction of the Atlantic Forest offers some insightful chapters on the 1950s and 1960s in Brazil (Dean 1987). The country was growing economically, companies were widely exploiting its vast natural resources, and new frontiers of deforestation were being opened with hardly any governmental restrictions or groups within civil society and the media willing to report on the environmental abuses that were taking place.

Until the 1970s, the fate of the indigenous populations produced even less political concern than the destruction of the forests. Some writers lamented, with a certain fatalism, the disappearance of these old forest peoples. Anthropologists tried to study them and some defended their rights to survival. The first political initiative by the federal state was taken in 1910 with the creation of the Indian Protection Service (later to be renamed the National Foundation for the Indian). But state policies aimed at keeping indigenous groups alive through assimilation. The main idea was to concentrate them in places controlled by religious missionaries or public officials (Cunha 1992). However, by the middle of the 20th century, the Brazilian state began to promote more intensive occupation of the

Center-West and the Amazon region. During the so-called "march towards the west," expeditions and state officials discovered indigenous villages and territories of the kind that had disappeared from the Atlantic coastal regions.

In this context, an innovative institutional initiative emerged in 1961 with the creation of the Xingu National Park, an area of 27,000 square kilometers in the Southern Amazon region. The idea was that in this park, indigenous societies would be able to live according to their traditional ways of life, protected from the approaching economic frontier. The model of large protected areas where indigenous communities could continue to practice a traditional way of life remained nonetheless marginal until the 1990s, when it started to embody an important aspect of the new "conservation imperative." In fact, even in 1961, the official political justification for the creation of the Xingu Park emphasized the protection of fauna and flora over indigenous cultural survival, even if the latter goal was crucial for the anthropologists and state officials who promoted the project (Garfield 2004).

Beginning in the 1970s, the military regime began an intense process of economic advancement in the Amazon region. In addition to the subsidies and tax benefits for the occupation of large tracts of land by big companies, colonization projects were implemented that would, according to the jargon of the day, "bring men without land to a land without men." In fact, one of the main features of that period was the legal concept known as the "value of bare land," through which land ownership was granted by the National Institute for Colonization and Agrarian Reform if the occupant was able to show that the forest had been destroyed (Hecht and Cockburn 2010). Many agribusiness companies and different kinds of economic adventurers moved to the region in search of "empty" land. The perspective taking shape at the time was therefore that the destruction of the Amazon rainforest would increase and be uninterrupted. The "conversion imperative" would be gradually implemented. The difference was that the process would be faster, taking into account the greater availability of capital and industrial equipment, such as trucks and chainsaws. The total destruction of the native forest would require no more than a few decades.

Some social actors in the present day Amazon region still try to follow this logic. However, their actions have clashed with the rise of new political perceptions about the value of tropical forests that have been gaining momentum since the 1990s. These new perceptions seem to converge toward the "conservation imperative." Their main component is the recognition by key sectors of society and the state, more strongly at the federal than at the regional level, that the large forest areas in northern Brazil represent a major geopolitical asset, which is currently endangered and should be preserved through concrete legislation and policy actions. They also incorporate the notion that the state should guarantee big tracts of forest land for Indian groups, rubber tappers and other forest peoples in the form of reserved areas.

This conservation turn manifested itself, for example, through the significant increase in the number of protected areas created in Brazil beginning in the 1980s (Drummond et al. 2011). The pace accelerated in the last decade, with Brazil accounting for around 74% of the protected lands created worldwide after 2003 (Jenkins and Joppa 2009). At the end of 2010, an area of around 2,197 million square kilometers (43.9% of the region) was formalized as protected reserves in the Amazon forest. Half of this total (around 22.2% of the region) was defined as different types of Conservation Units and the other half (around 21.7% of the region) as Indian Lands (Veríssimo et al. 2011). The creation of these huge corridors of protected areas, together with other federal policies, helps to explain the strong reduction in deforestation after 2004. There had been reductions in the rate of deforestation in the Amazon in the past. But they had never reached those recent levels, and were related to moments of national economic recession. By contrast, the most recent process has taken place in a period of economic growth, suggesting the perspective of a decoupling between economic growth and deforestation in the Amazon (Pádua 2013).

The making of the conservation turn is linked to the ideas, actions and fights of different social actors that converged in a quite complex way. Local communities of indigenous and forest peoples started to fight systematically against the destruction of their environments during the 1980s. They were moved by a particular sense of "danger" – the danger that their livelihoods and their very existence as social groups would be destroyed. In their case, community survival was linked to the forest resources they used for sustenance and economic life. An important example was the action of the rubber tappers' unions in the western Amazon state of Acre to defend themselves and their forest through the organizations of *empates*, or stand-offs – public demonstrations where workers and their families stayed in front of the trucks in order to halt the advance of deforestation. The assassination in 1988 of Chico Mendes, the movement's leader, was in fact just one of hundreds of murders of local peasant and Indian leaders in the region. In any case, the fight of the rubber tappers became a symbol of what was later called "environmentalism of the poor" or "livelihood ecology" (Martinez-Alier 2003).

On a parallel course, in the 1970s, Brazilian democratic and leftist political forces started to criticize the model of occupation imposed by the military regime in the Amazon region, and to bring to light its environmental and social consequences. They were inspired by and established links of solidarity with poor local communities against the expansion of business projects in the forests. It was a moment of political excitement, with the tangible prospect of defeating the authoritarian regime. New NGOs and social movements were being created for the defense of the environment, human rights and other demands in both rural and urban areas. Scientists and conservationists started to defend the ecological value of the tropical forest based on the rising concept of biodiversity. At the same time, a new wave of concern over the fate of the rainforests grew in the mass media and public opinions of many countries The theme of global warming was gaining momentum. In the context of an increasingly globalized world, this complex set of social actors were able to communicate with each other, learn from each other, and establish rich networks of dialogue and activism (Pádua 2012).

Since the fall of the military dictatorship in Brazil in 1984, new policies toward the Amazon have tried to change the chaotic heritage of the military years and create better conditions for ecological conservation and human development. The Brazilian Constitution of 1988 established innovative rules for environmental protection and human rights. It ordered, for example, the demarcation and registration of all indigenous lands. A crucial initiative was undertaken in 1996 by the social-democratic government of Fernando Henrique Cardoso: the proportion of any private property in the Amazon that for environmental reasons cannot be cleared (the so-called "legal reserve") was increased to 80%. Since 2003, a new coalition of leftist and center parties, led by the Workers Party, has managed to implement more extensively than ever before the constitutional rules and policy proposals for the conservation of tropical forests that had been under discussion since the end of the military regime. However, this process has been far from homogeneous through time, even considering the relative permanence of governments lead by the Workers Party. There has been a strong reduction in the creation of protected areas, including indigenous lands, in the period of President Dilma Roussef after 2011. However, the creation of these areas during the previous eight years' governments of President Lula da Silva was so intense that Brazil's recent historical advances in this issue are still quite impressive.

Some observers have suggested that the results achieved in forest conservation after 2003 have a temporary political underpinning, namely Marina Silva as environment minister for nearly seven years during Lula da Silva's two presidential terms. Marina Silva, a well-known politician and environmental leader from the state of Acre, spent her childhood inside the forest, where her father worked as a rubber tapper, and she was a close political partner of Chico Mendes. Her political will, without question, contributed to the progress in controlling deforestation. But the fact is that some of the institutional instruments for that purpose had been put into place prior to the Lula administration. For example, the Amazon Surveillance System (SIVAM) and the Amazon Protection System (SIPAM), a network of satellites, radars and watching stations developed, among other goals, to detect deforestation movements, were officially inaugurated in 2002. Since 1988, however, the National Institute for Space Research (INPE) had been using satellites to produce more precise data on the Amazon deforestation. The new system called DETER (Detection of Deforestation in Real Time), established by the INPE in 2004, is considered an essential tool for carrying out such controls (Mello 2006).

These increasingly integrated initiatives embody the rise of the "conservation imperative" in Brazilian political culture. After Marina Silva left the ministry and the governing Workers Party in 2008, the rate of deforestation continued to decline until 2012. She resigned due to discontent with former president Lula's lack of support for stronger environmental policies. When in 2010 she ran for president as the candidate of the minor Green Party, she received almost 20 million votes, nearly 20% of the total. In 2014, as candidate of the Brazilian Socialist Party, she received around 22 million votes (nearly 21.3% of the total). This entirely new level of votes for a candidate running on an environmentalist platform demonstrates, in part at least, the penetration of environmental ideas in Brazilian society.

There is a strong possibility, therefore, which history has yet to confirm, that there has been a shift in Brazilian political culture with respect to tropical forests, now mainly associated with Amazon endangerment and protection. There seems to have emerged a new political consensus at the federal state level, which is largely shared by the legislative and the judiciary, of no longer accepting the destruction of the Amazon rainforest, at least not at the levels seen in the late 20th century. In 2013, for example, deforestation in the Amazon was 5,891 square kilometers, an increase of almost 29% in relation to the historically low level of 2012 (4,571 square kilometers). This increase set off warning lights. Many social voices, inside and outside Brazil, including opposition parties, blamed President Roussef for relaxing the controls over the Amazon deforestation. The government, on the other hand, promised to renew its fight against deforestation. These moves helped to enforce my point about the building of a new political consensus against the destruction of the Amazon forest. Any fluctuation in the yearly data is becoming a tough political issue and a return to the previous rates of deforestation would cause a political scandal. The complacent attitude of previous administrations seems to no longer be possible.

The causes for such historical change are necessarily multiple. In the specific case of deforestation in the Amazon, the emergence of strong international discussions beginning in the 1970s at the scientific, diplomatic and general public levels made a significant contribution to Brazilian discourses. Over the years, numerous reports, films, books and public demonstrations have taken place in Europe, North America and Asia with respect to the problem. The circulation of information through the Internet and globalized media made these demonstrations immediately accessible in Brazil and feed the debate at the local level. However, given that there were also countless public demonstrations, political debates and scientific, intellectual and artistic works produced within Brazil, exogenous pressure does not suffice to explain the change (Hochstetler and Keck 2007; Pádua 2012).

The key is rather to be found in new geoeconomic and geopolitical factors. The Amazon has turned out to be an enormous reserve of natural resources of growing value at the global level. Such is the case with biodiversity, whose economic potential became increasingly clear with the prospects of biotechnology. The same could be said for Brazil's fresh water supplies, especially from the perspective of expected water scarcity in the world. The Amazon's complex hydrological system, which accounts for approximately 7% of the fresh water available on the planet, is dependent upon the conservation of the forest. Other factors could also be cited, such as solar radiation, high levels of biomass reproduction and the forest's carbon absorption (essential to the prevention of further global warming). It is in any case clear that the Amazon rainforest is rich in some of the elements that could become crucial to the new global economy.

It has therefore become evident to Brazilian political elites, including sectors of the armed forces, that the economic and national security potential of the standing forest is much greater than that of bare land, particularly in relation to livestock grazing, which has occupied more than 60% of the deforested land in the region. This is the case even though its value is not immediately realized, but instead functions as a reserve for the future. It has also become clear that the growth of Brazilian agribusiness does not require the destruction of the Amazon. The 2 million square kilometers of the Cerrado are more than sufficient to create a huge new frontier for crops production. In fact, the Cerrado is also being heavily destroyed – its total conversion to other forms of land use, especially export-driven agriculture, is already at the level of 50%. Since it lacks the charisma of tropical forests, it could be read as a zone of sacrifice for the preservation of the Amazon, almost without protest from Brazil or from abroad.

Economics does not entirely explain the rise of the "conservation imperative." Other factors have played a role, including the renewal of the traditional symbolic significance of the forest for Brazilian national identity. Some of the political and social actors that belong to the ruling Workers Party coalition have a tradition of defending the Amazon forest. They represent a heritage of social and democratic struggle, active since the period of the military dictatorship that is often embodied in environmentalism, as well as in the defense of social and cultural rights. However, the rise of increasingly pragmatic politics in Brazil's left has given more weight to economic and geopolitical factors; and equally pragmatic (largely economic) reasons have also led some traditionally conservative groups to accept the new "conservation imperative."

The debate regarding the classification of tropical forests as "endangered places" has had concrete historical consequences. The situation in Brazil, which contains the largest reserves of these forests, has driven what may become a major change in political culture with respect to forests, promoting real results for their conservation. The political options connected to the debate about deforestation as a form of endangerment are infused with values based upon ecological, economic and political considerations, which gain new forms as they travel through different geographies and spheres of social existence.

Acknowledgments

The final version of this chapter was written in Munich, where I was living during the first half of 2014 as a fellow of the Rachel Carson Center for Environment and Society. I wish to thank the members of the Work in Progress Seminar for their valuable suggestions. I am especially grateful for the detailed comments of Nicole Seymour. I also want to thank the editors of this book, Fernando Vidal and Nélia Dias, for their accurate suggestions.

Note

1 Political culture is an intangible and difficult term to define. I take it here as a synthetic image of the set of assumptions, beliefs and guidelines that have a strong, recurrent and dominant presence in the political life of a society. For the theoretical foundations of the concept, see Brint 1991.

Bibliography

Alemão, F. F., 1961 [1851]. Carta a Augustin de Saint-Hilaire em 23 de Novembro de 1851. *Anais da Biblioteca Nacional*, 81.

Antonil, A. J., 1976 [1711]. *Cultura e Opulência do Brasil por suas Drogas e Minas.* Melhoramentos, São Paulo.

Assunção, P. D., 2001. *A Terra dos Brasis: A Natureza da América Portuguesa vista pelos Primeiros Jesuítas.* Annablume, São Paulo.

Brint, M. A., 1991. *Genealogy of Political Culture.* Westview Press, Boulder.

Buckle, H., 1930 [1857]. *History of Civilization in England*, 1. Watts, London.

Caminha, P. V. de, 1963 [1500]. *Carta à El Rei Dom Manuel.* Dominus, São Paulo.

Costa Lima, L., 1989. *Control of the Imaginary.* University of Minnesota Press, Minneapolis.

Couto, J. V., 1848. Memória sobre a Capitania de Minas Gerais. *Revista do Instituto Histórico e Geográfico Brasileiro*, 11.

Cowell, A., 1990. *The Decade of Destruction: The Crusade to Save the Amazon Forest.* Henry Holt and Company, New York.

Cunha, M. M. Carneiro da, org., 1992. *História dos Índios no Brasil.* Companhia das Letras, São Paulo.

Dean, W., 1987. *Brazil and the Struggle for Rubber.* Cambridge University Press, Cambridge.

Dean, W., 1995. *With Broadax and Firebrand: The Destruction of the Brazilian Atlantic Forest.* University of California Press, Berkeley.

Drummond, J., Franco, J. L. and Oliveira, D., 2011. Uma Análise sobre a História e a Situação das Unidades de Conservação no Brasil. In: Ganem, T., org., *Conservação da Biodiversidade-Legislação e Políticas Públicas.* Edições Câmara dos Deputados, Brasília.

Enright, K., 2012. *The Maximum of Wilderness: The Jungle in the American Imagination.* University of Virginia Press, Charlottesville.

Freyre, G., 2003 [1936]. *Nordeste.* Global, São Paulo.

Fundação SOS Mata Atlântica and Instituto Nacional de Pesquisas Espaciais, 2013. *Atlas dos Remanescentes Florestais da Mata Atlântica.* SOS Mata Atlântica, São Paulo.

Garfield, S., 2004. A Nationalist Environment: Indians, Nature, and the Construction of the Xingu National Park in Brazil. *Luso-Brazilian Review*, 41 (1), 139–167.

Gondim, N., 2007. *A Invenção da Amazônia.* Valer, Manaus.

Goodland, R. and Irvin, H., 1975. *Amazon Jungle: Green Hell to Red Desert?* Elsevier Scientific Publishing, New York.

Grove, R., 1995. *Green Imperialism: Colonial Expansion, Tropical Island Edens and the Origins of Environmentalism.* Cambridge University Press, Cambridge.

Gusdorf, G., 1985. *Le Savoir Romantique de la Nature.* Payot, Paris.

Hecht, S. and Cockburn, A., 2010. *The Fate Of The Forest: Developers, Destroyers and Defenders of the Amazon.* Chicago University Press, Chicago.

Hochstetler, K. and Keck, M., 2007. *Greening Brazil: Environmental Activism in State and Society.* Duke University Press, Durham.

Holanda, S. B. de, 1968. *Visão do Paraíso.* Editora Nacional, São Paulo.

INPE/PRODES, 2014. *Monitoramento da Floresta Amazônica por Satélite* (http://www.obt.inpe.br/prodes/index.php). Accessed 5 December 2014.

Jenkins, N. and Joppa, L., 2009. Expansion of the Global Terrestrial Protected Area System. *Conservation Biology*, 142, 10.

Martinez-Alier, J., 2003. *The Environmentalism of the Poor: A Study of Ecological Conflicts and Valuation.* Edward Elgar, Cheltenham.

Mello, N. A. de, 2006. *Políticas Territoriais na Amazônia.* Annablume, São Paulo.

Nash, R., 1982. *Wilderness and the American Mind*. Yale University Press, New Haven.

Pádua, J. A., 1997. Biosphere, History and Conjuncture in the Analysis of the Amazon Problem. In: Redclift, M. and Woodgate, G., eds., *The International Handbook of Environmental Sociology*. Edward Elgar, London.

Pádua, J. A., 2002. *Um Sopro de Destruição: Pensamento Político e Crítica Ambiental no Brasil Escravista (1786–1888)*. Jorge Zahar Editor, Rio de Janeiro.

Pádua, J. A., 2004. Nature Conservation and Nation Building in the Thought of a Brazilian Founding Father: José Bonifácio (1763–1838). *CBS Working Paper*, 53. Center for Brazilian Studies, University of Oxford.

Pádua, J. A., 2010. European Colonialism and Tropical Forest Destruction in Brazil: Environment Beyond Economic History. In: McNeill, J., Pádua, J. and Rangarajan, M., eds., *Environmental History – As If Nature Existed*. Oxford University Press, New Delhi.

Pádua, J. A., 2011. Down by Blind Greed: The Historical Origins of Criticism Regarding the Destruction of the Amazon River Natural Resources. In: Tvedt, T., Chapman, G. and Hagen, R., eds., *A History of Water*, II 3. I. B. Tauris, London.

Pádua, J. A., 2012. Environmentalism in Brazil: An Historical Perspective. In: McNeill, J. and Stewart, E., eds., *A Companion to Global Environmental History*. Wiley-Blackwell, Oxford.

Pádua, J. A., 2013. The Politics of Forest Conservation in Brazil: a Historical View. *Nova Acta Leopoldina*, 114, 390.

Rebouças, A., 1876. *Excursão ao Salto do Guaíra: O Parque Nacional*. Typographia Nacional, Rio de Janeiro.

Schwarcz, L., 2004. *The Emperor's Beard: Dom Pedro II and His Tropical Monarchy in Brazil*. Hill and Wang, New York.

Silva, J. B. de Andrada e, 1973 [1825]. Representação à Assembléia Geral Constituinte e Legislativa do Império do Brasil sobre a Escravatura. *Obra Política de José Bonifácio*. Senado Federal, Brasília.

Slater, C., 1995. Amazonia as Edenic Narrative. In: Cronon, W., ed., *Uncommon Ground: Rethinking the Human Place in Nature*. W. W. Norton, New York.

Slater, C., 2002. *Entangled Edens: Visions of the Amazon*. University of California Press, Berkeley.

Thomas, K., 1983. *Man and the Natural World: A History of Modern Sensibility*. Pantheon, New York.

Tsing, A., 2005. *Friction: An Ethnography of Global Connection*. Princeton University Press, Princeton.

Valle, C. P., 2005. *Risonhos Lindos Campos: Natureza Tropical, Imagem Nacional e Identidade Brasileira*. Senai, Rio de Janeiro.

Varnhagen, F., 1975 [1857]. *História Geral do Brasil*. Melhoramentos, São Paulo.

Veríssimo, A. and Nussbaum, R., 2012. *Um Resumo do Status das Florestas em Países Selecionados*. Imazon/Proforest Initiative, Belém.

Veríssimo, A., Rola, A., Vedoveto, M. and Futada, S., 2011. *Áreas Protegidas na Amazônia Brasileira: Avanços e Desafios*. Instituto Socioambiental, São Paulo.

Worster, D., 1994. *Nature's Economy: A History of Ecological Ideas*. Cambridge University Press, Cambridge.

Part III

Technologies of preservation

In its 1984 definition of the conservator-restorer, the International Council of Museums characterized preservation as "action taken to retard or prevent deterioration of or damage to cultural properties by control of their environment and/or treatment of their structure in order to maintain them as nearly as possible in an unchanging state." Though applied to material objects, such a definition can, *mutatis mutandis*, be extrapolated to all entities considered endangered. Before any action can be taken, the entity has to be classified as threatened; and the very act of classifying it thus implies the obligation to protect it. Every step of the process involves professional communities and specialized technical tools – from conservation biologists measuring extinction rates to heritage experts assessing degrees of degradation, and in all cases, organizations drawing up lists – both at the national and the international level.

The multiplication of those communities and tools mirrors the expanding number of entities considered endangered – from species of dwindling numbers whose genetic material is frozen or whose remaining specimens are radio-tracked, to "intangible" cultural practices whose makers are encouraged not to adapt to circumstances that might lead them to abandon them. The chapters in this section, dealing with the California condor (Etienne Benson), the UNESCO lists of tangible and intangible heritage (Rodney Harrison) and cryopreserved biodiversity collections (Joanna Radin), demonstrate, on the one hand, how technologies embody preservation as the characteristic imperative of the endangerment regime. On the other hand, they illustrate those technologies' ontological impact – how they transform the entities they aim to safeguard and how, as Geoffrey C. Bowker noted in connection with biodiversity databases, they ultimately shape the world in their image. Labeling entities as endangered prompts diverse technical and bureaucratic interventions, and turns management into a necessary condition of their existence. This includes, among others, and as in the cases studied here, reintroduction of a species in the wild, captive breeding and cloning, and re-describing the producers of a cultural practice as its "transmitters."

The technologies of preservation aim at acting in the present for the sake of a foreseeable future. They manifest trust in experts' tools for making risk manageable. This confident vision coexists with a gloomier one: the future they anticipate (and want to prevent) is one in which present species or monuments, genes or cultural practices, might no longer be. *Might no longer be*: this anxious but generally well-founded vision is constitutive of the endangerment sensibility. But it also expresses a certain faith in a future in which the positive values of that sensibility survive, and in which the entities today under threat may still contribute their inherent and their instrumental value to making a biologically and culturally diverse world.

7 Endangered birds and epistemic concerns

The California condor

Etienne Benson

The condor as an endangered object

The designation of any particular object or category of object as endangered can have both positive and negative consequences from the perspective of a scientist interested in studying it. On the one hand, it tends to focus attention and muster resources for research; on the other, it heightens external scrutiny and constrains the actions of those who seek to study, and thereby perhaps to change irrevocably, what may be the last few, highly vulnerable instances of the category or object in question. Research on endangered objects is thus epistemologically and ethically fraught in ways that research on more common objects – atoms, sparrows, or friendships, for example – is not, even if research on those objects may, of course, be fraught in other ways. A broad range of tangible and intangible objects across the nature-culture divide falls into this zone of special concern, including artworks and archaeological artifacts, oral traditions and folkways, and biological species and ecosystems.

Here, I am concerned with the conditions for producing new knowledge about a set of objects that at first glance may appear solidly situated on the nature side of the divide, although their position becomes less clear on closer inspection: endangered species, and particularly endangered species of birds. I argue that varying conceptualizations of endangerment have shaped not only the way that ornithological knowledge is produced and enters into the scientific archive, but also the very objects – that is, the birds themselves – that are the focus of ornithologists' concern. As my primary example, I use the California condor (*Gymnogyps californianus*), which with a wingspan of up to about 3 meters (about 10 feet) and a population nadir of fewer than two dozen individuals in the 1980s is North America's largest vulture and one of the world's most endangered bird species. As of February 2012, there were estimated to be 173 California condors living in captivity and 213 living in the wild; the species is categorized as "endangered" under U.S. law and as "critically endangered" by the International Union for Conservation of Nature (BirdLife International 2013).

The story of the California condor's discovery, near-extinction, and still-tenuous survival is part of the broader history of concern about the extinction of species. In the United States, the country within which the present-day range of the condor lies, this history began in the eighteenth and nineteenth centuries with the recognition of the very possibility of anthropogenic extinction, continued in the late nineteenth century and early twentieth century with the emergence of the conservation and environmental movements, and entered its most recent phase with the identification of biodiversity loss as a global crisis in the 1970s and 1980s. The concept of biological extinction and the effort to prevent it in the United States have been closely intertwined with changing imaginations of wilderness, the frontier, and the pre-European history of North America (Alagona 2004; 2013, 122–148; Barrow 2009; Farnham 2007; Heise 2010; Takacs 1996).

Naturalists and scientists have played central roles throughout this several-centuries-long history. Only in the twentieth century, however, did the possibility of the extinction of the California condor and other species come to serve as the basis for the development of a distinctive set of practices, theories, ideas, and relationships that together might be considered a science of endangered species. Although the history of research on and conservation of the California condor is unique in ways that reflect both the species' biology and the historical conditions of its endangerment, it serves here as an example of the interrelation of ontologies and epistemologies produced through the concept of biological endangerment since the nineteenth century.

To show how this interrelation developed over time, I draw on the writings of condor researchers and conservationists as well as on existing histories of condor conservation by both condor conservationists and historians (Alagona 2004; 2013, 122–148; Alvarez 1993; Barrow 2009, 283–299; Darlington 1987; Koford 1953; McMillan 1968; Moir 2006; Nielson 2006; Snyder and Snyder 2000). I argue that the designation of a biological species as endangered not only transforms the conditions under which the species comes to be known but also – inasmuch as science is a mode of relating to the world that transforms its objects in order to make them knowable – the species itself. The ontological and epistemological conditions of endangerment are thus connected: the threat of non-existence leads to new ways of knowing at the same time that ways of knowing reconstitute their objects of inquiry. In other words, the endangerment sensibility facilitates certain kinds of relationships and hinders others. These relationships are simultaneously scientific and deeply laden with values and affects (Koelle 2012; Whitney 2013).

For the sake of clarity, albeit at some risk of overemphasizing historical discontinuities, I divide the history of research on and conservation of the California condor into four phases. First, *discovery*, from roughly the late eighteenth century to the late nineteenth century, when naturalists described the condor for the purposes of natural history, relying heavily on Native American informants to do so. Second, *salvage*, from the 1890s to the 1920s, when ornithologists recognized the condor as endangered by human activities, believed that its extinction was all but inevitable, and attempted to collect as many traces of the vanishing bird as they could. Third, *preservation*, from the 1930s to the 1970s, when conservationists made their first significant efforts to prevent human activities from threatening the condor's survival. Finally, *management*, from the 1980s to the present, when conservation biologists began to breed condors in captivity and intensively manage them in the wild. In each of these phases, the archive of condor data and specimens – including living specimens – was reimagined and reconstituted. By way of conclusion, I consider the continuities and discontinuities between these phases, the extent to which the condor can be used to understand the epistemo-ontological paths of other endangered species, and some possible future twists and turns in this species' relationship with humans.

Discovery

The California condor has been known for thousands of years by the indigenous peoples of western North America, who established various relationships with it, including ritual uses of condor feathers and bones (Bates, Hamber and Lee 1993; Snyder and Snyder 2000, 30–45). From a European perspective, recognizable written descriptions of the California condor – or the "California vulture," as it was sometimes called – can be found as early as the beginning of the seventeenth century. However, the first modern scientific reports and specimens date only to the late eighteenth century, when European museums acquired their first specimens (Harris 1941, 9; Snyder and Snyder 2000, 31). For most of the nineteenth century, the California condor was understood to be rare, but its rarity did not necessarily imply that it was at risk of extinction. Naturalists recognized that species varied naturally in abundance. The biological and ecological peculiarities of the condor, a large-bodied, long-lived, slow-breeding bird that fed on widely dispersed carrion, suggested that it had never been present in particularly large numbers. This rarity, along with the fact that condor populations were located far from the metropolitan centers of scientific research of the time and the absence of any particularly strong incentives to study the species, had epistemological consequences. There were few direct conflicts between condors and humans, no particular economic or cultural values attributed to the condor by the European-American settlers who began arriving in droves during the gold rush of the 1840s and 1850s, and numerous logistical obstacles in the way of any research expedition. As a result, the production of knowledge about the species during this period was largely opportunistic and often based on second-hand observations.

In the unstable swirl of conquest and commerce that characterized nineteenth-century California, naturalists were often forced to rely on an extended chain of trust to establish even the most basic of facts about the condor. Unlike better-established areas of science in which trust relations come to be obscured by the language of disinterestedness and objectivity (Daston and Galison 2007; Shapin 1994), the social relations upon which condor research depended remained visible and often problematic. Even when such a chain of trust had been established, crucial details often remained unclear. In the description of the condor for the 1839 *Ornithological Biography*, for example, John James Audubon relied on letters from John Kirk Townsend, who had traveled in the Columbia River region at the northern edge of the condor's range (Audubon 1839, 240–243; Branch 2008; Jobanek and Marshall 1992). From his own first-hand experience, Townsend could report that condors were rare, quite shy, and could be most easily observed when they gathered to feed on annual salmon runs. Nonetheless, much of his knowledge of the condor's behavior was based on the claims of his Native American informants. These indigenous reports, Townsend assured Audubon, "may generally be relied upon." From them, he had learned that condors used vision rather than smell to find carrion, that they tended to breed on swampy ground, and that the "Wahlamet" (i.e. Willamette) Mountains, which he had never visited, were their "favorite place of resort" (240). Audubon's description in the *Ornithology Biography* was thus partly second-hand, based on Townsend's direct observations, and partly third-hand, based on Townsend's reports of Native Americans' reports.

This form of opportunistic knowledge-at-a-distance left significant gaps of which early and mid-nineteenth century ornithologists were painfully aware. On the subject of the eggs of the condor, for example, Townsend had nothing to report, either from his own experience or from that of his informants, forcing Audubon to rely instead on the dubious claim by the English naturalist David Douglas that condors incubated two eggs that were "nearly spherical, about the size of those of a goose," and "jet black" (Audubon 1839, 242). It might be thought that collections of skins, skeletons, and eggs would have allowed opportunities for extended study of the condor's anatomy even if the living animals were difficult to observe in the field. However, the rarity and remoteness of the condor from metropolitan centers made the collection of specimens and their transport back to private collections and public museums of natural history extremely difficult, exacerbating problems also encountered in studies of more common species (Barrow 2000, 9–45; Farber 2000; Jardine, Secord, and Spary 1996; Kohler 2006; Prince 2003). Even the large size of the condor, which contributed to its status as one of the most spectacular North American birds, magnified the difficulty of transporting and storing specimens (Lewis 2012, 129).

By 1859, it was possible to view a specimen of a California condor in collections in Washington, D.C., San Francisco, Philadelphia, London, and Paris, but even for those who were able to gain access to those collections, the specimens provided limited information (Taylor 1859, 63; Wilbur 1974, 71–72). None of these institutions held eggs of the species and few had more than one specimen, which was often in poor condition. The lack of egg specimens led to the persistent repetition of Douglas' claim that the condor's egg was black, even though, as one zoologist noted acerbically in 1867, the claim must have seemed to any experienced in the study of bird eggs as "conflicting with all inference by analogy" (Brewer 1867, 114). Given the condor's rarity, even such seemingly outlandish claims could neither be dismissed out of hand nor easily corroborated.

My point is not merely that the rarity of the condor in an objective or factual sense made it difficult to settle scientific questions about the species, but also that naturalists' consciousness of the species' rarity shaped their expectations about what could be known and about what sorts of evidence were acceptable.

Second-hand reports that would have been dismissed as unacceptably flimsy evidence with regard to more common and accessible species were grudgingly accepted as the best available under the circumstances. The result was continued speculation, admissions of ignorance, and reliance on travelers' reports into the late nineteenth century. The California condor remained rare in the specimen archives of natural history museums, as it was in the wild, and the kind of knowledge that could be gathered about it was correspondingly limited. The information and artifacts that resulted from largely opportunistic research were highly valued, but were also viewed with skepticism because of the difficulty of corroboration. California condors played only a small part in the lives of California's Anglo-American settlers in the nineteenth century, however, and beyond a few curious ornithologists, the absence of detailed studies of the bird caused little concern.

Salvage

As California was occupied and developed by colonists from the United States in the second half of the nineteenth century, some of the constraints that had previously hindered ornithologists' study of the species began to loosen. By the 1890s, it was clear to many naturalists that the California condor's numbers had been severely reduced, even taking into account the species' naturally low numbers, and that this reduction was largely due to the very developments that had made the species easier to observe. The first ornithologist to sound the alarm was James G. Cooper, who argued in the journal *Zoe* in 1890 that the condor was, as his title made clear, "A Doomed Bird" (Cooper 1890; see Barrow 2009, 285). Although Cooper's and others' warnings of the species' imminent extinction during this period were sometimes accompanied by pleas for protective measures, these pleas had little effect. The condor was nominally protected by California state law in 1905, for example, but the law was rarely enforced and, in any case, did nothing to protect the condor's habitat (Snyder and Snyder 2000, 55). Few expected such legal measures, by themselves, to reverse the condor's decline. The dominant mood in texts about the condor from the late nineteenth century through the early twentieth century was one of resignation, even fatalism. The only outstanding questions seemed to have been how long the process of extinction would take and how much could be learned about the species before it met its inevitable fate.

Research on the condor during the period was thus what might be called, in analogy to anthropological and archaeological studies, "salvage biology" (Janzen 1986): an attempt to preserve as many observations and material traces as possible for later research before a species' presumably inevitable disappearance. The salvage biology of this period was informed by the idea that the advance of civilization, while itself desirable, would inevitably lead to the extinction of numerous species, the loss of which might be lamentable for aesthetic and scientific reasons. In comparison to the period when the condor was simply understood as rare, the growing awareness of the condor's apparently imminent extinction led to an intensification of research effort, an increasingly wide circulation of specimens, and, ultimately, a significant expansion of the scope and reliability of knowledge about the species.

One of the consequences of this fatalism was a boom in interest in collecting condors and their eggs before the species vanished completely, which probably contributed to the decline in the species' numbers. In the 1890s, there was a veritable race to collect specimens, which some later researchers have argued significantly depleted the population. An inventory of all known condor specimens in the early 1970s, which undoubtedly failed to account for some that were never reported or did not end up in collections, uncovered "185 mounted birds and study skins, 51 skeletons, and 55 eggs" (Wilbur 1974, 71). Given that the total population of California condors probably numbered around 200 birds in 1900 and that more than half of the specimens collected were taken between 1881 and 1910, these are significant numbers (Barrow 2009, 285; Snyder and Snyder 2000, 56). Fatalism about the endangerment of the condor thus had both epistemological and ontological consequences, generating new knowledge about the species even as it contributed to its decimation.

Extraordinary attempts were made to acquire specimens of the condor during this period, both by those who were confident the species could survive and by those who believed it would inevitably go extinct. In 1895, just five years after Cooper warned of the condor's extinction, Frank Stephens submitted a more optimistic report to *The Auk*, the journal of the American Ornithological Society (Stephens 1895). Stephens described having seen twenty-six birds in a single flock while traveling near the southern end of the Sierra Nevada mountain range. After having expended "two charges of buckshot in the futile hope that a stray buckshot might strike one" (81), he had ringed the carcass of the horse on which the condors had been feeding with a dozen steel traps in anticipation of their return. This was an "unfair method" (81), he admitted, but one he thought few ornithologists would not have attempted under the circumstances. Ultimately, the attempt was unsuccessful, and Stephens did not see condors again during his months of travel in the region. Nonetheless, he had no worries about the species extinction, explaining the absence of further encounters by the condors' increasing wariness. He was convinced that "[u]nless an epidemic or some other disaster overtakes the species its extermination will not occur in our day" (82). That natural history collecting might be the very disaster that extinguished the species does not seem to have occurred to him.

Even those who were more concerned than Stephens about the condor's survival did not hesitate to collect specimens when the opportunity provided itself. A particularly striking example is William Finley, who helped inspire a turn toward photographic studies of birds and away from specimen collecting around the turn of the twentieth century (Barrow 2009, 286; Mathewson 1986; Mitman 1999, 96–99; Snyder and Snyder 2000, 49–58; Wilson 2010, 48–51). In a landmark series of articles between 1906 and 1910, Finley published photographs and detailed descriptions of the nesting behavior of California condors, particularly a mating pair with a hatchling discovered in the spring of 1906 in the mountains north of Los Angeles (see Figure 7.1; Finley 1906). The initial article in the series

Figure 7.1 A photograph of William Finley with a condor adult and young at a nest site near Los Angeles in 1906; part of a hand-painted series by Finley and his partner Herman Bohlmann. Digital scan contributed by the Audubon Society of Portland to the National Digital Library of the U.S. Fish and Wildlife Service, http://digitalmedia.fws.gov/, Item ID: Finley and Bohlman Slides009.jpg.

contained a harrowing description of the near-death of a chick temporarily taken out of the nest for photography. It is clear from the text that Finley was concerned about the possibility of the chick's death, but not because he was worried about contributing to the condor's extinction. Rather, what seems to have concerned him most was wasting a unique research opportunity. In July 1906, about three and a half months after the hatching of the egg, Finley captured the two-thirds grown chick, transported it to Portland, Oregon, and gave it a name. A few months later, "General" was sent to New York, where he joined the collection of the Bronx Zoo (Finley 1910). Finley believed that the species was on the verge of extinction, fought for protections for threatened bird species, and helped establish the use of photography in bird studies – a technique that seemingly posed a less direct threat to the condor than the collection of skins, skeletons, and eggs. But he expressed no particular concern that his removal of a healthy chick from the wild population would hasten the condor's extinction.

The growth in photography and observational field studies of condors at this time should therefore be seen as an elaboration rather than a replacement of salvage biology. It was motivated as much by the desire to completely document the species before it vanished as by any desire to prevent the practice of science itself from hastening that disappearance (Haraway 1984–1985; MacKenzie 1988, 306–307). It provided information about the condor and its evolutionary and ecological position that could not be derived from field notes or from examining skins, skeletons, and eggs. As Finley had argued in the journal *The Condor* a few years before photographing and eventually capturing General, "mere collecting" continued to be "essential," even if photography "shows a much higher development in a man's love for nature" (Finley 1901, 137). The focus on photography in combination with other methods was congruent with a broader turn in biology from anatomical and taxonomic studies toward physiology and behavior (Kohler 2006, 1–46; Nyhart 2009). In this context, salvage biologists sought to construct archives of photographs, films, texts, and biological remains so that future generations of scientists would have a fuller picture of the history of life. At a time when the survival of non-domesticated animals of all kinds seemed increasingly unlikely and that of a large, rare vulture particularly so, scientists related to the condor primarily in terms of the material traces it would leave behind.

It is difficult to precisely date the end of the salvage period for condors. Some ornithologists continued to practice and to advocate this form of research even after most had adopted a less fatalistic perspective on the relationship between biology and conservation. In 1941, for example, Harry Harris noted in the journal *The Condor* the "remarkable and singular fact that not until the California Condor . . . was on the very threshold of extinction has an adequately equipped and properly organized effort been made to secure a comprehensive body of data on its life history" (Harris 1941, 3). Harris gave no indication that studies of the condor's life history could aid in the prevention of the species' extinction or that certain research methods might themselves be considered as threats. Rather, he lauded one fellow researcher for "taking the fullest advantage of what opportunity remains to secure a definitive life history of the doomed raptor," and he speculated that the time was not far off when another writer would be able to summarize the history of research on the species in its entirety and add to it a "fitting epitaph" (55). Harris's vision of a definitive endpoint to condor research suggests the link between endangered objects and the sciences that focus on them: the end of the condor also meant the end of a certain kind of ornithology. In the face of inevitable and imminent extinction, many scientists felt authorized to use any and all research methods available to create an archive that would serve science beyond the extinction of the species, even when the very methods they chose hastened the arrival of the end.

Preservation

The transition from fatalism about the extinction of the California condor to the belief that its decline could be reversed by the implementation of appropriate conservation measures was gradual and incomplete. By the 1930s, however, the tone of the discussion had begun to change, as had the conditions for condor research. Whereas ornithologists around the turn of the twentieth century had been eager to collect skins and eggs of the species before it vanished from the face of the Earth, many within the next few decades came to realize not only that such collecting was one of the human activities that were hastening the condor's and other species' demise but also that the extinction might not be inevitable after all (Barrow 2000, 150–153). At the same time, research seemed more essential than ever. Instead of being seen as the production of records for the archive of a dying species, it could now be understood as providing crucial information for the implementation of measures that might save the species from extinction. The result of this shift in perspective was a newly intensified research effort that was shaped by conservation ideals in both its goals and its methods.

Among the most influential of those who argued that the condor's fate was a matter of human choice rather than inevitable destiny was Carl Koford. Under the supervision of Joseph Grinnell, director of the Museum of Vertebrate Zoology at the University of California at Berkeley, and with financial support from the National Audubon Society, Koford began intensive observations of condors in 1939 (Barrow 2009, 565–566; Moir 2006, 43–54; Snyder and Snyder 2000, 59–72). Delayed by the outbreak of war, Koford's book-length study of the species eventually appeared in 1953. The diversity of hands-on and hands-off research methods he used included photography, close inspections of nests and feeding sites, banding of individuals, and distant observation with binoculars. Eventually, Koford came to the conclusion that human disturbance of any kind, including for the purpose of scientific research, was the major threat to the condor's survival. What the species needed to survive, he came to believe, was to be left alone. It is not a coincidence that the period when Koford was carrying out his preservationist form of condor science was also the period when the American wilderness movement was launched, with an emphasis on protecting "untrammeled" pieces of land within a colonized and industrialized landscape (Sutter 2002; Turner 2013). Clinging to existence in the face of progress, condors stood for the nature that had been lost in the United States' heedless rush toward development, as well as for the possibility of redemption through the preservation of remaining species and "pristine" spaces. While disagreements emerged between activists most concerned with protecting the species and those most concerned with protecting its wilderness habitat, the two causes became closely intertwined in mid-twentieth-century California (Alagona 2004, 568; Alagona 2013, 128–131).

Even photography, a seemingly non-intrusive method that had been used to great effect by advocates of wilderness and wildlife protection, became suspect in this moment of wilderness enthusiasm. Among the "Major Mortality and Welfare Factors" Koford included in his report were the close encounters necessary for high-quality photography, which had the potential to disturb what Koford believed were extraordinarily sensitive creatures (Alagona 2004, 570; Koford 1953). Although in retrospect it might seem strange to group photography along with poisoning, shooting, and habitat loss, at least some readers of Koford's text found his concerns convincing. Their willingness to reject close-up photography was perhaps made easier by the extensive photographic records that already existed. One reviewer of Koford's book for the *Wilson Bulletin* noted that since virtually every aspect of condor nesting behavior had already been photographed and filmed "and since it is impossible to photograph the nesting birds without disturbing them there is no justification for further photography" at nest sites or at feeding and watering areas. The same reviewer stressed that despite claims since 1890 that the species was a relict of the Pleistocene with "one wing in the grave," the condor was "not doomed to extinction" (Marshall 1954, 75–76). Together, the existence of an already-established archive and renewed hope for the species' survival undermined the apparent need for further research, at least of a certain kind.

The possibility that the condor might survive had also been stressed in the preface to Koford's book by Alden H. Miller, Grinnell's successor at Berkeley, who focused on the potential benefits of Koford's work to ornithological science. Only secondarily, Miller argued, should Koford's study be seen as a contribution to an archival record of the condor that would survive the species itself. Such a record Miller compared to life insurance: something not without value, but nonetheless "a cold and wholly inadequate payment for a life" (Miller 1953, viii). In Miller's view, as in Koford's, it made little sense to harm the bird for the sake of the data one might thereby produce. More broadly, the fact that the condor's survival was no longer a hopeless cause placed scientists in a new position of responsibility. They were still responsible for preserving an archive of texts, images, and artifacts of the condor, but they now also had to consider means of saving the living condor itself within a landscape divided into areas of industrial development and a few lingering enclaves of apparently pristine nature, where a representative sample of a rapidly vanishing natural world could be preserved.

Management

Between the 1960s and 1980s, ornithologists and conservationists gradually shifted from the preservationism advocated by Koford, Miller, and others to a more managerial approach to condor conservation that combined habitat protection with intensive management of the condor population, including assisted reproduction. Whereas the preservationist approach had emphasized the protection of condors and condor habitat from human harms such as shooting, poisoning, and disturbance (including close-up photography), the managerial approach focused on captive breeding and hands-on monitoring and manipulation of reintroduced wild populations to ensure maximum population growth and genetic diversity. This approach necessitated a shift toward research on condor breeding in captivity and authorized increasingly interventionist research methods on free-living condors, such as radio-tracking (Benson 2010) and regular medical examination of nesting birds. These methods were described by their advocates as essential to producing the knowledge needed to manage the lives of condors in their smallest details, without which the species seemed likely to go extinct despite even the strongest protective measures.

The condor was neither the first nor the only species to be subjected to this newly interventionist approach to endangered species conservation, which had its roots in the years after World War II (Mitman 1996). The whooping crane, for example, another large, impressive, and highly endangered North American bird, had been the subject of a captive breeding program since the 1950s (Barrow 2009, 301–306). However, in the wake of the environmental movement and particularly the passage of the Endangered Species Act of 1973, which gave the U.S. federal government wide-ranging authority and responsibility for endangered species protection, this managerial turn became increasingly apparent for a wide range of species. Observers at the time recognized that a major transition was in progress. In 1976, for example, David R. Zimmerman, author of *To Save a Bird in Peril*, described in the journal *Science* a shift from the "classic conservation techniques" of preserving habitat and restricting hunting, trapping, and poisoning to a new approach in which "[b]iologists intercede directly in the afflicted birds' life cycles, either by moving or manipulating the birds themselves, or by manipulating their immediate breeding or feeding habitat" (Zimmerman 1976, 876; see also Zimmerman 1975). The hands-off policies advocated by Koford and by many wilderness activists were thus challenged by a new enthusiasm for active management.

The very act of creating a list was an important part of this managerial turn. One of the primary inspirations for the U.S. endangered species list, which was first developed in the early 1960s and was formalized in federal legislation in 1966, was the Red List developed by the International Union for Conservation of Nature. While the latter would eventually develop a set of gradations of endangerment from "Least Concern" to "Extinct," the U.S. list as defined by the Endangered Species Act consisted of only two categories: "endangered" and "threatened" (Barrow 2009, 301–344; Bean and Rowland 1997, 193–282; Heise 2011; International Union for Conservation of Nature 2001). Inclusion in the former category indicated that a species was imminently threatened by extinction; inclusion in the latter indicated that the species was likely to become endangered in the foreseeable future. A number of what might be called implicit or negative lists were also generated by the act, including the lists of species that were unlisted, of species that had been proposed for listing, and of species that had been delisted. The act of listing depended both on the often-contested science of the present and on visions of the foreseeable future, and it had the power to mobilize a massive if not always effective legal and institutional apparatus around a particular species. As with the designation of wilderness areas (Cronon 1996), the listing of endangered species reinforced a division between those parts of the nonhuman world that deserved special care and those parts that did not.

As one of the first species listed as endangered by the U.S. federal government, the condor would soon become a prominent and controversial example of the new managerial approach. In 1978, a committee convened by the National Audubon Society and the American Ornithological Union concluded that captive propagation, which had first been proposed in the late 1940s, offered the only hope of preventing the species' extinction (Snyder and Snyder 2000, 67, 87). The committee's report and the writings of other supporters of captive breeding recast earlier preservationism as a failed strategy for both conservation and science. A typical statement came from Stanley Temple, one of the members of the committee. In a 1979 review of a report on condor research, Temple lamented the fact that the species "may have received too much protection" since "early admonitions about disturbing condors have limited recent research on the birds so much that essential information about condors has not been collected" (Temple 1979, 291). The result of this protectionism, Temple argued, was that no one could say exactly why the population continued to decline or what needed to be done to save it.

Within the small community of endangered species scientists and the even smaller group of those who had studied condors, such criticisms were necessarily personal as well as programmatic. Although Temple did not single out Koford by name, it was Koford's style of research and advocacy for strict protection of the condor from all those who might harm it – including those who thought they were helping it – that he was attacking. Koford, in turn, in one of his last statements on condors before his death in 1979, assailed the committee on which Temple had served for its willingness to secure the survival of the condor through the sacrifice of its wildness – the very characteristic, in his opinion, that made it worth saving (Phillips and Nash 1981, 67–98). For Koford, captive propagation was a misguided effort to save the biological species as such, rather than preserving the condor and its ecological and cultural significance *in situ* (Alagona 2004, 570–571; Snyder and Snyder 2000, 91).

Despite the opposition of Koford and a number of prominent scientists and wilderness advocates who temporarily gained the upper hand when a condor chick died while being examined at a nesting site in 1980, advocates of intensive management and captive breeding eventually won the debate (Moir 2006, 65–81, 121–137; Nielson 2006, 171–189; Snyder and Snyder 2000, 99–101, 278–306, 316–338). Some opponents of captive breeding were convinced the condor could be saved through intensified protection, others that the species was doomed regardless of what conservationists did. The essence of the winning argument in favor of captive breeding was that mortality from a variety of causes, including lead poisoning, could not be reduced in time to save the condor from extinction. These sources of mortality, rather than the kinds of disturbance that Koford had sought to prevent, were identified as the primary cause of the condor's decline. In 1987, the last free-living condor was taken into captivity; five years later, the first captive-bred condor was released to the wild. For the five intervening years, the California condor was "extinct in the wild," with the world's entire population living in zoos and breeding facilities (Moir 2006, 121–137; Nielson 2006, 171–189; Snyder and Snyder 2000, 278–306, 316–338).

From the 1980s onward, the shift from a preservationist to a managerial paradigm for condor conservation led to research in two related areas that had hitherto received little attention. The first was the reproductive behavior and biology of condors, which was essential to expanding the captive condor population as quickly as possible without the loss of genetic diversity. This had never before been a major concern of condor researchers, with perhaps the exception of a few zoo biologists. That Koford had only reported witnessing copulation between condors once in his 1953 report indicates both the perceived unimportance of mating behavior and the difficulty of studying condor reproduction using the methods common in the mid-twentieth century (Wilbur and Borneman 1972). As of 1980, no condor had ever hatched in captivity. In the early 1980s, it was still unclear whether mating pairs would "double clutch," laying a second egg in the same season if the first was removed, which was a crucial piece of information for the success of a captive breeding program (Harrison and Kiff 1980; Holden 1980). The first successful hatching of an egg in captivity, taken from a breeding pair in the wild, took place in 1983 at the San Diego Zoo (Miller 1983). It marked the beginning of an intensive period of research into the breeding and developmental biology of condors with an eye toward raising a replacement population in captivity (Snyder and Snyder 2000, 278–306, 316–338).

The second new research area concerned the behavior of adult condors in the wild. This was a topic that had preoccupied Koford and others from the 1930s to the 1970s, but the new generation of researchers pursued it in a very different manner. Concerned less with establishing the typical life history of the species than with establishing precise demographic measures and determining causes of mortality, this program depended on the use of numbered wing-tags and radio transmitters to identify individual birds and to monitor them regularly in the field (see Figure 7.2). It also relied on the development of techniques that would allow condors to be safely, effectively, and repeatedly captured and measured (Snyder

Figure 7.2 One of the first captive-bred condors to be released in the wild sunning its wings in December 1993. This condor, designated Y89, died the following June after colliding with a power pole (Sommer 1994). Public domain photograph from the National Digital Library of the U.S. Fish and Wildlife Service, http://digitalmedia.fws.gov/, Item ID: WVCon04-condor.

and Snyder 2000, 339–363). This kind of regular monitoring and handling had been unnecessary when the solution to condor endangerment was strict protection of the population as a whole within designated wilderness areas. It only made sense in the context of a decision to protect individual condors against specific threats and to intervene when a bird's life was threatened within a landscape assumed to be thoroughly influenced by humans. Such constant monitoring made it possible, for example, to bring an individual in for chelation treatment when it became apparent that it had ingested lead pellets from a gut pile left behind by a deer hunter (Cade 2007), or to recapture and relocate a condor that had strayed into an area of dense human population.

Since the 1980s, research on and conservation of the California condor has continued to be dominated by this managerial approach. A recent assessment concludes that "the condor's apparent recovery is solely because of intensive ongoing management," without which the condor would soon be extinct (Finkelstein et al. 2012, 1; see also Walters et al. 2010). In 2000, the husband-and-wife team of Noel and Helen Snyder, who had been prominent advocates of captive breeding and intensive research during the particularly contentious period of the early 1980s, published a large, lavishly illustrated book about the condor research and conservation program. Their *The California Condor* provided a comprehensive and sympathetic view of the managerial turn in endangered species conservation, arguing that controversies over condor conservation in the late twentieth century revealed the peril to endangered species that was posed by "anti-research attitudes" and "factions that . . . oppose all intensive activities on the grounds that the greatest threat to the species is research itself, not to mention evil researchers bent on torturing the species" (Snyder and Snyder 2000, 371).

This is, of course, a caricature. While those who opposed the Snyders' proposals may sometimes have accused them of tormenting the animals they studied, they were not "anti-research" but rather simply opposed to the kind of research that they thought would do more harm than good. Nonetheless, the Snyders' characterization contained a grain of truth. Preservationists were opposed to certain kinds of research that they deemed both harmful and unnecessary, as were the advocates of management. A program of managerial endangered species conservation simply demanded different kinds of knowledge and authorized different kinds of interventions for the sake of producing it.

The shift from preservation to active caretaking therefore represented more than merely a continued expansion of the use of science, technology, and government authority to address problems of species conservation. It also represented a reorientation of these forces toward new goals and a reconceptualization of the condor as an object of inquiry. Under the preservationist approach, it had been possible to imagine human activities, including scientific research, as brief interventions into the essentially natural lives of condors, even if those condors now fed on the carcasses of domesticated cattle rather than of the wild megafauna that had sustained their ancestors during the Pleistocene. Scientists and conservationists had attempted to minimize those interventions even while recognizing that they could not be eliminated entirely. Now, in contrast, human management was the fundamental condition of condor lives. Henceforth, research on condors would necessarily be research on captive condors and their descendants. The result of this development can be seen as a reconciliation of wilderness preservation and wildlife management (Alagona 2004; 2013), but it also reflected a deeper shift in the understanding of nature.

After this shift, proposed research procedures could be reasonably criticized as ineffective or unsafe but not as unnatural or intrusive. To resume the archival metaphor, one might say that the archive of science and the archive of nature had been collapsed into a single postnatural archive that encompassed living things in "the wild" and in captivity as well as specimens and records. One consequence was that the collection of condors in the traditional natural-historical sense was no longer truly possible. All condors, even those living outside of captivity in California or Arizona, were already part of a collection that was being carefully curated by conservationists and biologists. In the wild, as in the zoo, endangered condors were now the subject of a sustained exercise of biopower aimed at ensuring the reproduction of the species into the future (Braverman 2012; Friese 2013). Trapping or shooting one of these free-living condors, even for the purposes of science, bore less resemblance to collecting in the style of late-nineteenth-century naturalists and egg enthusiasts than to stealing a specimen from a museum drawer or a living animal from a zoo exhibit. This is not to say that the entire world had been turned into a museum or a zoo, but only that in regard to certain closely managed species such as the condor, the lines between zoo, museum, nature park, and wilderness area were no longer as discernible as they had once been.

Conclusion

By showing how changing understandings of the species' endangerment led to major shifts in research methods and goals, this chapter has emphasized the discontinuities in the history of California condor research and conservation since the nineteenth century. A wide gap separates the late-nineteenth-century conviction that the advance of civilization inevitably meant the doom of certain wild animals from the late-twentieth-century conviction that the application of science and technology could save vulnerable biological species even as human populations and levels of consumption soared. Equally far apart are the collecting free-for-all of late-nineteenth-century zoologists and the highly regulated interventions of condor managers since the 1980s. Nonetheless, the continuities in this story should not be ignored. Since Cooper first sounded the alarm in 1890, the production of knowledge about the California condor has been pervasively and persistently shaped by the idea that human activities are threatening the species with extinction. The human activities in question have been as broad as the advance of civilization and as narrow as the use of lead shot by recreational hunters, but the fundamental idea of humanity as a threat to the condor and to other endangered species – and ultimately, perhaps, to nature on a much grander scale – has persisted. In trivial and not-so-trivial ways, the threat of extinction has now been shaping biologists' view of the condor and of natural variety for more than a century.

Although each of the phases presented here represents assumptions and visions that were widely shared at the time, the sequence cannot be applied unchanged to all endangered bird species within the United States, let alone to all endangered species. The passenger pigeon (*Ectopistes migratorius*), for example, which astounded European settlers in North America by its sky-darkening flocks of millions, was never rare in the sense that the condor was understood to be rare in the early and mid-nineteenth century. It became rare only after it was recognized as threatened, and its rarity was understood not as its natural condition but as the penultimate step on its road to extinction in the early twentieth century (Price 1999, 1–55). It did not survive long enough to become subject either to the preservationism of the mid-twentieth century or the managerialism of the late twentieth century. If it had, its distinctive biology and significance to Americans as both popular meat source and unpopular crop pest would undoubtedly have led researchers and conservationists down a very different path than that traced out by the condor. Similarly contingent variations would almost certainly also be seen in any detailed study of other extinct or endangered North American birds: the heath hen, the Carolina parakeet, the ivory billed woodpecker, the whooping crane, Kirtland's warbler, the bald eagle, and many others. Despite their commonalities and the many ideas, practices, tools, and institutions that have been transferred from one endangered species to another, each endangered species has ultimately been endangered in its own way.

Even the California condor may yet become endangered in new ways, or perhaps, though it is harder to imagine, escape the category of endangerment altogether. The current managerial approach has been moderately effective, leading to a global population of about 400 condors, about half of them living outside of captivity. But it is both costly and unproven in the face of new threats that could yet alter environmental conditions crucial to the condor in ways that are beyond the control of even the most well-funded and technologically advanced recovery effort. From here, it is possible to imagine a variety of potential paths. The species might recover to the point where intensive management can be replaced by habitat protection, as in the mid-twentieth century, or new threats and population declines might lead to a return to an entirely captive population, as in the late 1980s, but this time with little hope of a successful reintroduction to the wild. Or it may be that the very idea of endangered species and endangered species protection will lose its appeal and efforts to protect the condor will be abandoned, effectively returning – albeit under radically changed circumstances and perhaps only for a moment before the species' extinction – to the "pre-endangerment" approach of most of the nineteenth century. Under such circumstances, new ways of knowing the condor might emerge from under the shadow of endangerment. When they do, they will be dependent on new social forms and a network of trust, just as Audubon was dependent on Townsend's translation of Native American informants' claims in the period of the condor's scientific discovery.

Whether the endangerment sensibility should be seen as shadow or light – that is, as a help or a hindrance to the development of knowledge about the condor and the preservation of the species – has not been the primary focus of this chapter, which instead has focused on the ways that the threat of extinction has colored, for better or worse, the way knowledge about condors is produced. But by way of conclusion, the question may be worth considering. For all of the changes in the endangerment sensibility that have taken place from the late nineteenth century, one thing has remained constant: it is the human, rather than the condor, who has been seen as responsible for the condor's fate. Humans are the ones whose civilization, cattle ranches, hunting regulations, wilderness preserves, lead shot, breeding programs, power lines, and radio-tags have been seen as having the power either to save the condor or to doom it. In this light, the designation of the species as endangered appears both as an acknowledgment of human power and a recognition of human culpability.

However morally admirable that acknowledgment may be, it remains profoundly anthropocentric. Little room is left in this all-encompassing vision for condors to innovate, experiment, or adapt to changing circumstances; they remain trapped in the Pleistocene past, so to speak, even as the world is transformed around them. That condors clearly did fail to adapt to a landscape poisoned with lead shot and other novel threats quickly enough to prevent their precipitous decline in the nineteenth and twentieth centuries should not, I think, be taken to mean that they do not have the capacity to do so in the future, with a little time and some human facilitation. But viewing condors through the lens of endangerment does not encourage providing those opportunities or developing the kinds of knowledge that would make them possible. On the contrary, the endangerment sensibility in all of its varieties seems inextricably tied to a vision of species as frozen in time and of conservation as an effort to insulate them against change. Perhaps thinking of condors not as somehow essentially endangered but simply as in danger, like many other creatures on this planet if perhaps often in distinctive and particularly perilous ways, would open up new possibilities for their flourishing over the long term.

Bibliography

Alagona, P.S., 2004. Biography of a 'Feathered Pig': The California Condor Conservation Controversy. *Journal of the History of Biology*, 37 (3), 557–583.

Alagona, P.S., 2013. *After the Grizzly: Endangered Species and the Politics of Place in California*. University of California Press, Berkeley.

Alvarez, K., 1993. *Twilight of the Panther: Biology, Bureaucracy and Failure in an Endangered Species Program*. Myakka River, Sarasota, FL.

Audubon, J.J., 1839. *Ornithological Biography, or an Account of the Habits of the Birds of the United States of America*, Vol. 5. Adams & Charles Black, Edinburgh.

Barrow, M.V. Jr., 2000. *A Passion for Birds: American Ornithology after Audubon*. Princeton University Press, Princeton.

Barrow, M.V. Jr., 2009. *Nature's Ghosts: Confronting Extinction from the Age of Jefferson to the Age of Ecology*. University of Chicago Press, Chicago.

Bates, C. D., Hamber, J. A. and Lee, M. J., 1993. The California Condor and California Indians. *American Indian Art Magazine*, 19 (1), 41–47.

Bean, M. J. and Rowland, M. J., 1997. *The Evolution of National Wildlife Law*, 3rd ed. Praeger, Westport, CT.

Benson, E., 2010. *Wired Wilderness: Technologies of Tracking and the Making of Modern Wildlife*. Johns Hopkins University Press, Baltimore.

BirdLife International, 2013. Gymnogyps californianus. *The IUCN Red List of Threatened Species*, Version 2014.2 (www.iucnredlist.org). Accessed 2 August 2014.

Branch, M. P., 2008. John Kirk Townsend. In: Patterson, D., ed., *Early American Nature Writers: A Biographical Encyclopedia*. Greenwood, Westport, CT., 373–380.

Braverman, I., 2012. *Zooland: The Institution of Captivity*. Stanford University Press, Stanford, CA.

Brewer, T. M., 1867. Some Errors Regarding the Habits of our Birds. *American Naturalist*, 1 (3), 13–23.

Cade, T. J., 2007. Exposure of California Condors to Lead From Spent Ammunition. *Journal of Wildlife Management*, 71 (7), 2125–2133.

Cooper, J. G., 1890. A Doomed Bird. *Zoe*, 1 (8), 248–249.

Cronon, W., 1996. The Trouble with Wilderness: Or, Getting Back to the Wrong Nature. *Environmental History*, 1 (1), 7–28.

Darlington, D., 1987. *In Condor Country*. Houghton Mifflin, Boston.

Daston, L. and Galison, P., 2007. *Objectivity*. Zone, Brooklyn, NY.

Farber, P. L., 2000. *Finding Order in Nature: The Naturalist Tradition from Linnaeus to E. O. Wilson*. Johns Hopkins University Press, Baltimore.

Farnham, T. J., 2007. *Saving Nature's Legacy: Origins of the Idea of Biological Diversity*. Yale University Press, New Haven, CT.

Finkelstein, M. E., Doak, D. F., George, D., Burnett, J., Brandt, J., Church, M., Grantham, J. and Smith, D. R., 2012. Lead Poisoning and the Deceptive Recovery of the Critically Endangered California Condor. *Proceedings of the National Academies of Science of the United State of America*. Vol. 109, No. 28 (2012): 11449–11454 (http://www.pnas.org/content/109/28/11449). Accessed 2 August 2014.

Finley, W. L., 1901. Catching Birds with a Camera. *Condor*, 3 (6), 137–139.

Finley, W. L., 1906. Life History of the California Condor. Part I. Finding a Condor's Nest. *Condor*, 8 (6), 135–142.

Finley, W. L., 1910. Life History of the California Condor. Part IV. The Young Condor in Captivity. *Condor*, 12 (1), 4–11.

Friese, C., 2013. *Cloning Wild Life: Zoos, Captivity, and the Future of Endangered Animals*. New York University Press, New York.

Haraway, D., 1984–1985. Teddy Bear Patriarchy: Taxidermy in the Garden of Eden, New York City, 1908–1936. *Social Text*, 11 (Winter), 20–64.

Harris, H., 1941. The Annals of Gymnogymps to 1900. *Condor*, 43 (1), 1–55.

Harrison, E. N. and Kiff, L. F., 1980. Apparent Replacement Clutch Laid by Wild California Condor. *Condor*, 82 (3), 351–352.

Heise, U. K., 2010. *Nach der Natur: Das Artensterben und die moderne Kultur*. Suhrkamp, Berlin.

Heise, U. K., 2011. Lost Dogs, Last Birds, and Listed Species: Cultures of Extinction. *Configurations*, 18, 39–62.

Holden, C., 1980. Condor Flap in California. *Science*, 209 (4457), 670–672.

International Union for Conservation of Nature, 2001. *IUCN Red List Categories and Criteria Version 3.1* (http://www.iucnredlist.org/technical-documents/categories-and-criteria/2001-categories-criteria). Accessed 20 September 2012.

Janzen, D.H., 1986. Science is Forever. *Oikos*, 46 (3), 281–283.

Jardine, N., Secord, J.A. and Spary, E.C., eds., 1996. *Cultures of Natural History*. Cambridge University Press, Cambridge.

Jobanek, G.A. and Marshall, D.B., 1992. John K. Townsend's 1836 Report of the Birds of the Lower Columbia River Region, Oregon and Washington. *Northwestern Naturalist*, 73 (1), 1–14.

Koelle, A., 2012. Intimate Bureaucracies: Roadkill, Policy, and Fieldwork on the Shoulder. *Hypatia*, 27 (3), 651–669.

Koford, C.B., 1953. *The California Condor*. Dover, New York.

Kohler, R.E., 2006. *All Creatures: Naturalists, Collectors, and Biodiversity, 1850–1950*. Princeton University Press, Princeton.

Lewis, D., 2012. *The Feathery Tribe: Robert Ridgway and the Modern Study of Birds*. Yale University Press, New Haven, CT.

MacKenzie, J.M., 1988. *The Empire of Nature: Hunting, Conservation and British Imperialism*. Manchester University Press, Manchester.

Marshall, J.T., 1954. Review of "The California Condor" by Carl B. Koford. *Wilson Bulletin*, 66 (1), 75–76.

Mathewson, W., 1986. *William L. Finley: Pioneer Wildlife Photographer*. University Press, Corvalis, OR.

McMillan, I., 1968. *Man and the California Condor: The Embattled History and Uncertain Future of North America's Largest Free-living Bird*. E.P. Dutton, New York.

Miller, A., 1953. Preface to *The California Condor* by Carl B. Koford. Dover, New York, vii–viii.

Miller, J.A., 1983. New Condor Chicks from Captive Eggs. *Science News*, 123 (15), 229.

Mitman, G., 1996. When Nature is the Zoo: Vision and Power in the Art and Science of Natural History. *Osiris*, 11, 117–143.

Mitman, G., 1999. *Reel Nature: America's Romance with Wildlife on Film*. Harvard University Press, Cambridge, MA.

Moir, J., 2006. *Return of the Condor: The Race to Save our Largest Bird from Extinction*. Lyons Press, Guilford, CT.

Nielson, J., 2006. *Condor: To the brink and back – The Life and Times of one Giant Bird*. Harper Collins, New York.

Nyhart, L., 2009. *Modern Nature: The Rise of the Biological Perspective in Germany*. University of Chicago Press, Chicago.

Phillips, D. and Nash, H., 1981. *The Condor Question: Captive or Forever Free?* Friends of the Earth, San Francisco.

Price, J., 1999. *Flight Maps: Adventures with Nature in Modern America*. Basic Books, New York.

Prince, S.A., ed., 2003. *Stuffing Birds, Pressing Plants, Shaping Knowledge: Natural History in North America 1730–1860*. American Philosophical Society, Philadelphia.

Shapin, S., 1994. *A Social History of Truth: Civility and Science in Seventeenth-century England*. University of Chicago Press, Chicago.

Snyder, N.F.R. and Snyder, H., 2000. *The California Condor: A Saga of Natural History and Conservation*. Academic Press, San Diego.

Sommer, C., 1994. California Condor dies at L.A. Zoo. *Los Angeles Times*, 2 August (http://articles.latimes.com/1994-08-02/local/me-22783_1_california-condor). Accessed 3 August 2014.

Stephens, F., 1895. Notes on the California Vulture. *Auk*, 12 (1), 81–82.

Sutter, P.S., 2002. *Driven Wild: How the Fight against Automobiles Launched the Modern Wilderness Movement*. University of Washington Press, Seattle.

Takacs, D., 1996. *The Idea of Biodiversity: Philosophies of Paradise*. Johns Hopkins University Press, Baltimore.

Taylor, A. S., 1859. The Great Condor of California. *Hutchings' Illustrated California Magazine*, 4 (2), 61–64.

Temple, S. A., 1979. Review of "The California Condor, 1966–1976: A Look at Its Past and Future" by S. R. Wilbur. *Bird-Banding*, 50 (3), 190–191.

Turner, J. M., 2013. *The Promise of Wilderness: American Environmental Politics since 1964*. University of Washington Press, Seattle.

Walters, J. R., Derrickson, S. R., Fry, D. M., Haig, S. M., Marzluff, J. M. and Wunderle, J. M. Jr., 2010. Status of the California Condor (Gymnogyps californianus) and Efforts to Achieve its Recovery. *Auk*, 127 (4), 969–1001.

Whitney, K., 2013. Tangled up in Knots: An Emotional Ecology of Field Science. *Emotion, Space and Society*, 6, 100–107.

Wilbur, S. R., 1974. California Condor Specimens in Collections. *Wilson Bulletin*, 86 (1), 71–72.

Wilbur, S. R. and Borneman, J. C., 1972. Copulation by California Condors. *Auk*, 89 (2), 444–445.

Wilson, R. M., 2010. *Seeking Refuge: Birds and Landscapes of the Pacific Flyway*. University of Washington Press, Seattle.

Zimmerman, D. R., 1975. *To Save a Bird in Peril*. Coward, McCann, and Geoghegan, New York.

Zimmerman, D. R., 1976. Endangered Bird Species: Habitat Manipulation Methods. *Science*, 192 (4242), 876–878.

8 World Heritage listing and the globalization of the endangerment sensibility

Rodney Harrison

> Nothing seems easier than to draw up a list, in actual fact it's far more complicated than it appears; you always forget something, you are tempted to write, etc., but an inventory is when you don't write, etc.
>
> (Georges Perec 'Notes Concerning the Objects that are on my Work-table' [1976] in *Species of Spaces and Other Pieces* 2008, 146)

Introduction

The aim of this chapter is to consider the origin and subsequent history of the World Heritage concept as an attempt to develop a universal system for the specification and definition of heritage 'at risk' within the context of the development of a late twentieth/early twenty-first century 'endangerment sensibility' (Vidal and Dias this volume). In particular, it explores the practical implications of the establishment and operation of the World Heritage List and the Lists of Intangible Cultural Heritage, their history and development at the institutional and conceptual level, and the ways in which such lists have come to operate as key 'apparatuses' for the identification, categorization and designation of objects, places and practices of heritage that simultaneously specify and operationalize global (and *globalizing*) categories of endangerment and their management (Rico 2014).[1] In using the term 'apparatus', I draw on Michel Foucault's definition of a 'heterogeneous ensemble consisting of discourses, institutions, architectural forms, regulatory decisions, laws, administrative measures, scientific statements, philosophical, moral and philanthropic propositions' (1980, 194); a device that specifies (and hence helps to create) a subject so that it might control, distribute and/or manage it (see further discussion in Rabinow 2003, 49ff.). Importantly, the Foucauldian *dispositif* is a 'formation which has as its major function at a given historical moment that of responding to an *urgent need*' (Foucault 1980, 195, original emphasis). It has a certain sense of historical 'inertia' and functions in such way so as to promote its own continued existence and self-replication, even if the conditions under which it arose subsequently change. Giorgio Agamben further defines an apparatus as 'anything that has in some way the capacity to capture, orient, determine, intercept, model, control, or secure the gestures, behaviours, opinions, or discourses of living beings' (2009, 14) (and as Foucault emphasizes, the *system of relations between them*).

I argue that one of the important implications of heritage lists as apparatuses is that they instigated and have continued to facilitate forms of government 'at a distance' (after Foucault 2011). They have also made possible the expansion of museological forms of management, from endangered monuments and buildings *in situ* to imperiled traditions and customs, and thence to cultural practices labeled 'intangible' heritage. To put this another way, as soon as heritage (as an 'official' object of knowledge and management) was specified and collected in the shape of a list, it was possible to detach it from its material presence and thus for it to become spatially and temporally *simultaneous*. The list could be present in one place but operative in many other places at the same time. Heritage itself was no longer shackled to the presence of physical objects and buildings but could be assembled virtually as a part of a list and thus understood to transcend national borders. It thereby became mobile, intangible, virtual and global. This chapter argues that such lists are intimately connected with the development of a contemporary endangerment sensibility. Far from being neutral reflections of an attempt to codify and protect 'universally' cherished aspects of human culture, they can be understood to have certain strategic objectives (discussed further below) that require globalizing the perception of sources of endangerment, as well as the systems generated to respond to them. They therefore have wide reaching epistemological, political, institutional and material implications.

Some notes on heritage, modernity, classification and endangerment

This chapter concerns itself exclusively with what I have elsewhere (Harrison 2010; 2013a) termed 'official' heritage. I use this term to refer to a set of professional practices that are authorized by the state and motivated by some form of legislation or written charter. This represents an operational definition of heritage as the series of mechanisms by which objects, buildings and landscapes are set apart from the everyday and conserved for their aesthetic, historic, scientific, social, spiritual or recreational values. By contrast, I would suggest that 'unofficial' heritage should be understood as the lived experience and cultural work that individuals and groups undertake to keep the past alive in the present. This encompasses a broad range of practices that may be represented using the language of heritage, but that are not formally recognized or protected by forms of legislation and/or authorized by the state. What constitutes official and unofficial heritage is not fixed, and ideas about what constitutes heritage in one arena significantly influence those in the other. The expansion of the range of categories of objects, places and practices considered to constitute heritage globally, which occurred over the decades following the ratification of the World Heritage Convention, has come about as a direct result of ideas concerning heritage in the unofficial realm becoming officially recognized. On the other hand, ideas about what heritage *is* and *does* that circulate within official heritage can also significantly influence what people believe constitutes their own (unofficial) heritage, whether it is recognized by the state or not. So these categories are such that each influences the definition of the other.

It has long been observed that 'heritage' is generally defined in the context of some sort of threat to objects, places or practices that are perceived to hold a form of collective value (Davison 2008 [2000]; Dicks 2000; Lowenthal 1985; Smith 2006). This threat might manifest itself as the demolition of historic buildings, the loss of cultural objects, the erosion of social values, or the impact of urban development on natural landscapes. In such contexts, physical destruction or decay is perceived to injure not only the *object, place* or cultural *practice* in question, but also the group of *people* who hold it as part of their heritage (Harrison 2013a). Threat implies risk and uncertainty, and one way in which modern societies manage them is through placing increased trust in 'experts' and abstract 'expert systems' over local forms of knowledge (Giddens 1991, 29–32; see also Beck 1992; Douglas 1992). Experts define risk by means of statistics and data that make it *calculable* and hence *manageable*. Identifying and classifying risk are integral to the process of managing it. And such a process, as Paul Rabinow (1989) has highlighted, simultaneously *specifies* and *produces* its own target. These strategies for the care and management of heritage 'risk' are closely connected to the increasing bureaucratization and professionalization of heritage that has occurred since the development of the World Heritage Convention and other allied developments that occurred in the 1970s (Harrison 2013a).

Indeed, as Geoffrey Bowker (2005; Bowker and Star 2000) has noted, classificatory systems are not only structures for the organization of information, but they are also 'memory practices': structures that, by producing and maintaining knowledge systems, shape the way in which we perceive the past and present. In *Organizing Modernity*, John Law (1994) shows that 'modes of ordering' the world not only characterize, but also define and hence generate, the qualities of their objects. Hence the way in which we perceive and experience the subject of a classificatory system results, in part at least, from the classificatory system itself. In *Purity and Danger* (1966), Mary Douglas observes that taxonomic anomalies are treated with distrust because they represent potential sources of social disorder and threat. One way of dealing with them is to purify them by rendering them to the realm of myth. Another is to build more elaborate systems that can take account of them. So classificatory systems should not be considered as fixed, and the process of classification should be understood as involving mutually constructive 'looping effects' (see Hacking 1986; 1995) between the taxonomic system and its objects.

Bruno Latour has shown how the history of the modern sciences is, in part, a history of the mobilization of objects and forms of data that can be reassembled in the laboratory (1987, 224–225). His model describes the construction of scientific knowledge by way of the production and accumulation of 'immutable and combinable mobiles' – objects, specimens, charts, maps, tables, field notebooks and other recorded observations – collected from the peripheries (or 'field') and brought to a center (such as a laboratory) where they may be combined and interpreted. This allows laboratories to 'act at a distance' (Latour 1987, 229) on the fields of collection. Tony Bennett has applied this model to the operation of museums (e.g. 2005; 2009; 2013), and I have elsewhere discussed its potential utility to understand the role of heritage lists and registers (Harrison 2013b). While there are differences between the operations of museums and heritage registers, the World Heritage List has had the effect of making fixed places 'mobile' and susceptible to administration in sites far removed from their physical locations. In many ways, the documents that describe and collect these places, and that are the focus of administrative work, become more instrumental than the 'actual' places themselves. It is after all the descriptions of these places' value as defined by World Heritage criteria that determine how they are understood and managed, as well as how they are represented across various forms of media. Furthermore, like museums, UNESCO's List and its associated guidelines have come to play a key role in attributing value by way of increasingly refined criteria and the forms of expertise that are defined in relation to them.

Clearly, the practices of classification, ordering and cataloguing have a whole range of implications for heritage in its official guise. Central to the museum are processes of assembling, categorization, comparison, classification, ordering and reassembling (e.g. Baudrillard 1994; Bennett 1995; 2004; Harrison 2013b; 2013c; Pearce 1995) – processes that relate to modern scientific practices more generally (Bowker 2005; Hopwood et al. 2010; Latour 1993; Schlanger 2010). Similarly, the listing of heritage sites on various registers, and the classificatory schemes that accompany these processes, can also be seen to be among the phenomena that enter into the management of endangerment and risk. Bowker and Star (2000, 137ff.), drawing on the work of Foucault (1970), Latour (1987), Tort (1989) and others, show how lists of various kinds have figured within the history and sociology of science as technologies not only for hierarchical ordering, but also for the bureaucratic management of work and labor. They also suggest that lists should be seen as modes of ordering that articulate with other lists to form regulatory systems that require standardization, coordination and collaboration across varied administrative technologies and bureaucratic structures. In their expansive cross-referencing and complicated 'looping effects', lists generate their own worlds that are articulated across a heterogeneous series of sites, objects, actors and practices, maintained and held together by a robust set of supporting administrative and governmental (and inter- and trans-governmental) infrastructures.

Entwined histories: World Heritage and the endangerment sensibility

In 2012, UNESCO celebrated the fortieth anniversary of the adoption of the *Convention Concerning the Protection of the World Cultural and Natural Heritage* ('World Heritage Convention') by the General Conference of the United Nations Educational, Scientific and Cultural Organization (UNESCO) meeting in Paris at its seventeenth session, held from October 17 to November 21, 1972. The Convention, forged in the crucible of the post-war internationalism characteristic of the various organizations that emerged following the 1944 Bretton Woods United Nations Monetary and Financial Conference, announced its arrival with a dire warning regarding the threats to heritage in the modern world:

> the cultural heritage and the natural heritage are increasingly *threatened* with destruction not only by the traditional causes of decay, but also by changing social and economic conditions which aggravate the situation with even more formidable phenomena of damage or destruction . . . deterioration or disappearance of any item of the cultural or natural heritage constitutes a harmful impoverishment of the heritage of all the nations of the world.
>
> (UNESCO 1972, my emphasis)

The idea of international collaboration on safeguarding cultural heritage, first discussed in 1931 at the Athens Conference on the Restoration of Historic Buildings organized by the International Museums Office (IMO), led to the drafting of the Athens Charter; however, its recommendations were not realized until well after the end of World War II. The new sense of global responsibility for cultural heritage, rooted in the experience of the two world wars, was initially embodied in the 1954 United Nations Convention for the Protection of Cultural Property in the Event of Armed Conflict (known as 'Hague Convention'). The Hague Convention specified that signatories must refrain from damaging cultural properties in their own or other countries' territories during times of armed conflict, and made any act directed by way of reprisals against cultural property a violation of the convention. The sense of international responsibility for heritage was tested the same year when the Egyptian government announced its plans to construct the Aswan High Dam, which would require the flooding of a valley containing ancient Egyptian monuments, including the Abu Simbel temples. An appeal was subsequently launched by UNESCO Director General Vittorino Veronese on March 8, 1960, to undertake 'a task without parallel in history' (cited in Hassan 2007, 80): a global campaign to save the antiquities of Egypt and Sudan (Säve-Söderberg 1987). The international safeguarding campaign, which would run for twenty years (the construction of the dam itself was completed by 1970), involved a large-scale archaeological excavation and recording program, as well as the relocation and reconstruction of the Abu Simbel and Philae temples and other monuments from the valley. Over half of the estimated $80 million cost of the project was raised from forty-seven donor countries. A series of influential and wealthy individuals formed an 'Honorary Committee of Patrons' to lobby governments on UNESCO's behalf, while an exhibition of Tutankhamen's treasures toured the UK, Europe and North America between 1972 and 1979 to help enlist private support. The bulk of the financial support came from the USA, France, Italy and the Federal Republic of Germany, while private contributions in excess of $7 million were received. A tourist tax levied on visitors to Egypt raised almost $2 million (Hassan 2007, 84). As part of the safeguarding campaign, twenty-three temples were documented and relocated (see Figure 8.1).

Figure 8.1 Relocation of the temples of Ramses II at Abu Simbel in progress, 1965.
(© UNESCO)

While the first UNESCO safeguarding campaign was under way, the Second
Congress of Architects and Specialists of Historic Buildings met in Venice in 1964
and adopted a number of resolutions. The first created the International Charter
on the Conservation and Restoration of Monuments and Sites, or 'Venice Charter'. The Venice Charter provided an international framework for the preservation
and restoration of historic monuments and buildings. A subsequent resolution,

put forward by UNESCO, created the International Council of Monuments and Sites (ICOMOS) to oversee its implementation. ICOMOS was founded in 1965, and in this same year, a White House conference called for a 'World Heritage Trust' to preserve the world's natural and scenic areas and historic sites 'for the present and the future of the entire world citizenry'. The flooding of Venice in November 1966 (see Figure 8.2) and the subsequent development of a second international safeguarding campaign appeared to underline the need for global collaboration on heritage issues. Images of a flooded Venice conveyed the vulnerability of global heritage and gave urgency to these institutional developments. The International Union for Conservation of Nature (IUCN), formed in 1948 as the International Union for the Protection of Nature (IUPN), echoed the 1964 Venice resolutions. In 1972, the UN conference on Human Environment that

Figure 8.2 View of women knee deep in water in Rialto Square in Venice during the floods of November 1966. The floods stimulated a second international safeguarding campaign and were instrumental in rallying support for the development of a World Heritage Convention.

(Photograph by A.F.I. Venise. © UNESCO/A.F.I. Venise.)

took place in Stockholm drafted a Convention Concerning the Protection of the World Cultural and Natural Heritage, which was adopted by the General Conference of UNESCO that same year (Bandarin 2007; UNESCO 1972).

The 1972 Convention created a World Heritage Committee, which would be advised by ICOMOS, IUCN and the International Centre for the Study of the Preservation and Restoration of Cultural Property (ICCROM). The World Heritage Committee would administer the nomination of places to a World Heritage List, including 'properties forming part of the cultural heritage and natural heritage . . . which it considers as having outstanding universal value in terms of such criteria as it shall have established'. It placed the question of the identification and management of heritage squarely within the context of the circumstances of late-modern life by appealing to the idea of threat, and suggesting that the threat of the loss of heritage was an issue for the concern of all humanity.

The Convention defined 'cultural heritage' and 'natural heritage' separately, using different criteria. Cultural heritage was defined as 'monuments', 'groups of buildings' or 'sites', employing language that emphasized its essentially material, or tangible, quality. Natural heritage was defined as 'natural features', 'geological and physiographical formations' or 'natural sites' in a way that mirrored the structure of types of cultural heritage, but that also made a clear distinction between 'works of man [sic]' and works of nature. This classificatory system implicitly defined heritage as something that is endangered and remote from contemporary everyday life (both in the sense in which natural heritage as 'wilderness' excludes the presence of humans, and cultural heritage occupies a past temporal dimension considered to be separate from the present). But perhaps the most novel and defining aspect of the 1972 Convention text was its concept of 'universal heritage value':

> parts of the cultural or natural heritage are of outstanding interest and therefore need to be preserved as part of the World Heritage of mankind as a whole . . . in view of the magnitude and gravity of the new dangers threatening them, it is incumbent on the international community as a whole to participate in the protection of the cultural and natural heritage of outstanding universal value, by the granting of collective assistance which, although not taking the place of action by the State concerned, will serve as an efficient complement thereto . . . it is essential for this purpose to adopt new provisions in the form of a convention establishing an effective system of collective protection of the cultural and natural heritage of outstanding universal value, organized on a permanent basis and in accordance with modern scientific methods.

Such a move from the notion of collective to universal values implied a corresponding shift from the idea of heritage as part of a nation's patrimony to one that was not contained by national borders. This idea had a clear lineage from the safeguarding campaigns in Egypt/Sudan and Venice. The suggestion that there existed places that should be equally cherished by all humans and whose management was therefore a matter of international responsibility has had profound and far-reaching consequences both for the notion of 'heritage' and for how it is preserved, managed and exhibited. The creation of the List was not an end in itself, but an instrument to involve each Convention member in the 'duty' of 'ensuring the identification, protection, conservation, presentation and transmission to future generations of the cultural and natural heritage'.

This duty would involve the establishment not only of a World Heritage List, but also a range of subsidiary state- and regional-level instruments, lists, policies, and administrative and bureaucratic processes that replicated specific hierarchies of value, implicating objects, places and, subsequently through the Intangible Heritage Convention, practices in a series of lists that functioned together to contain and codify the heritage of humanity. In this way, the List both produced and reproduced a global hierarchy of value (cf. Herzfeld 2004), generating a world in its own image, simultaneously specifying and operationalizing global categories of endangerment and the appropriate means for their management. The World Heritage List, as an apparatus of endangerment, not only structured and represented but also defined and, hence, generated its subject – *heritage at risk* (Rico 2014). In addition to the World Heritage List, the 1972 Convention announced the creation of a subsidiary 'List of World Heritage in Danger', which would be

> a list of the property appearing in the World Heritage List for the conservation of which major operations are necessary and for which assistance has been requested under this Convention. This list shall contain an estimate of the cost of such operations. The list may include only such property forming part of the cultural and natural heritage as is threatened by serious and specific dangers, such as the threat of disappearance caused by accelerated deterioration, large-scale public or private projects or rapid urban or tourist development projects; destruction caused by changes in the use or ownership of the land; major alterations due to unknown causes; abandonment for any reason whatsoever; the outbreak or the threat of an armed conflict; calamities and cataclysms; serious fires, earthquakes, landslides; volcanic eruptions; changes in water level, floods and tidal waves.

The long catalog of sources of threat, including both natural and cultural dangers and calamities, suggested that the sources of endangerment of heritage had become increasingly uncertain and hence unable to be easily specified. In this, the 1972 World Heritage Convention contrasted strongly with earlier international conventions, in which the threats to heritage were explicitly identified. In the case of the 1931 Athens Charter and 1964 Venice Charter, the threats to heritage are narrowly focused on new conservation and restoration techniques considered incompatible with the aesthetic, historic or scientific values of the monuments or their setting. In drawing attention to the many and uncertain sources of endangerment of natural and cultural heritage, and initiating a system of listing and ranking of heritage at risk, the Convention articulated with other contemporary international instruments of the late twentieth century endangerment sensibility, in particular those being produced for the management of 'natural heritage' such as the IUCN Red List of Threatened Species, established in 1963 as a global inventory of the international conservation status of biological species, and UNESCO's own Man and the Biosphere program established in 1971 (Bargheer this volume).

Of course, these lists were themselves also arranged in elaborate hierarchies of international, national and regional systems for defining and managing various endangered entities, from species (Benson this volume) to languages (Muehlmann this volume), and to the whole spectrum of 'biocultural diversity'. In addition to common perception of sources of endangerment, these lists and conventions were linked through shared networks of governmental and intergovernmental bodies, expert committees and associated organizations, cross-referential codes of practice and groups of practitioners, each of which reinforced and further defined a developing 'endangerment sensibility' during the later part of the twentieth century.

The World Heritage List grew rapidly. The first inscriptions to the List were made in 1978, and by 1980, eighty-two World Heritage sites had been inscribed in thirty-seven countries, and fifty-five countries had ratified the Convention. These first inscriptions included many familiar sites that had already been subject to intensive state management and utilized in the construction of national identities, such as Yellowstone National Park in the USA, and Chartres Cathedral in France. New sites were added regularly to the List, which grew to number 335 by 1990, and had more than doubled again in the following decade. Similarly, the number of states ratifying the Convention continued to grow. By its thirtieth anniversary in 2002, the World Heritage List numbered 730 sites, and 175 signatories had ratified the Convention.

The expansion of an endangerment sensibility was further explicitly facilitated by a number of 'heritage at risk' registers established during this period. I have already mentioned that the 1972 World Heritage Convention established the principles by which the World Heritage Committee could inscribe on a 'List of World Heritage in Danger'. The first site was inscribed on the List of World Heritage in Danger in 1979, and the List grew slowly through the 1980s and 1990s. Other international and national lists of threatened objects and places were subsequently established by various international NGOs following the UNESCO model. For example, the World Monuments Fund began to publish a biannual 'List of 100 Most Endangered Sites' in 1996, and ICOMOS launched a similar 'Heritage@ Risk' list in 1999, with occasional thematic foci (e.g. special reports on underwater heritage and European modernist buildings in 2006). Comparable lists, intended to highlight the vulnerability of particular heritage objects, sites and landscapes, also proliferated at the national level among states and heritage interest groups. These not only established a hierarchy of value and vulnerability, but also criteria by which gradients of endangerment could be determined. In England, for example, English Heritage began to keep a 'Buildings at Risk Register' in 1998, listing Grade I and Grade II buildings and Scheduled Ancient Monuments believed to be at risk due to neglect or external threat. These grades are applied to listed buildings according to statutory criteria. Grade I sites are of 'exceptional interest'; Grade II sites are 'particularly important, of more than special interest'; and Grade II sites are 'of special interest, warranting every effort to preserve them'. Such gradients not only establish and articulate with other systems of valuation, but are strategic in the sense in which they imply different actions (or regimes of management) that should be taken if sites are threatened. I have noted elsewhere that sites are very rarely 'de-accessioned' from such lists, many of which do not have any formal mechanism for de-accessioning (Harrison 2013d). This means that with time, the lists continue to grow. Once again, this initiates looping effects by which endangerment is also perceived to increase exponentially with the passage of time, and the need for such lists is perceived as increasingly urgent.

The rise of 'intangible heritage'

Almost as soon as the World Heritage Convention had been ratified, various challenges to its definition as a 'universal' system forced it to redefine and adopt new categories and forms of heritage, not only to expand its own field of governance, but also to maintain its claims to universality (see Harrison 2013a; 2013b). Foucault (1980, 194–196) notes that apparatuses function through two linked processes of *functional overdetermination*, in which each effect of the apparatus is the result of multiple, potentially contradictory causes, and *strategic elaboration*, the reworking of negative effects produced by such contradiction into positively functioning ones within the apparatus. These also operate as interrelated 'looping effects'. We can see such functional overdetermination and strategic readjustment and elaboration in operation in relation to the concept of intangible heritage.

As early as 1973, Bolivia suggested to UNESCO that a protocol might be added to the Universal Copyright Convention in order to protect aspects of folklore (Aikawa 2004). In 1982 at the Mondiacult World Conference on Cultural Policies, held in Mexico City, UNESCO officially acknowledged these concerns regarding the importance of intangible cultural heritage and included it in its new definition of cultural heritage (UNESCO 2012). It subsequently established a Committee of Experts on the Safeguarding of Folklore, and in 1989, it adopted the Recommendation on the Safeguarding of Traditional Culture and Folklore (Kurin 2004). The Recommendation called on member states to develop a national inventory of institutions concerned with folklore, with a view to their inclusion in regional and global registers, and to establish national archives where collected materials could be stored and made available. Moreover, the Committee made it clear that

> the preservation of folklore is concerned with protection of folk traditions *and those who are the transmitters*, having regard to the fact that each people has a right to its own culture and that its adherence to that culture is often eroded by the impact of the industrialized culture purveyed by the mass media.
> (UNESCO 1989, my emphasis)

It recommended that 'measures must be taken to guarantee the status of and economic support for folk traditions both in the communities which produce them and beyond'. In using the term 'folklore', it seems clear that UNESCO's initiative was aimed primarily at the preservation of the non-material aspects of indigenous/first nations and developing peoples. Two important points are worth noting here. The first is the way in which this Recommendation opened a problem area that continued into the Intangible Heritage Lists regarding the ultimate 'materiality' of the entities to be preserved. While the List itself helps to 'materialize' supposedly 'intangible' folkloric practices, it is not possible to conserve intangible heritage without also redefining its 'producers' as 'transmitters' of something abstract, which is perhaps ultimately independent of them. Although very precisely localized, culturally and geographically, the practitioners of 'intangible heritage' are rather like the instruments by which it is held and passed on. Such a redefinition is an eminently political move: while purporting to conserve the (copy)rights and interests of the people who produce intangible heritage, it actually puts distance between them and their cultural productions, and transforms them into the vessels by which it is 'transmitted'.

In 1996, the Report of the World Commission on Culture and Development, *Our Creative Diversity*, noted that the World Heritage Convention is not appropriate for celebrating and protecting intangible expressions of cultural heritage such as dance or oral traditions, and called for the development of other forms of recognition (besides World Heritage listing) (UNESCO 2012). An underlying principle of the report was that the diversity of world cultures must be protected and nurtured, but its emphasis on the threats to 'traditional' cultures seems to reflect UNESCO's broadening, more anthropological focus on living cultures over this period. However, as Schmitt (2005; 2008) and Skounti (2009) have noted, it was not until significant external pressure developed around a specific place and local circumstances that there was sufficient impetus to get the development of a convention onto UNESCO's agenda. The place was Jemaa el Fna Square (see Figure 8.3), and the nature of the heritage it embodied, as well as the work of campaigners (in particular, the Spanish writer Juan Goytisolo) to highlight its significance, had an important influence on the final shape of the convention that would be finally adopted in 2003.

Figure 8.3 Jemaa el Fna Square in Marrakech, Morocco.

(Photograph by Jane Wright, 1992. © UNESCO/Jane Wright.)

Jemaa el Fna is a square in the old city of Marrakech in Morocco. As Schmitt (2005; 2008) notes, although it is sited on the edge of the medina (historic old city) of Marrakech (itself a World Heritage Site), its layout owes as much to modern developments as to its medieval origins. As one of the main squares in the city, it hosts a changing suite of entertainers, drink sellers and restaurateurs. Snake charmers, monkey trainers, musicians and magicians perform for audiences in the open air, while herbalists, dentists and charm sellers set up temporary stalls on tables or carpets. In the evenings, the square becomes more crowded, as the snake charmers and monkey trainers depart, and the square is populated first by Berber or Arabic *halaiqui* (story-tellers), dancers and magicians, and, subsequently, with food stalls that are popular with locals and tourists alike. Juan Goytisolo, who left Spain under Franco and has lived much of his life in self-imposed exile in Marrakech and France, had featured the square in a number of his novels and short stories. In the mid-1990s, he became concerned about local authority plans to construct a tower block with a glass façade and an underground car park immediately adjacent to Jemaa el Fna. He believed this would pose a serious threat to the tradition of oral story-telling by *halaiqui* in the square, and after failing to convince local authorities of his concerns, he wrote to Federico Mayor, Director General of UNESCO, through his publisher, Hans Meinke. In his letter, dated January 26, 1996, he proposed that the square be placed under the protection of UNESCO as part of the '*patrimonio oral de humanidad*' (oral heritage of humanity), outlining the threats to the square as he perceived them. Schmitt translates and reproduces the relevant part of the letter:

Dear Federico,

. . . yesterday your ears must have been burning, for Juan Goytisolo came to talk [to me] about taking action on behalf of the famous Jemaa el Fna Square in Marrakech, and about the possibility of persuading UNESCO to declare it as 'oral heritage of humanity'. According to Goytisolo, this square, with its storytellers and reciters, is the only place in the Arab world where the tradition of oral literature is still cultivated. Apparently it is advisable to propose its protection by UNESCO in order to avoid it being destroyed by speculation.

(cited in Schmitt 2008, 98–99)

Meanwhile, Goytisolo began a campaign to save Jemaa el Fna, which gained local support and spread to the international press, bringing worldwide attention to the issue.

Subsequent discussions around the formulation of a Convention for the Safeguarding of the Intangible Cultural Heritage were strongly influenced by the particular circumstances of Jemaa el Fna, especially through the emphasis on oral story-telling and on the 'cultural spaces' in which such story-telling occurs (Schmitt 2008) (see Figure 8.4). This was notably the case when, in June 1997, the UNESCO Cultural Heritage Division and the Moroccan National Commission organized an international consultation on the preservation of popular

Figure 8.4 Street performers at Jemaa el Fna Square in Marrakech, Morocco.
(Photograph by Jane Wright, 1992. © UNESCO/Jane Wright.)

cultural spaces in Marrakech, which was integral to discussion of the creation of an international convention to protect outstanding examples of folklore and popular culture. While the meeting agreed that defined 'cultural spaces' such as that of Jemaa el Fna were actually quite exceptional, and in the majority of cases the transmission and performance of oral culture was not confined to specific delimited spaces, the idea of the 'cultural space of transmission' as a category was maintained in subsequent discussions and represented in UNESCO's final definition of 'intangible heritage' (see below). However, as Schmitt (2005; 2008) points out, while Goytisolo and other Moroccan heritage experts played an important role, the local *halaiqui* were not themselves involved in the discussions. Thus, the definition of intangible heritage that derives from Jemaa el Fna was very much filtered through the lens of the writer and Moroccan conservation professionals. This is consistent with the view of people not as producers and makers, but as 'transmitters' of heritage, and, instantiating 'governing at distance', legitimates the external experts' role.

In 1999, UNESCO and the Smithsonian Institution jointly organized a conference in Washington, DC, titled 'A Global Assessment of the 1989 Recommendation on the Safeguarding of Traditional Culture and Folklore: Local Empowerment and International Cooperation'. This was followed in May 2001 by the First Proclamation of nineteen 'Masterpieces of the Oral and Intangible Heritage of Humanity', which included Jemaa el Fna. Goytisolo himself acted as Chairman of the International Jury at the announcement, and made specific reference to the square in his speech.

> As I have learned from my custom of listening in the Square of Marrakesh, the halakis (story-tellers) perform in the framework of a changing society anxious for instruction and constantly looking over its shoulder at those who – outside an education almost exclusively linked to the practice of the competitive norms ruling the Global Village – preserve and memorise for the future the narratives of the past. . . . the holders of oral knowledge can be and at times are more cultured than their compatriots versed only in the use of audio-visual and computing techniques. But in a world subjugated by these ubiquitous techniques, oral culture, whether primary or hybrid, is seriously endangered and warrants an international mobilisation to save it from gradual extinction.
>
> (Goytisolo 2001)

In October 2003, the General Conference adopted the Convention for the Safeguarding of the Intangible Cultural Heritage and added twenty-eight new 'Masterpieces of the Oral and Intangible Heritage of Humanity'. A further forty-three were inscribed in 2005 (UNESCO 2012). The ninety items on the list of 'Masterpieces' were extremely diverse, ranging from the language, dance and music of the Garifuna in Belize, Guatemala, Honduras and Nicaragua to sand drawing traditions in Vanuatu, and the *Nyngyo Johruri Bunraku* puppet tradition of Japan (see Figures 8.5 and 8.6). Many of these were expressions and cultural practices of small-scaled, non-monumental societies, which had previously been largely outside of the interests of UNESCO. The process thus represented an expansion of the official definition of world heritage, and, perhaps more importantly, an extension of the governance of the World Heritage Committee beyond objects, places and buildings to languages, festivals and cultural practices, as well as to the spaces in which they are enacted and the persons who perform them.

Figure 8.5 Traditional Japanese puppet theatre, Nyngyo Johruri Bunraku, performed here at the UNESCO headquarters in Paris in 2004, was originally proclaimed as one of the Masterpieces of the Oral and Intangible Heritage of Humanity in 2003 and subsequently inscribed on the Representative List of the Intangible Cultural Heritage of Humanity in 2008.

(Photograph by Michel Ravassard, 2004. © UNESCO/Michel Ravassard.)

Figure 8.6 Alasita Bolivian ritual symbolizing prosperity and wealth, performed at the UNESCO headquarters in Paris in 2011. The Andean cosmovision of the Kallawaya was originally proclaimed as one of the Masterpieces of the Oral and Intangible Heritage of Humanity in 2003 and subsequently inscribed on the Representative List of the Intangible Cultural Heritage of Humanity in 2008.

(Photograph by Michel Ravassard, 2011. © UNESCO/Michel Ravassard.)

The 2003 Convention for the Safeguarding of the Intangible Cultural Heritage defined intangible cultural heritage as

> the practices, representations, expressions, knowledge, skills – *as well as the instruments, objects, artefacts and cultural spaces associated therewith* – that communities, groups and, in some cases, individuals recognize as part of their cultural heritage. This intangible cultural heritage, transmitted from generation to generation, is constantly recreated by communities and groups in response to their environment, their interaction with nature and their history, and provides them with a sense of identity and continuity, thus promoting respect for cultural diversity and human creativity.
>
> (UNESCO 2003, my emphasis)

Earlier documents had defined 'intangible' heritage in material terms, and this was underlined in the new Convention. While much broader than that which emerged from early discussions around Jemaa el Fna, its reference to 'cultural spaces' as integral to the conservation of intangible heritage bears the mark of the original campaign to save the square. Indeed, the very notion of 'intangible' heritage, with its connotation of 'immateriality' and hence 'vulnerability', is itself exposed here as a problematic concept. In fact, the conservation targets of the Intangible Heritage Convention were not fundamentally different from those specified within the World Heritage Convention. Rather, they manifest strategic elaborations and readjustments, which responded to the inherent dynamics of the heritage institutions.

Barbara Kirshenblatt-Gimblett (2004) remarks on this process when she observes that, although the categories of tangible and intangible heritage appear to distinguish between 'things' and 'events', even 'things' can be 'events'. For instance, the Japanese Shinto shrine Ise Jingū has been regularly rebuilt sixty-one times since it was first erected in 690. She argues that this is an extreme example of a quality embodied within heritage conservation processes themselves: objects of conservation are the sum of ongoing processes of selection, definition, maintenance, management and curation, which work together to make up a heritage 'site'. That is why the Intangible Heritage Convention must make reference not only to cultural practices, but also to the instruments, objects, artifacts and cultural spaces associated with them. Indeed, in doing so, the Convention explicitly acknowledges the falseness of the dualism it establishes and the impossibility of separating intangible from tangible heritage, since events are also objects and objects, events. Hence, the requirement to preserve not only *halaiqui* story-tellers' *practices*, but also Jemaa el Fna as their *setting* and the bodies of those *halaiqui* story-tellers and their audiences as their tangible containers.

As in the case of the World Heritage Convention, the Intangible Heritage Convention ensured its own perpetuation and growth by urging member states to establish parallel inventories of intangible cultural heritage in their territories and to ensure recognition and respect for intangible cultural heritage. Such a process involved further looping effects between the various lists of endangered cultural heritage. It also directed UNESCO to develop a 'Representative List of the Intangible Cultural Heritage of Humanity . . . [to] ensure better visibility of the intangible cultural heritage and awareness of its significance, and to encourage dialogue which respects cultural diversity'. Accordingly, in 2005, the list of Masterpieces was closed, to be replaced in 2008 by a Representative List of the Intangible Cultural Heritage of Humanity. At this time, all of the existing 'Masterpieces' were moved on to the Representative List. This appears to be a profoundly significant move. Masterpieces were general and seen as representing and, hence, belonging to humanity as a whole. In contrast, 'representative' entities are by definition specific and can only manifest (and hence stand for) 'local' heritage values. But a strange twist of logic nonetheless attributes to them 'universal' value by virtue of their being included in a 'World' Heritage List. This is partly accomplished through a sort of Platonic move that, as discussed above, turns heritage into an intangible entity that is not produced, but merely transmitted. Moreover, as Kirshenblatt-Gimblett (2006) has noted, the attribution of universality distinguishes the (local) transmitters of cultural heritage, and the (global) humanity to which that heritage belongs.

Listing, categorizing and governing at a distance

What are the epistemological, political, institutional and ontological implications of this move from the conservation of endangered monuments and buildings *in situ* to 'intangible' heritage? I would like focus here on the operations of these new lists themselves, as apparatuses that accommodated a broader definition of heritage while simultaneously maintaining the 'universal' principles of the World Heritage Convention. The criteria for inclusion on the Representative List, which were revised in June 2010, acknowledge explicitly the role of risk and threat by requiring safeguarding measures to be outlined in any nomination. Similarly, the process of listing and categorizing was maintained, despite the multiplication of the categories of heritage that are recognized to deal with the anomalies that its claims to universality have exposed. In recognizing intangible heritage as a specific category, the Convention separates objects, buildings and places from the practices and traditions associated with them. It thereby simply reorganizes the universal categories of World Heritage, rather than operate a fundamental revision of its classificatory system – in Georges Perec's vision, this would be an attempt to resist writing 'etc.,' under circumstances where to do so would draw attention to the inventories' very deficiencies (Perec 2008, 146).

The invention of the notion of 'cultural landscape' provides another example of the strategic elaboration and readjustment of putatively 'universal' categories. The notion is often considered as a revision of World Heritage classificatory systems. Elsewhere (Harrison 2013a; 2013b), I have discussed in detail the example of Uluru-Kata Tjuta National Park World Heritage Site, which was originally listed as a 'natural' site in 1987 and relisted as 'cultural landscape' in 1994. The outcry surrounding the nomination under the natural heritage category, which was seen as effectively denying the ongoing social and religious importance of the place to Anangu people (Layton and Titchen 1995), had a strong influence on the development of the 'cultural landscape' concept and its deployment by UNESCO. Against its expected effect, the concept reinforces the culture/nature dualism, to the extent that it makes a distinction between authentic nature and nature as 'cultural', i.e. as shaped by humans. 'Cultural landscape' is therefore not an entirely new category, but an amplification of that of 'cultural' heritage (see further discussion in Harrison 2013a; 2013c). Such separation of the 'natural' and the 'cultural' also ignores the ways in which heritage listing itself produces 'natural' landscapes as 'cultural' ones, while sustaining the myth that there are wild places somehow completely separate from humans (see further discussion in Harrison 2004, 12–13; Meskell 2012; 2013; Rose 1996).

The Intangible Heritage Lists have also contributed to the extension of the endangerment sensibility through the specification of new potentially threat-ened entities, expanding and spreading to culture itself the sense of vulnerability and risk that was previously focused on its material products. In doing so, as already pointed out, the Lists have fashioned 'cultural heritage' as something that is separate from, yet somehow contained within, the minds and bodies of those who 'transmit' it. While the global, late-modern development of an endanger-ment sensibility has not been unidirectional, as it clearly encompasses various modes and techniques of protection, in the case of cultural heritage, there does appear to have been an exponential growth in the kinds of objects considered to be at risk. This broad historical trend involves a movement from the acquisi-tion and collection of objects in museums to the application of museological conservation to monuments and buildings *in situ*, and a subsequent obsession with categorizing and listing as 'intangible' heritage cultural practices and the people who perform them. While these various shifts have been motivated by attempts to incorporate an increasingly diverse and 'representative' repertoire of forms of heritage within the World Heritage Convention's universal categories, the process required strategic readjustments at the institutional, conceptual and administrative levels. The subsequent specification of new targets documents the extraordinary growth of the late-modern endangerment sensibility in rela-tion to natural and cultural heritage.

Conclusion

My intention has been to show the ways in which heritage lists of various kinds have emerged as central devices in contemporary global processes. The World Heritage List and Intangible Heritage Lists are connected closely with other international instruments of the endangerment sensibility in initiating a rank-ing system that operates as one of a series of apparatuses for redescribing objects, places and practices as heritage 'at risk'. These lists now form vast assemblages graded according to status and value – from the global, national and regional to the local – and categorized according to an ever-expanding list of criteria, each one reflecting the perceived risk of potential loss of 'culture'. It is not only the existence of these lists themselves that is important, but their hierarchical structure and the looping effects produced within the networks that connect them. Further, listing on such registers comes with the expectation of particular forms of action (Byrne 1991) – conservation works, adaptation for re-use, public access or even reconstruction, depending on the specifics of the listing and the nature of the list.

I have argued that one of the important implications of World Heritage listing is that it enables forms of government 'at a distance' – a system of management in which national authorities and local bureaucrats effectively 'self-govern' by complying with regulations and standards understood as involving a universal 'duty' to be carried out for the benefit of the global 'community'. This process has made possible a shift from the conservation of endangered monuments and buildings to the listing of forms of imperiled traditions and customs as part of the expansion of museological forms of management. The Intangible Heritage Lists, in particular, have facilitated the extension and expansion of the endangerment sensibility through the specification of new objects of endangerment (and hence targets for management), incorporating cultural agents and performances into a sense of vulnerability and risk that was previously focused only on the material products of culture. These shifts have been motivated by attempts to include an increasingly diverse and 'representative' repertoire of forms of heritage within the 'universal' categories of World Heritage. They have not only reorganized and nourished these lists themselves, but also broadened the range of targets perceived to be 'at risk', as well as the practices by which risk is documented, inscribed and managed in the contemporary world. Such an expansion has taken place by way of mutually amplifying looping effects – from list to list, from institution to institution, and across various levels, from local initiatives to the international organizations – such that they ultimately generate their own endangered worlds.

Note

1 I use this term to collectively denote the *List of Masterpieces of the Oral and Intangible Heritage of Humanity,* which was established with the first *Proclamation of Masterpieces of the Oral and Intangible Heritage of Humanity* in 2001 and expanded as a result of two subsequent proclamations in 2003 and 2005 respectively, and the *Representative List of the Intangible Cultural Heritage of Humanity* and *List of Intangible Cultural Heritage in Need of Urgent Safeguarding,* which were established when the *Convention for the Safeguarding of Intangible Cultural Heritage* came into effect in 2008 and into which the original list of Masterpieces was incorporated.

Bibliography

Agamben, G., 2009. *What is an Apparatus? And Other Essays.* Stanford University Press, Stanford.

Aikawa, N., 2004. An Historical Overview of the Preparation of the UNESCO International Convention for the Safeguarding of the Intangible Cultural Heritage. *Museum International,* 56 (1–2), 137–149.

Bandarin, F., ed., 2007. *World Heritage: Challenges for the Millennium.* UNESCO World Heritage Centre, Paris.

Baudrillard, J., 1994. The System of Collecting. In: Elsner, J. and Cardinal, R., eds., *Cultures of Collecting*. Reaktion, London, 7–24.

Beck, U., 1992. *Risk Society: Towards a New Modernity*. Sage, London.

Bennett, T., 1995. *The Birth of the Museum: History, Theory, Politics*. Routledge, London/New York.

Bennett, T., 2004. *Pasts Beyond Memory: Evolution, Museums, Colonialism*. Routledge, London/New York.

Bennett, T., 2005. Civic laboratories: Museums, Cultural Objecthood and the Governance of the Social. *Cultural Studies*, 19 (5), 521–547.

Bennett, T., 2009. Museum, Field, Colony: Colonial Governmentality and the Circulation of Reference. *Journal of Cultural Economy*, 2 (1–2), 99–116.

Bennett, T., 2013. *Making Culture, Changing Society*. Routledge, Abingdon/New York.

Bowker, G.C., 2005. *Memory Practices in the Sciences*. MIT Press, Cambridge, MA.

Bowker, G.C. and Star, S.L., 2000. *Sorting Things Out: Classification and its Consequences*. MIT Press, Cambridge, MA.

Byrne, D., 1991. Western Hegemony in Archaeological Heritage Management. *History and Anthropology*, 5 (2), 269–276.

Davison, G., 2008 [2000]. Heritage: From Pastiche to Patrimony. In: Fairclough, G., Harrison, R., Jameson, J.H. Jr. and Schofield, J., eds., *The Heritage Reader*. Routledge, Abingdon/New York, 31–41.

Dicks, B., 2000. *Heritage, Place and Community*. University of Wales Press, Cardiff.

Douglas, M., 1966. *Purity and Danger: An Analysis of Concepts of Pollution and Taboo*. Routledge and Kegan Paul, London.

Douglas, M., 1992. *Risk and Blame: Essays in Cultural Theory*. Routledge, London.

Foucault, M., 1970. *The Order of Things: An Archaeology of the Human Sciences*. Tavistock Publications, London.

Foucault, M., 1980. *Power/Knowledge: Selected Interviews and Other Writings, 1972–1977*. Longman, Harlow.

Foucault, M., 2011. *The Government of Self and Others: Lectures at the College de France, 1982–1983*. Picador, New York.

Giddens, A., 1991. *Modernity and Self-Identity: Self and Society in the Late Modern Age*. Polity Press, Cambridge.

Goytisolo, J., 2001. Defending Threatened Cultures. Speech delivered at the opening of the meeting of the Jury, 15 May 2001 (http://www.unesco.org/bpi/intangible_heritage/goytisoloe.htm). Accessed 12 October 2012.

Hacking, I., 1986. Making Up People. In: Heller, T.C., Sosna, M. and Wellbery, D.E., eds., *Reconstructing Individualism: Autonomy, Individuality, and the Self in Western Thought*. Stanford University Press, Stanford, 222–236.

Hacking, I., 1995. The Looping Effect of Human Kinds. In: Sperber, D., Premack, D. and Premack, A.J., eds., *Causal Cognition: A Multidisciplinary Debate*. Oxford University Press, Oxford/New York, 351–383.

Harrison, R., 2004. *Shared Landscapes*. UNSW Press, Sydney.

Harrison, R., 2010. What is heritage? In: Harrison, R., ed., *Understanding the politics of heritage*. Manchester University Press/Open University, Manchester/Milton Keynes, 5–42.

Harrison, R., 2013a. *Heritage: Critical Approaches*. Routledge, Abingdon/New York.

Harrison, R., 2013b. Assembling and Governing Cultures 'At Risk': Centers of Collection and Calculation, from Ethnographic Museums to UNESCO World Heritage List. In: Harrison, R., Byrne, S. and Clarke, A., eds., *Reassembling the Collection: Ethnographic Museums and Indigenous Agency*. SAR Press, Santa Fe, 89–114.

Harrison, R., 2013c. Reassembling Ethnographic Museum Collections. In: Harrison, R., Byrne, S. and Clarke, A., eds., *Reassembling the Collection: Ethnographic Museums and Indigenous Agency*. SAR Press, Santa Fe, 3–36.

Harrison, R., 2013d. Forgetting to Remember, Remembering to Forget: Late Modern Heritage Practices, Sustainability and the 'Crisis' of Accumulation of the Past. *International Journal of Heritage Studies*, 19 (6), 579–595.

Hassan, F. A., 2007. The Aswan High Dam and the International Rescue Nubia Campaign. *African Archaeological Review*, 24, 73–94.

Herzfeld, M., 2004. *The Body Impolitic: Artisans and Artifice in the Global Hierarchy of Value*. University of Chicago Press, Chicago/London.

Hopwood, N., Schaffer, S. and Secord, J., eds., 2010. Seriality and Scientific Objects in the Nineteenth Century. *History of Science*, 46 (3–4).

Kirshenblatt-Gimblett, B., 2004. Intangible Heritage as Metacultural Production. *Museum International*, 56 (1–2), 52–64.

Kirshenblatt-Gimblett, B., 2006. World Heritage and Cultural Economics. In: Karp, I., Kratz, C. A., Szwaja, L. and Ybarra-Frausto, T., eds., *Museum Frictions: Public Cultures/Global Transformations*. Duke University Press, Durham/London, 161–202.

Kurin, R., 2004. Safeguarding Intangible Cultural Heritage in the 2003 UNESCO Convention: a critical appraisal. *Museum International*, 56 (1–2), 66–77.

Latour, B., 1987. *Science in Action: How to Follow Scientists and Engineers through Society*. Harvard University Press, Cambridge.

Latour, B., 1993. *We Have Never Been Modern*. Harvard University Press, Cambridge.

Law, J., 1994. *Organizing Modernity: Social Order and Social Theory*. Blackwell, Oxford.

Layton, R. and Titchen, S., 1995. Uluru: An Outstanding Australian Aboriginal Cultural Landscape. In: von Droste, B., Plachter, H. and Rossler, M., eds., *Cultural Landscapes of Universal Value*. Gustav Fischer Verlag Jena with UNESCO, Stuttgart/New York, 174–181.

Lowenthal, D., 1985. *The Past is a Foreign Country*. Cambridge University Press, Cambridge.

Meskell, L., 2012. *The Nature of Heritage: The New South Africa*. Wiley-Blackwell, Maldon/Oxford/Chichester.

Meskell, L., 2013. A Thoroughly Modern Park: Mapungubwe, UNESCO and Indigenous Heritage. In: González-Ruibal, A., ed., *Reclaiming Archaeology: Beyond the Tropes of Modernity*. Routledge, Abingdon/New York, 244–257.

Pearce, S., 1995. *On Collecting: An Investigation into Collecting in the European Tradition*. Routledge, London/New York.

Perec, G., 2008. *Species of Spaces and Other Pieces*. Penguin Books, London.

Rabinow, P., 1989. *French Modern: Norms and Forms of the Social Environment*. MIT Press, Cambridge, MA.

Rabinow, P., 2003. *Anthropos Today: Reflections on Modern Equipment*. Princeton University Press, Princeton.

Rico, T., 2014. The limits of a 'heritage at risk' framework: The construction of post-disaster cultural heritage in Banda Aceh, Indonesia. *Journal of Social Archaeology*, 14 (2), 157–176.

Rose, D. B., 1996. *Nourishing Terrains: Australian Aboriginal Views of Landscape and Wilderness*. Australian Heritage Commission, Canberra.

Säve-Söderberg, T., 1987. *Temples and Tombs of Ancient Nubia*. Thames and Hudson, London.

Schlanger, N., 2010. Series in Progress: Antiquities of Nature, Numismatics and Stone Implements in the Emergence of Prehistoric Archaeology. *History of Science*, 48 (3–4), 343–369.

Schmitt, T., 2005. Jemaa el Fna Square in Marrakech: Changes to a Social Space and to a UNESCO Masterpiece of the Oral and Intangible Heritage of Humanity as a Result of Global Influences. *The Arab World Geographer*, 8 (4), 173–195.

Schmitt, T., 2008. The UNESCO Concept of Safeguarding Intangible Cultural Heritage: Its Background and *Marrakchi* roots. *International Journal of Heritage Studies*, 14 (2), 95–111.

Skounti, A., 2009. The Authentic Illusion: Humanity's Intangible Cultural Heritage, the Moroccan Experience. In: Smith, L. and Akagawa, N., eds., *Intangible Heritage*. Routledge, Abingdon/New York, 74–92.

Smith, L., 2006. *Uses of Heritage*. Routledge, Abingdon/New York.

Tort, P., 1989. *La Raison Classificatoire: les Complexes discursifs – Quinze Etudes*. Aubier, Paris.

UNESCO, 1972. *Convention Concerning the Protection of the World Cultural and Natural Heritage* ('World Heritage Convention'). UNESCO, Paris.

UNESCO, 1989. *Recommendation on the Safeguarding of Traditional Culture and Folklore*. UNESCO, Paris.

UNESCO, 2003. *Convention for the Safeguarding of the Intangible Cultural Heritage*. UNESCO, Paris.

UNESCO, 2012. *Brief history of the Convention for the Safeguarding of the Intangible Cultural Heritage* (http://www. unesco.org/culture/ich/index.php?pg=00007). Accessed 12 October 2012.

9 Planning for the past

Cryopreservation at the farm, zoo, and museum[1]

Joanna Radin

Introduction

Historians of the life sciences have long appreciated the farm, the zoo, and the museum as important sites of technical and social innovation (e.g. Findlen 1994; Kimmelman 1983; Livingstone 2003; Outram 1996; Rader and Cain 2008; Star and Griesemer 1989). In the late 20th century, representatives of these institutions adopted liquid nitrogen-based cryopreservation systems to help establish repositories of frozen blood, tissue, sperm, and eggs. Anxieties about the perceived endangerment of biodiversity found expression in efforts to preserve these bits of animal bodies for future uses. That such frozen collections are often referred to as "arks" invokes the dominant Euro-American narrative about the end of the world: the tale of Noah and the Flood (e.g. Wojcik 1997). That they have also been described as biodiversity banks, or "sources of living biological material for an essentially limitless range of research projects" (Volobouev 2009, 132) underscores their positioning as materials that are potentially generative of knowledge for remaking the world (Watson and Holt 2001). The relationship between these facets of genetic salvage projects gives texture to what, in the Introduction to this volume, Nélia Dias and Fernando Vidal have termed the "endangerment sensibility."

The interplay between apocalyptic prophecies – linked to Christian soteriology – and rational prognoses – linked to Enlightenment rationality – in descriptions of genetic salvage projects demonstrate how distinctive articulations of the past and visions of the future meet in the freezer to orient action in the present. Here, I am inspired by historian Reinhart Koselleck's comments on the "space of experience" and "horizon of expectations" that are opened by modernity. Koselleck presents the embrace of rational prognosis as a historical development that marks a break with prophecy and that has consequences for the lived experience of time. He writes,

> [p]rognosis produces the time within which and out of which it weaves, whereas apocalyptic prophecy destroys time through its fixation on the End. From the point of view of prophecy, events are merely symbols of what is already known . . . Rational prognosis assigns itself to intrinsic possibilities, but through this produces an excess of potential controls on the world.
>
> (Koselleck 2004 [1979], 19)

Paying attention to how time is conceptualized and instrumentalized in cryo-preservation projects framed in terms of endangerment at the farm, the zoo, and the museum reveals that prophetic orientations toward the future persist and are perhaps necessary for being able to make rational prognoses or predictions.

The beginnings of technoscientific interest in cryopreservation can be situated in the 1930s when Basile Luyet, a biophysicist and Catholic priest, trained his focus on a mode of existence between life and death. For Luyet, who is often referred to as the "Father of Cryobiology," examining latency – the state of supposed suspended animation wherein an organism is neither fully alive nor fully dead – yielded important insights about the nature of life itself. Gradually, as I will describe, the ability to harness latency through practices of freezing allowed cryopreservation to be recognized also as a strategy for hedging against the risks posed by an unknown future.

The contours of this kind of hedge were described in a 1984 report to the U.S. National Science Foundation, based on a special workshop sponsored by the Association of Systematics Collections. The goal of the workshop was to take stock of existing, *ad hoc* collections of frozen non-human tissues that had been made for specific projects and attempt to streamline their coordination for use in the future. Two contributors to this report called for a strategy of "planned hindsight." They wrote that while "samples collected to explore one problem may prove to be of historical value for investigating many future problems in distantly related areas . . . their value in retrospective studies should increase dramatically with carefully planned sampling programs" (Dessauer and Hafner 1984, 10–11).

The *Oxford English Dictionary* defines "hindsight" as seeing what has happened, and what ought to have been done, after the event; perception gained by looking backward. Planned hindsight, then, is an orientation to the future that attempts to prepare for the need to look back to the past. The paradox of planned hindsight generates meaning in the desire to articulate a pragmatic compromise between the awareness that it is impossible to perfectly plan for the future and a belief that hindsight is 20/20. If cryopreservation, through its harnessing of life in a state of suspended animation, has promised to bring a slice of the living present into the future, planned hindsight has been the attempt to guide how that slice should be made.

This paradox is deepened, however, when we recognize complex temporality or the open-ended and multiple dimensions of time (Asdal 2012; Bowker 2005; Schrader 2010). In the present of planned hindsight, frozen collections are envisioned as material manifestations of what Geoffrey Bowker has described as a bureaucratic databasing impulse. Such practices of storing life on ice attempt to flatten or decontextualize both bodies and time even as they promise novel ways of using bodies *in* time (Bowker 2005). In other words, by situating tissue from frozen organisms as particular kinds of biodiversity resources for the future, planned hindsight – by definition – partakes of a restriction of possibilities for future life, even as it generates new possibilities. The impulse to plan is one that seeks to exert deliberate influence on futures that cannot be known. It is a project that simultaneously recognizes its own impossibility even as it acts to shape what is understood as possible.

At its base, planned hindsight is also a response to a problem resulting from past actions for which the fix is elusive, and believed to reside in an unknowable yet singular unfolding future. If the problem is endangerment of biodiversity, then cryopreservation – a technology for harnessing the latent potential of life – is an action that can be performed in the present. At stake is not only the desire to fix the problem, but absolution from guilt in having participated in its creation. Here, I am referring to a pervasive sense of the anthropogenic roots of contemporary endangerment of biodiversity (Sepkoski this volume). An important justification for the creation of collections of frozen biodiversity is the belief that those who might make use of these materials in the future – scientists who will have more knowledge about or need for such variation than those working today – will look back with gratitude for the time and energy that was invested in creating stockpiles of variation. Those who endeavor to engineer planned hindsight must, then, inspire both guilt and hopefulness in those whose financial and emotional investments are necessary for its success. Often, as I will show, the way in which this is accomplished is by recourse to more prophetic discussion of apocalypse, resurrection, and redemption. This makes it important to attend to both the epistemic and affective as well as the prognostic and prophetic orientations of planned hindsight. It is through their interplay that possibilities and constraints on action in the present are negotiated.

I begin this chapter by summarizing the technical conditions of possibility that supported the uptake of liquid nitrogen as a resource for freezing life. Then I present examples of three kinds of contemporary "frozen" biodiversity collections. At each institution – farm, zoo, and museum – I am especially interested in how ideas about endangerment are mobilized to encourage public participation as well as financial and emotional investment in efforts to maintain frozen bits of biodiversity. In the final section, I draw out certain blind spots in planned hindsight that are made visible when frozen collections of non-human biodiversity are placed alongside efforts to freeze tissues from supposedly endangered human groups. In each of these cases, planned hindsight strains against its architects' own best intentions to control the unruliness of an infinite array of possible futures by deferring, via the act of freezing, problem solving in the present.

A brief and selective history of cryopreservation

For centuries, scientists have struggled with how to maintain tissues at low temperature without destroying life processes (e.g., Boyle 1683; Kavaler 1970). Basile Luyet, in his efforts to understand latent life, catalogued several centuries of studies in what has come to be regarded as the founding text of the field of cryobiology, *Life and Death at Low Temperatures* (Keilin 1959; Luyet and Gehenio 1940; Smith 1961). Yet it was not until 1949 that British reproductive biologists Chris Polge, Audrey Smith, and Alan Parkes made the accidental discovery that adding glycerol to fowl sperm could preserve viability of cellular material after freezing to $-79°C$ (the temperature of dry ice) (Polge et al. 1949; Smith 1950). The revelation that a simple sugar substance could safely protect cells during freezing and thawing processes was a boon to researchers (Coriell et al. 1964; Landecker 2005; Parry 2004).

Several years later, an American entrepreneur named Rockefeller Prentice learned of Smith and Polge's success in freezing sperm and saw this as a development of great potential for his cattle breeding business (Foote 2002; Polge and Lovelock 1952). Prentice had founded American Breeders Service in 1941 and would grow it into one of the world's most influential industrial agriculture concerns. He realized that if he could find a more efficient way to preserve bull semen, he could use it to artificially inseminate herds scattered across the country. Shipping a vial of semen would surely be more efficient than shipping an entire bull. In 1953, Prentice hired Chris Polge as a consultant to work on the fundamentals that would enable more robust commercial application of cryopreservation. That year, the first calf – appropriately dubbed "Frosty" – was born in North America from semen frozen by Prentice's staff at American Breeders Service.[2]

As they refined methods for freezing and defrosting semen, American Breeders Service also established connections with the Linde Division of chemical company Union Carbide, based in Tonawanda, New York. In the late 19th century, Carl von Linde, a German scientist and brewer, had established the first commercial facility for producing liquid gases. Under Prentice's instruction, investigations were begun into the potential use of liquid nitrogen to maintain low temperatures as semen was moved between lab and field. Not only could it provide extremely low temperatures (-196°C), it was inert and nonflammable, making it safe for the biologics it was protecting as well as the field technicians handling it.

In collaboration with the Linde Division, American Breeders Service succeeded in producing a portable unit that combined both a new insulation material and a vacuum to support liquid nitrogen. It allowed for a two-week holding period that enabled it to be "transported from farm to farm as the local inseminator made his rounds."[3] Pleased with the success of linking field and lab across the American Midwest, American Breeders Service converted to this liquid nitrogen-based system completely in 1958. The company did not seek a patent, allowing competitors to gain access to it through Linde. The result was a transformation of both industries – cattle breeding and liquid nitrogen sales.

A 1962 ad for the Denver-based company Cryenco expressed the power of this new ability to harness latency with an image of a tilted hourglass and a message that celebrated the extension of basic biophysics research to industrial application:

> Suspended animation – stopping and starting the biological clock at will – has been one of man's age old dreams . . . including Jules Verne. Today, through Cryobiology, scientists are slowing down and even theoretically stopping the chemistry of life processes.
>
> (AIBS Bulletin 1962)

The text went on to note, "Prolonging cell life is finding practical application in the use of frozen bull semen for artificial insemination of cattle" (AIBS

Bulletin 1962). Cattle breeders intent on reducing genetic variation – in partnership with companies like Linde – had become a resource for knowledge on practical considerations of working with cryopreservation technology.

An article in the *Science Newsletter,* also published in 1962, seized upon the potential to apply this technology to the perceived problem of wildlife endangerment. The article was titled, "Live cells frozen alive: The science of freezing cells at extremely low temperatures is expected to lead to preserving samples of species about to become extinct." This particular application had been selected from a list of potential future uses provided to the author by Dr. Joseph F. Saunders, Head of the Medicine and Dentistry Branch of the Office of Naval Research, which included:

- Preserving such large specimens as the brains of vertebrates for study.
- Long-term banking of antibiotic producing organisms, eliminating today's necessity of 'starting again from scratch'.
- Arresting chemical and other life processes in disease as well as in health.
- Studying now unidentifiable but important intermediates in biochemical reactions.
- Preserving today's generations of organisms for later comparison with future generations, and also tissues affected by today's diseases for comparison with diseased tissues several years from now.
- And far in the future . . . the preservation of live plants and animals to prevent extinction of a species.

"Such an application," said Saunders, "cannot be dispelled as 'being impossible or fantastic,' because, theoretically, life virtually does experience a temporary halt at temperatures around and below the liquid nitrogen range" (Ewing 1962, 246).

Today, whole vertebrate organisms have yet to be successfully frozen and defrosted in ways that approximate Saunders' vision, but by the late 1970s, gametes and other genetic materials were beginning to colonize the freezers of those explicitly concerned with issues of loss of biodiversity (Dessauer and Hafner 1984).[4] The technical ability to preserve biodiversity through cryopreservation was made possible, in part, through efforts put into developing the technology to reduce genetic variation for economic purposes: to standardize cattle for mass production. The former future of cryopreservation as beneficial agent of high-modern agriculture was recast as one of extremely restricted genetic possibility (Koselleck 2004 [1979], xix). In other words, under the auspices of planned hindsight, cryopreservation is revealed to have been both an agent of endangerment and a means of forestalling its further encroachment.

The Swiss Village Farm Foundation (Newport, Rhode Island)

The first example that I want to discuss is the Swiss Village Farm Foundation (SVF).[5] I have chosen to begin with SVF because its mission – to preserve semen, eggs, and embryos from rare and endangered breeds of food and fiber livestock – is a direct response to the genetic homogenization that has resulted from the success of companies like American Breeders' Service (Estabrook 2010).

SVF was founded in 1999 in Newport, Rhode Island, by a philanthropist and heiress to the Campbell's Soup fortune in partnership with Tufts School of Veterinary Medicine. Most of the existing protocols for freezing gametes were standardized for major industrial herds used for commercial purposes. SVF technicians quickly learned that each species' and even each breed's gametes require a unique recipe of extenders and buffers to ensure functionality upon defrosting. In this sense, cryopreservation can be construed as part of the craft of planned hindsight (Roosth 2010).

According to the organization's website, "Collecting 200 embryos and 3,000 straws of semen per breed, SVF will be able to reawaken a breed, with its full genetic diversity." According to SVF's website, there are many reasons why one might want to restore a breed that has become "severely endangered":

> Rare or heritage breeds of livestock carry valuable and irreplaceable traits such as resistance to disease and parasites, heat tolerance, mothering ability, forage utilization, and unique flavor and texture qualities. A particular breed that now dominates the marketplace may find its future jeopardized for any number of reasons. Recall the Irish Potato Famine . . . With the lack of diversity in today's animal agriculture, we are at tremendous risk.[6]

Here, risk-based discourses of population health and food security are the future goals to which cryopreservation-enabled planned hindsight claims to be oriented. But, as an SVF spokesperson admitted to me, "We really don't know where it's going to take us, we just know that these breeds are going extinct and if we don't collect them now, the opportunity is going to be gone."[7] The prognosis for how such endangered genes will be used as resources for the future, which guides their collection and curation, comes also with the admission of the impossibility of predicting a future threatened by the loss of biodiversity. In this case, as in those that follow, I am interested in the interplay between such discourses of prognosis and prophecy.

While cryopreservation has its roots in agricultural breeding, repurposing the technology to exploit a perceived weakness or vulnerability in contemporary modes of food security has involved recourse to endangerment discourses generated in conjunction with wildlife biodiversity. SVF says that, "We want to educate people and make them aware that different breeds of goats, sheep, and cattle can be endangered just like endangered species can."[8]

Unlike those conservation genetic organizations that focus on rare and endangered wildlife, SVF's strategy involves encouraging people to eat the very animals they are asked to save. A tote bag sold through their online gift shop encourages one to "put diversity on your plate."[9] SVF's marketing strategy is to argue that exploiting an "economically viable niche in today's market" will also lead to breed preservation. It is a strategy in which consuming diversity is a means of overcoming endangerment. This is in keeping with Donna Haraway's critique that in the world of purebred dogs, "breeds become like endangered species, inviting the apparatus of apocalyptic wildlife biology . . . [the] prominent role given to Species Survival Plans invites a reproductive tie between natural species and purebred dogs," or, in the case of SVF, heritage cows, sheep, and pigs. Either way, "In this mongrelizing tie, the natural and the technical keep close company, semiotically and materially" (Haraway 2003, 210).

Making this formula work has meant that SVF has had to establish itself as a tastemaker. For SVF, endangerment provides a rhetorical framework through which little-known breeds can achieve a higher profile. No one knows which breeds will prove to be the most palatable and, therefore, profitable, but they must all be kept in play. In the process, SVF has tapped into circulating aesthetics of heritage and reclamation to fashion facilities for maintaining animals "on the hoof" as well as on ice. Although the "farm" is only open to visitors one day per year (for biosecurity reasons), the foundation's website devotes a lengthy description to the aesthetic reawakening of the 45-acre property's pastoral glory. During its renovation:

> There were no details left to the imagination. Leaded glass and vaulted ceilings adorn the cottages, pecky cypress was used in the construction of the exterior doors and trim, and hand-hewn beams were used throughout.[10]

There is a steampunk flavor to this endeavor, in which a high-modernist "stainless steel" project of cryopreservation is mediated by 19th-century "brass" of a working farm (Kasson 1976). The explicit denial of imagination, of reference to historical specificity, betrays the extent to which the entire enterprise is itself saturated by a creative nostalgia for a specific kind of past.

Steampunk is an aesthetic movement with ties to Luddism that represents a reaction against modernism and its tendency to "black box" or conceal the workings of machines and knowledge systems (Lears 1989; Sale 1995). According to one commentator, "steampunks seek less to recreate specific technologies of the 19th century than to re-access what they see as the affective value of the material world" of the Victorian period (Onion 2008). Steampunk devotees view the 19th century as a time of "innocence in technological and scientific knowledge" (Onion 2008, 142). The Victorian era was also a time characterized by an intensification of efforts to master the order of nature (Ritvo 1997; Stocking 1987). The embrace of the Euro-American aesthetic of a period in which machines were imagined to be more accessible to the everyday person could be seen as a strategy for making the hard-to-visualize workings of reproductive physiology and cryopreservation familiar and acceptable.

It is in this sense that SVF's effort at planned hindsight borrows from the more explicit retro-topic imaginary of steampunk. While steampunk may be more post-modern than anti-modern in the way it "picks and chooses from previously existing styles of physical technology and ideological modes of technological engagement," the "risk of misinterpretation of an aesthetic movement as simple aesthetics leaves the deeper relationship between human and object unexamined," as is the epistemological and moral position of steampunk theorists (Onion

2008, 142–155). It looks toward the future with a specific, 19th-century vision of the past – what can go wrong (potato famines) as well as what can go right (humans living in greater harmony with their machines).

SVF's recent alliance with the "Fair Food" stand in Philadelphia's Reading Terminal Market is another example of how a nostalgic steampunk aesthetic frames the action toward which this particular agricultural cryopreservation project is oriented. Built in a former train station, the Market (established 1893) is one of the oldest in the United States. It is a symbolic and actual point of contact to witness the contradictions posed by consumers looking to reconcile their urban lifestyles with a hunger for the country. It is also a way of mediating the cryopreservation project's inherent values of planned hindsight with those of a movement inspired by a desire to see both machines and humans as capable of breaking down and of dying (Tresch 2012).

The Frozen Zoo (Escondido, California)

Zoos generally condemn the eating of endangered organisms. In protecting endangered species, zoos also see themselves as committed to preserving habitats. With linkages to empire and territorial expansion, in the 21st century, the zoo has fashioned itself as an institution dedicated to conducting science in the name of habitat conservation; the best way to save endangered species is to preserve or reclaim the habitats to which they are adapted (Baratay and Hardouin-Fugier 2004; Rothfels 2002). In the meantime, having access to their genetic material may be useful for raising awareness about the extent of homogenization (or loss of hybrid vigor) in these wild populations. This is the official position of the "Frozen Zoo®," a pioneer among efforts to create frozen repositories of tissue for purposes of genetic research and conservation (Chemnick et al. 2009; Kumamoto 1998).

At the time I visited in 2009, The Frozen Zoo® was located in the Arnold and Mabel Beckman Center for Conservation and Research for Endangered Species (CRES), adjacent to the "Wild Animal Park" arm of the San Diego Zoo in Escondido, California (Bruns 1983). The "Wild Animal Park" has since been rebranded as the "San Diego Zoo Safari Park." At the Frozen Zoo®, viable fibroblast cell lines and tissue pieces were then frozen from 7,000 individual mammals, birds, and reptiles. It also maintained sperm and embryos from a number of those species (Chemnick et al. 2009). Of major importance are animals listed as endangered and threatened and, of the approximately 500 species and subspecies in the collection, more than 200 were listed in the IUCN Red List of Threatened Species (Chemnick et al. 2009, 127). An equally important concern from the outset was the interaction of conservation efforts with those of the Species Survival Plan groups under the auspices of the American Zoo and Aquarium Association.

The concept for a frozen zoo was conceived of and implemented by Dr. Kurt Benirschke in the mid-1970s. Benirschke, a German émigré, began his career with no interest in conservation, though his students would later honor him as "a person who can charm the birds from the trees" (Gotze 1984). He obtained his medical degree from the University of Hamburg and came to the U.S. for his

internship, where he met a nurse who would become his wife. He considered pursuing a career as a surgeon but was persuaded to apply for a pathology residency
at Harvard. It was there, through rotations at Boston Lying-in Hospital, that he
became interested in placenta pathology and, later, cytology. He sought animal
models with which he could explore the phenomenon of multiple births, chime-
rism, and freemartinism (which occurs in cattle when a female co-twin to a male
is born sterile) (Gotze 1984).

First, he looked for research materials at Chicago's slaughterhouses, which were
efficiently rendering American Breeders Service cows (e.g. Cronon 1991). But it
was a visit with his children to the Catskill Game Farm in New York in the early
1960s that initiated a long project of comparative mammalian cytogenetics (Hsu
and Benirschke 1967). The owner, Roland Lindemann, also from Germany, had
begun the park in 1933 (Rothfels 2002). In exchange for assistance with a techni-
cian's pet project on the genetics of camels, Lindemann supplied Benirschke with
the initial materials for what would become the Frozen Zoo® (Benirschke 2007).
When Benirschke was hired as Professor of Reproductive Medicine and Pathology
at U.C. San Diego, his vast knowledge of mammalian reproduction made an affili-
ation with the famed San Diego Zoo obvious. He brought the modest collection of
frozen materials he had begun earlier with him (Benirschke and Kumamoto 1991).

Shortly after arriving in San Diego, Benirschke encountered an ambitious grad-
uate student in biology. This student, Oliver Ryder, was among the first to think
about the role molecular studies could play in conservation. Ryder has reflected
on his own career, "I was a product of the culture; it was save the whales, save
the mountain gorillas, and wildlife ecology. I was a geneticist though, and there
was no obvious interface."[11] Together, Benirschke and Ryder began to examine
what it would mean to use genetics for wildlife conservation purposes (Ryder
and Fleischer 1996). It was in the context of an emerging endangerment sensibility that Benirschke's interest in maintaining significant specimens, originally
begun in tradition of medical pathology collections, began to take on the logic of
planned hindsight. There are now a number of such wildlife repositories at zoos
around the world but the Frozen Zoo® remains a thought leader in the public and
policy realms (Ryder et al. 2000).

In February 2009, I traveled to Escondido to learn more about the facility.
There, I met Ryder, who continues as the current steward of the collection. As
he led me to the freezers, he attempted to lower my expectations. He confessed,
"One problem with the Frozen Zoo® is that it isn't much to look at." We entered
a small, off-white room. It was Karen Knorr-Cetina (1999) who observed that
the lab is like a nursery, and this case was no exception. Six giant liquid nitrogen
tanks were surveilled by a blue and white baby monitor. The baby monitor, Ryder
explained, was a solution to a design flaw in the alarm system. Because few people
spend any significant amount of time with the materials, they cannot hear the
cries of a failing freezer. A brightly colored banner in a playful font announced
that the freezers cradle the Frozen Zoo®. Ryder told me how he had wanted to
install dramatic lighting that would spotlight each freezer, allowing it to be seen
beckoning as it was approached down the long hallway that led to the freezer

room. "Unfortunately," he lamented, "the people who designed the space didn't want to budget for that."

I asked him what kinds of people he imagined being impressed by that aesthetic. "Kids," he replied. From time to time, school groups visiting the Wild Animal Park are brought on a tour of the Frozen Zoo® facilities. Ostensibly, this is to reinforce and reproduce the zoo's longtime image as a place of science as well as entertainment. But it may also serve to inculcate a sense of the importance of planned hindsight, to introduce young people to the project of stewardship of frozen collections that are intended to grow in value as these children grow in age.

Around the time I visited the Frozen Zoo, Philadelphia's unfrozen zoo had launched a major campaign that underscores the emotions that are conjured by the endangerment sensibility and how they are oriented toward investment in planned hindsight. "X-tink-shun" is a multi-media presentation designed to get "zoo-goers to learn about endangered species."[12] The focal point is a series of puppets, such as Didi the dodo, that, in the terms of program designers, function to "reawaken" endangered or extinct species.[13] These figments of the past are made contemporaneous with the present through the illusion of puppetry, which performs the technical feats of reanimation of extinct species that science has not yet perfected. In a video clip provided on the X-tink-shun website, Didi tells prospective zoo-goers, "We can change the world for all of my fellow creatures, but we need to get started now! Get to work! Let's go, chop chop!" The urgent exhortation of this resurrected dodo represents a substitution of entertainment for a discussion of how science should figure in the politics of nature. Zoo-goers are encouraged to embrace the illusion and, with it, the immediacy of thinking ahead about the past without recognizing the consequences for the present. Viewers of this clip are also invited to "Learn more from Didi by visiting her at a Bank of America Eco-Stage at the Zoo."[14] The zoo in both its warm and cold states reveals itself as a space where endangerment is a form of edutainment that places science and fiction in tension, performed under terms provided by banking (Bowker 2005).

Back in Escondido, as liquid nitrogen swirled out of a tank that Ryder had opened, he remarked that he was pleased to have an inquiry from an academic, rather than a reporter looking for a sensational story. "Too often, people are incorrectly focused on a '*Jurassic Park*' inspired fantasy of the zoo, which is not," he emphasized, "at all what inspires his work . . . the focus is on conservation – finding ways to prevent organisms from going extinct in the first place." Ryder was invoking the 1990 Michael Crichton techno-thriller in which dinosaur genetic material preserved in amber is manipulated to bring dinosaurs back to life in a science-themed amusement park. In typical Crichton fashion, the hubris of scientists combines with the greed of investors with horrific consequences. *Jurassic Park* is a high-tech morality tale in which man's god-like aspirations wreak havoc (Haraway 1997). Anthropologist Sarah Franklin offers a compelling reading of the relevant stakes described in the fantasy of *Jurassic Park*:

> [It] collapses the trajectory of evolutionary time in a fantasy of recreation, in which we stand beside our extinct genealogical brethren in awe at the

miracle of (re)creation only to be reminded there was only one true Genesis which we mimic at our peril. But what it does best is to hold these two things together: the hubris of "playing God" alongside its equally profound appeal.

(Franklin 2000, 215)

Ryder's defensiveness about the association to *Jurassic Park* belied a fertile seed of truth that makes Franklin's interpretation of the novel relevant to the paradox presented by planned hindsight. Researchers, including Ryder, have engaged in captive breeding and cloning using materials maintained in the Frozen Zoo® (e.g. Friese 2009). Whereas Ryder sought to suppress the enterprise's kinship with Crichton's novel, the reproductive potentials of the Frozen Zoo were generative for Charis Thompson Cussins' 1999 fictional essay, "Confessions of a Bioterrorist." In that narrative, wherein Ryder is acknowledged by name in a note in the conclusion, Cussins presented a different kind of Christian-inflected reproductive fantasy, not of resurrection but of virgin birth. The ability to bank reproductive material on ice enabled Mary, a physiologist who collects gametes from endangered species and the story's protagonist – to impregnate herself with a bonobo embryo.

In Cussins' critical feminist account, this act of "bioterror" is both ironic and transgressive because it brings humans and non-humans into new kinds of relations of care and compassion. Her virgin Mary shares the wisdom that "you can't know which gestations will save the world." Via the Frozen Zoo®, resurrection might come from new kinds of assisted reproduction, made possible through the harnessing of latent life. This is one kind of prophecy upon which planned hindsight depends but also seeks to suppress. Nevertheless, such reproductive potential as well as countless other currently unfathomable forms are also latent in the freezer (Radin 2013).

Those engaged in planned hindsight are more comfortable in making prognoses about the future ability of frozen biodiversity to generate genetic information about populations. Improving knowledge about genetic diversity (or the lack thereof) beyond the zoo is used to strengthen claims for conservation (O'Brien 2003; O'Brien and Johnson 2005). Ryder, who contributed to the 1984 report in which planned hindsight was proposed, has dedicated a great deal of his time and energy to promoting the coordination of existing repositories of frozen materials. He and Benirschke contributed a chapter to the 1984 report titled, "The Value of Frozen Tissue Collections for Zoological Parks," in which they highlighted specific kinds of future uses to which these tissue could be put, including harvesting chromosomes and the investigation of infectious diseases and pathology (Dessauer and Hafner 1984, 7–8). These are very similar to the future uses explicitly advocated by SVF and the museum collection that I describe below.

The Ambrose Monell Cryo Collection (New York, New York)

Though the oldest frozen specimen in the Ambrose Monell Cryo Collection (AMCC) at the American Museum of Natural History in New York City is from 1989, the repository only became operational in 2001 and occupies a large suite of rooms in the basement of the museum.[15] Administratively, the facility is part of the museum's Sackler Institute for Comparative Genomics and supports the molecular research of a range of scientists within the institution and around the world. The scientists connected to the AMCC have done much to foreground contributions such frozen collections can make toward the future of systematics. However, like SVF and the Frozen Zoo®, a series of more fantastical prophetic possibilities vibrate in the background.

The concept of "planned hindsight" was originally named in the context of efforts to devise a coordinated system of tissue sharing that would elevate a distributed network of freezers into a "national resource" for systematists. From their vantage point in the museum world, the coordinators of the 1984 report, systematists Herbert Dessauer and Mark Hafner, argued that collections designated as frozen tissue depositories should both be managed in accordance with traditional curatorial procedures for systematic collections and shared by the home institution and the biological resource programs of the U.S. National Science Foundation and National Institutes of Health. They also encouraged others to view the tissues they had collected for a range of other reasons as important also for the future of conservation research, observing that, "Not many people realize the value of frozen tissue and those that do operate independently without oversight" (Dessauer and Hafner 1984).

In the 1984 report, Dessauer and Hafner linked their interest in the value of what they called "undenatured tissues" to that of the British pathologist George Nuttall and Alan Boyden. Boyden inherited Nuttall's collection of specimens in the late 1940s. Historian Bruno Strasser has argued that Boyden's recourse to serology as an objective method to understand the relationships among species, his creation of a serological museum for applying a comparative perspective, and his negotiations between natural historical and experimental traditions illustrate the rise of a new hybrid research culture in the mid-20th century. Boyden insisted, as would his intellectual descendants at the AMCC, that "the proteins of the bodies of organisms were just as worthy of preservation and conservation as their 'skins and skeletons' because they were just as characteristic" (Strasser 2010, 166).

In that moment, freezers were coming to assume an importance in the study of biological diversity as significant as the display cases and tables that had previously contained the materials for the study of the order of things (Foucault 1971). Much as Oliver Ryder's work at the zoo would no longer depend solely on animals in cages or habitat replicas, "Boyden did not need the 'great halls and showcases' of the traditional natural history museum in order to exhibit his objects of study, since 'bottles of sera look much like each other'; what he needed were 'adequate cold rooms for the preservation of these sera. A Rockefeller Foundation grant made possible the construction of such a room in 1951, in a former coal bin" (Strasser 2010, 168). Almost 50 years later, at the American Museum of Natural History, the AMCC was installed in a basement, next to the carpentry shop that continues to prepare dioramas.

If Boyden was engaged in building a "serological empire" at the dawn of the genomic age, then the AMCC represents its extension to frontiers both new and old (Strasser 2010). Not only are fresh materials being gathered and added to the collection, old materials are being rescued from the freezers of individual scientists who might otherwise dispose of them. Following the exhortations of Dessauer and Hafner (1984), the AMCC processes new material but also seeks to salvage old collections of biodiversity that were made decades earlier (e.g. Gruber 1970). This, the practice of incorporating older frozen collections, has helped it to become the fastest growing collection in the museum for the least amount of money and the only one that is fully digitized. It has a capacity of over 1 million specimens and defines the cutting edge in cryopreserved biodiversity collections (Hanner et al. 2009).

A huge part of the activities of support labs like AMCC involve getting people in other departments and institutions to transfer their materials into what is seen as a more secure environment. They are, as one human biologist said to me of his efforts to preserve old human blood samples, "salvaging science itself." And they do this with the full knowledge that tissues that may have been languishing unused may take on new and unanticipated kinds of value in the future. What, ironically, is less often appreciated is the way that choices about what was originally frozen and why influence how that future value is imagined. This is particularly true for instances wherein newer ideas about old frozen blood, such as endangerment, are invoked to rescue it from being forgotten or destroyed.

I spent several weeks at AMCC in 2010 learning about the history and practices involved in maintaining the AMCC in an era of planned hindsight. There, I spoke with Rob DeSalle, a leader in the field of comparative genomics who initiated the effort to get the museum to support a centralized repository for frozen tissues. Like Boyden had decades earlier, DeSalle recognized the importance of establishing an equivalency between this kind of frozen collection and older collections in the museum of fossils or animal skins. Part of what enabled him to speak authoritatively about the importance of salvaging old blood was that, in 1991, he had isolated what was then the oldest living DNA ever extracted from amber. This was the technique whereby the DNA for making dinosaurs was recovered in Michael Crichton's *Jurassic Park*. DeSalle, unlike Ryder, has allowed these kinds of prophecies to be articulated as inspiration for a project of planned hindsight. In his book *The Science of Jurassic Park and the Lost World*, he wrote, "although any number of these problems [of building a dinosaur from ancient DNA] might be insoluble now, it's not clear that any of them are, comparatively speaking, in their infancy. There's a huge amount that is, as scientists are fond of saying, 'not well understood,' yet" (DeSalle and Lindley 1997). On the subject of the AMCC, DeSalle told me,

> Cold is about thinking long term. Right now, most people are thinking about DNA, but without cold, you can't get at the proteins, parasites, or microbes. But you don't know what people will want in the future. The safest way to save stuff is to freeze it in large chunks so that you're never limited by technological innovation.[16]

George Amato, who was the current head of the Comparative Genomics program at the museum when I visited, sought to distance himself from discussions of cloning and resurrection. "It's just so obvious," he said, "that to understand transmission of diseases like SARS, WNV, and H5N1, it would make sense to go look at frozen bird samples from different regions." Concerns about biosecurity, then, were also coming to be used as a justification for planned hindsight in the case of SVF and the Frozen Zoo®.

Amato's stated personal interest in supporting the AMCC, however, is to produce a complete tree of life.[17] For Amato, assisted reproduction and cloning are not obviously valuable for endangered species; it is tantamount to taking the "eye off the prize." In other words, they are not the future for which he imagines frozen biodiversity collections should be used. Despite the evolutionary bent of the effort to construct a tree of life, it is a curiously synchronic one. As Geoffrey Bowker has argued:

> The tree of life maps the diversity of life by breaking the web into countable units which are assumed to be entities in the world. These entities are then aligned in a regular historical time in which there is no turning back and no speeding up. The unchanging species is mapped onto the flat time. Although temporality is thus doubly invoked, and is clearly central to all discourse, it is often invisible in discussions of the tree. And yet this folded temporality is precisely our effort to map the world, in all its complexity, onto a linear, featureless time.
>
> (Bowker 2005, 118)

Bowker's critique of the tree of life goes to the heart of the paradox expressed by planned hindsight. The kind of time that is valued in the tree of life is a singular one purged of its generative richness, its unpredictability, and its multiplicity. The double invocation of temporality vis-à-vis discourses of evolution and endangerment crowds out the space for an equally urgent consideration of the enactment of time in relation to markets and emotions like fear and hope, or what might be called "lively" time (Rajan 2012).

Along these lines, those affiliated with the collection sometimes employ a different set of metaphors. They call it a "Frozen Ark." Indeed, the museum is a leading member of an international consortium of the same name, as is the Frozen Zoo. The Frozen Ark Consortium is inspired to action by a perceived sense of urgency in the face of predictions of accelerating "loss of wild animal species across the globe." The Consortium had set a goal of collection and storage of samples from at least 10,000 species by 2015 and stated that:

> To achieve this we need urgently to expand the rate of sample collection and recruit the conservation community to join the museum, university, institute, zoo and aquaria consortium membership . . . As the renowned biologist Edward O Wilson said 'We should preserve every scrap of biodiversity as priceless while we learn to use it and come to understand what it means to humanity'.[18]

In addition to funding from the Ambrose Monell Foundation – connected to the chemical industry – the AMCC received significant startup support from NASA. According to DeSalle, this support was consistent with the space agency's interest in astrobiology, deep space travel, and arks. Indeed, astrobiologists have begun experimenting with re-animating "extinct" genes by inserting ancient genetic material into contemporary organisms (Schirber 2009; Schwarz et al. 2009). As DeSalle explained, "You would need to have a system in place if you were going to send life into outer space."[19]

AMCC curator Julie Feinstein, a passionate naturalist who was working toward a PhD in conservation genomics, drove this alternative vision of planned hindsight home when I asked her how visitors respond to the collection. (The AMCC is one of the main stops on tours given to trustees and potential donors.) While Ryder lamented that the Frozen Zoo®, which uses the same stainless steel cryovats, was not "much to look at," Feinstein confessed to seeing more possibility:

> It looks like a spaceship, doesn't it? The freezers are sort of science fiction generation ships, collecting and safekeeping for unknown future uses. Going into the AMCC evokes a science fiction state of mind. Maybe it's the blank-slateness of the tanks because, look at them, they're just waiting to be projected upon.[20]

Vanishing points

In the introduction to this chapter, I intimated that planned hindsight revealed certain blind spots when one considers the prospect of coordinating similar kinds of collections of frozen tissue from biodiverse human communities. In the late 1950s and 1960s, when cryopreservation technologies were first made available, certain indigenous human communities found themselves as subjects of genetic salvage. Anthropologists and population geneticists affiliated with the Cold War-era International Biological Program (IBP) collected and froze blood samples from individuals understood to live in a state of optimal adaptation and therefore whom were likely to be in possession of unique genetic signatures (Radin 2013). These human life scientists justified their salvage efforts as part of an urgent response to risks posed by the destructive and standardizing forces of modernity (Erickson et al. 2013).

For example, an American Breeders Service pamphlet on "Precision Processing" of bull semen, which outlined the techniques necessary for freezing life, was sent to a Harvard anthropologist named Albert Damon who sought to accumulate blood from geographically isolated, "primitive" human populations.[21] Though Damon and other scientists affiliated with the IBP did not use the phrase "planned hindsight," their efforts to collect tissues from such human communities were strikingly similar to those subsequently undertaken at SVF, the Frozen Zoo®, and the AMCC. At a time when scientists and members of the public were becoming aware of new globalizing forms of risk (Beck 1992), the vanishing indigene appeared as both a harbinger of doom for the species as a whole as well as a genetic resource for future understandings of adaptability to extreme environments. Those who collected and froze blood from such peoples invested great hope in future scientists' ability to harness life on ice to make this material generative of new knowledge "as yet unknown" (Radin 2013; 2014). They outlined the case for salvaging human genetic material in scientific journal articles as well as popular texts. Prophecy and prognosis reinforced each other in what was one of the first large-scale biodiversity cryopreservation projects.

Though it did not explicitly discuss cryopreservation, consider the contemporaneous coffee table book *The Horizon Book of Vanishing Primitive Man* (Severin 1973). This oversize volume, with a foreword by anthropologist Colin Turnbull (who helped population geneticist Luca Cavalli-Sforza, best known for proposing the Human Genome Diversity Project in the early 1990s, collect blood as part of the IBP during the Cold War), described the scientific importance of preserving specimens of human biodiversity. The book brought summaries of decades of work in anthropology together with *National Geographic*-style photomontages of members of indigenous communities in various poses evoking Rousseau's Noble Savage (Lutz and Collins 1993; Qureshi 2012). The final chapter, titled "Exodus," described them as "the people of the twilight, the tribes that are already so reduced in numbers as to be on the verge of extinction" (Severin 1973, 348). The author went on to synthesize perspectives shared by participants in then on-going IBP, noting that:

> These little-known primitive communities would stand a better chance of survival if they could be identified and counted, and the nature of their problems recognized and made public so that help could be given them. But no one knows exactly how many Stone Age hunter-gatherers and farmers are left, and thus *the problem of their preservation* is made even more acute, for it is double difficult to protect what one cannot identify or define.
>
> (Severin 1973, 351, emphasis added)

For scientists, the problem became one of how to act in the face of endangerment and inadequate information; this problem supplanted the needs and desires of the living community members themselves. The possibility of attempting to preserve the entire community, akin to efforts to make nature preserves for non-human biodiversity, was acknowledged to be impossible:

> An alternative solution, which is sometimes mooted, is that surviving primitive people be placed off limits . . . But such policies are no longer practicable. Pressure on land has grown too great, and there are qualms about creating 'human zoos', where man is condemned to a deliberate backwater.
>
> (Severin 1973, 352)

It was an endangerment sensibility coupled with a sense of impotence that led scientists to embrace cryopreservation as a way of deferring possibilities for action. If they could not engage directly in preserving the life ways of such groups in the face of an anthropogenic apocalypse brought about by incipient globalization, at least scientists might salvage some of their genetic material as a resource for the future. "Advanced societies, therefore, are left with the moral obligation of devising for primitive man the best transition from his aboriginal state to a condition in which he can survive alongside more modern cultures" (Severin 1973, 354). Or, at the very minimum, deferring such moral obligation by freezing bits of his (and her) body.

Compare this discourse with another coffee table book: *Vanishing Animals*, produced a decade later through a collaboration between Kurt Benirschke – founder of the Frozen Zoo® – and pop artist Andy Warhol. In the introduction, Benirschke explained,

> Extinction – the tragic and permanent loss of entire species of animals – should be a concern for everyone. This concern and a strong desire to take action toward preventing the loss of more animals has brought about an unusual collaboration between art and science. The result is this beautiful volume in which artist and scientist have joined efforts to inform and inspire others to take action.
>
> (Benirschke and Warhol 1986, i)

But what kind of action? In what may have more to do with his fame than his process, it is nonetheless worth mentioning, apropos the standardizing potential of cryopreservation, that Warhol, famous for his harnessing of mass production, was the artist chosen for the project of advertising the endangerment sensibility as it pertained to the preservation of biodiversity. When cryopreservation has been used to preserve biodiversity in projects guided by planned hindsight, the forms by which the future can be imagined are constrained, often – inadvertently – reproducing particular aspects of the present, even as unexpected dimensions may be emergent through time.

It is in this sense also of consequence that Benirschke quoted Seattle, Chief of Dwamish and Allied tribes of Puget Sound, to legitimate his project. Vanishing "primitive" man, whose body parts are already in the freezer, is made to speak for vanishing animals vis-à-vis the project of genetic salvage: "What is man without the beasts? If all the beasts were gone, men would die from a great loneliness of spirit. For whatever happens to the beasts, soon happens to man. All things are connected" (Benirschke and Warhol 1986 page 1).

Benirschke called up cinematic narratives of apocalypse and redemption in order to throw into relief the more realistic, but still not fully realized, possibilities of cryopreservation. Yet, even as he performed doubts about the redemptive potential of cryopreservation, he tacitly endorsed it as the self-described creator of the Frozen Zoo®. He wrote,

> I often wonder when I see *Star Wars*, and similar movie spectaculars, about how our children envisage that these warriors grow their plants or food animals to sustain their lives. Frequently, we extol that cryopreservation, the freezing away of seeds, embryos, and the like, will save existing genetic diversity from oblivion. How unrealistic are such hopes when we don't yet have a clue of how to get the semen, let alone the fertilized embryo, for safekeeping

from such forms as the blue whale, the bumblebee bat, or the many other animals and plants that are presently threatened . . . And we are in this together; no one escapes . . . As far as animals are concerned – and, after all, that is the topic of this book – we need to be mindful of the possibility that 'Tomorrow's Ark is by Invitation Only'.

(Benirschke and Warhol 1986, 6)

Such claims a mix of prophecy and prognosis, provoked profound confusion in geneticist Bentley Glass, who reviewed the book for a scientific journal, noting:

What is mainly at fault, at least in the opinion of this reviewer, is that the dramatic collages often obscure the naturalistic details that are required to distinguish an endangered species of animal from its near relatives who may be quite common . . . For a biologist, this lack of distinctive traits seems almost dishonest if the intention is really to marshal recruits for saving *endangered* species. At a price of 50 cents a page, what is the function of such a book?

(Glass 1987)

Apart from Glass' criticisms, not everyone wants to be on the ark or in the freezer. The vital legacies of the human specimens collected by Damon and others working during the IBP – which remain latent in freezers like those at SVF, the Frozen Zoo, and AMCC – have demonstrated that the future cannot be planned. In the mid-1990s, the Human Genome Diversity Project, a successor to the IBP, was derailed by indigenous peoples' resistance both to the idea that they were endangered and to the possibility of becoming resources for future biotechnological projects (M'Charek 2005; Reardon 2005). More recently, members of the Yanomami, whose blood samples were collected during the IBP and frozen, have demanded their repatriation (Couzin-Frankel 2010).

In this way, certain human subjects of salvage have come to voice resistance to the vision of planned hindsight that justified the freezing of their blood decades earlier. It is inaccurate to conclude that indigenous peoples do not care to partake in genomics or in science more broadly. But it is also a form of violence to formulate scientific projects predicated on understandings of indigenous peoples as existing outside of time (Fabian 1983; Wolf 1982). Unlike animals classed as endangered, these human communities can speak and express contemporaneous temporal perspectives that present epistemic and ethical challenges to the seemingly self-evident benefits of efforts to put life on ice (Harmon 2010; Kowal, Radin and Reardon 2013; Reardon and TallBear 2012).

Conclusion

In this chapter, I have attempted to add texture to the endangerment sensibility as it pertains to efforts to salvage biodiversity through practices of cryopreservation.

Though cryopreservation gained momentum through its demonstrated ability to standardize genetic diversity in the realm of industrial agriculture, I have also shown how it was deployed to counter homogenization of genetic diversity both within agriculturally relevant breeds as well as wildlife.

Explication of the logic of planned hindsight has been helpful in unpacking how these various cryo-collections have been made and described as worth continuing to maintain. Planned hindsight requires the imagination of the present as a future past. Endangerment serves as both a prophecy and a prognosis, which makes the project of freezing life appear worthy of investment even if its benefits cannot be clearly defined. The insistence on planning for a future that cannot be known but must be anticipated leads scientists to privilege the likelihood of using such frozen materials to address contemporary threats, such as biosecurity and famine, even as they barely suppress premonitions about how efforts to bring the past into the future will generate possibilities like interspecies kinships (in the case of Cussins' virgin Mary) or antagonisms (in the case of *Jurassic Park*), machines that die (in the case of steampunk), or subjects that ask to cease serving as resources for the future. All of these visions are preserved in the freezer.

Historian David Sepkoski (this volume) has emphasized the important links between the endangerment sensibility and the valuing of biodiversity. He notes that the endangerment sensibility sees biodiversity as worthy of protection because of its "inherent (i.e. non-instrumental) worth." In the cases I have described, reference to the "inherent" value of biodiversity might also be understood as a deferral of an ability to name the specific instrumental properties of that worth, which will ostensibly be realized at some future time. The official goal of all these initiatives is to preserve diversity. In this way, they are also kindred with projects of linguistic and cultural salvage. What makes these frozen collections distinctive from the more traditional instantiations of the farm, zoo, and natural history museum is that through participation in an endangerment sensibility, they displace processes of production of either food or species or knowledge in time, making the present a space of postponing action onto future generations rather than a space of reflection for what can be done today.

Notes

1 This chapter is derived in part from an article published in *Journal of Cultural Economy*, 8 (3), 2015. http://dx.doi.org/10.1080/17530350.2015.1039458
2 A brief corporate history is available at http://www.absglobal.com/.
3 http://usa.absglobal.com/company/history.phtml
4 This chapter focuses on the cryopreservation of animal life. However, there is also a rich story to be told about endangerment and preservation as it pertains to plant life.
5 http://svffoundation.org/
6 http://svffoundation.org/
7 Interview with SVF spokesperson, January 26, 2010.
8 Ibid.
9 http://svffoundation.org/
10 http://svffoundation.org/

11 From an author interview with Oliver Ryder, Escondido, CA, February 14, 2009.
12 http://www.xtinkshun.org/
13 Ibid.
14 http://www.xtinkshun.org/pages/didi.html
15 http://research.amnh.org/genomics/Facilities/AMCC
16 Interview with Rob DeSalle, NY, NY, January 19, 2010.
17 Interview with George Amato, NY, NY, January 14, 2010.
18 http://www.frozenark.org/consortium
19 Interview with Rob DeSalle, NY, NY, January 19, 2010.
20 Interview with Julie Feinstein, NY, NY, January 22, 2010.
21 Box 5 Harvard Solomon Island Expedition (1966–1972) A. Damon, Correspondence, Data. Solomons Corresp Misc 1966 Harvard Peabody Museum.

Bibliography

AIBS Bulletin 12 (5), 1962.

Asdal, K., 2012. Contexts in Action – and the Future of the Past in STS. *Science, Technology and Human Values*, 37 (4), 379–403.

Baratay, E. and Hardouin-Fugier, E., 2004. *Zoo: A History of Zoological Gardens in the West*. Reaktion Books, London.

Beck, U., 1992. *Risk society: Towards a New Modernity*. Sage Publications, Newbury Park/London.

Benirschke, K., 2007. *Animal Tales*. Trafford Publishing, Victoria, BC.

Benirschke, K. and Kumamoto, A. T., 1991. Mammalian Cytogenetics and Conservation of Species. *Journal of Heredity*, 82, 187–191.

Benirschke, K. and Warhol, A., 1986. *Vanishing Animals*. Springer, New York.

Bowker, G. C., 2005. Time, Money and Biodiversity. In: Collier, S. J. and Ong, A., eds., *Global Assemblages: Technology, Politics, and Ethics as Anthropological Problems*. Blackwell, Malden, MA.

Boyle, R., 1683. *New Experiments and Observations Touching Cold*. R. Davis, London.

Bruns, B., 1983. *A World of Animals: The San Diego Zoo and the Wild Animal Park*. Harry N. Abrams, Inc., New York.

Chemnick, L. G., Houck, M. L. and Ryder, O., 2009. Banking of Genetic Resources: The Frozen Zoo at the San Diego Zoo. In: Amato, G., DeSalle, R., Ryder, O. and Rosenbaum, H., eds., *Conservation Genetics in the Age of Genomics*. Columbia University Press, New York.

Coriell, L. L., Greene, A. E. and Silver, R K., 1964. Historical Development of Cell and Tissue Culture Freezing. *Cryobiology*, 1 (1), 72–79.

Couzin-Frankel, J., 2010. Researchers to Return Blood Samples to the Yanomamo. *Science*, 328 (5983), 1218.

Crichton, M., 1990. *Jurassic Park*. Knopf, New York.

Cronon, W., 1991. *Nature's Metropolis: Chicago and the Great West*. W.W. Norton, New York.

DeSalle, R. and Lindley, D., 1997. *The Science of Jurassic Park and the Lost World*. Basic Books, New York.

Dessauer, H. C. and Hafner, M. S., eds., 1984. *Collections of Frozen Tissues: Value, Management, Field and Laboratory Procedures, and Directory of Existing Collections*. Association of Systematics Collections.

Erickson, P., Klein, J.L., Daston, L., Lemov, R., Sturm, T. and Gordin, M., 2013. *How Reason Almost Lost Its Mind: The Strange Career of Cold War Rationality*. University of Chicago Press, Chicago.

Estabrook, B., 2010. Rare Breeds, Frozen in Time. *The New York Times*, 5 January.

Ewing, A., 1962. Live Cells Frozen Alive: The Science of Freezing Cells at Extremely Low Temperatures Is Expected to Lead to Preserving Samples of Species about to Become Extinct. *The Science News-Letter*, 81 (16), 246, 253.

Fabian, J., 1983. *Time and the Other: How Anthropology Makes Its Object*. Columbia University Press, New York.

Findlen, P., 1994. *Possessing Nature: Museums, Collecting, and Scientific Culture in Early Modern Italy*. University of California Press, Berkeley.

Foote, R.H., 2002. The History of Artificial Insemination: Selected Notes and Notables. *Journal of Animal Science*, 80, 1–10.

Foucault, M., 1971. *The Order of Things: An Archaeology of the Human Sciences*. Pantheon Books, New York.

Franklin, S., 2000. Life Itself: Global Nature and the Genetic Imaginary. In: Franklin, S., Lury, C. and Jackie, S., eds., *Global Nature, Global Culture*. Sage, London.

Friese, C., 2009. Methods of Cloning, Models for the Zoo: Rethinking the Sociological Significance of Cloned Animals. *BioSocieties*, 4, 367–390.

Glass, B., 1987. Vanishing Animals by Kurt Benirschke. *The Quarterly Review of Biology*, 62 (3), 306.

Gotze, H., 1984. Foreword. In: Ryder, O.A. and Byrd, M.L., eds., *One medicine: A Tribute to Kurt Benirschke, Director, Center for Reproduction of Endangered Species, Zoological Society of San Diego, and Professor of Pathology and Reproductive Medicine, University of California, San Diego from his Students and Colleagues*. Springer-Verlag, Berlin/New York.

Gruber, J., 1970. Ethnographic Salvage and the Shaping of Anthropology. *American Anthropologist*, 72 (6), 1289–1299.

Hanner, R., Corthals, A. and DeSalle, R., 2009. Biodiversity, Conservation, and Genetic Resources in Modern Museum and Herbarium Collections. In: Amato, G., DeSalle, R., Ryder, O. and Rosenbaum, H.C., eds., *Conservation Genetics in the Age of Genomics*. Columbia University Press, New York.

Haraway, D.J., 1997. *Modest-Witness@Second-Millennium.Femaleman-Meets-Oncomouse: Feminism and Technoscience*. Routledge, New York/London.

Haraway, D.J., 2003. Cloning Mutts, Saving Tigers: Ethical Emergents in Technocultural Dog Worlds. In: Franklin, S. and Lock, M.M., eds., *Remaking Life and Death: Toward an Anthropology of the Biosciences*. School of American Research Press, Santa Fe, 293–327.

Harmon, A., 2010. Where'd You Go with My DNA. *The New York Times*, 25 April.

Hsu, T.C. and Benirschke, K., 1967. *An Atlas of Mammalian Chromosomes*. Springer, New York.

Kasson, J.F., 1976. *Civilizing the Machine: Technology and Republican Values in America, 1776–1900*. Grossman Publishers, New York.

Kavaler, L., 1970. *Freezing Point: Cold as a Matter of Life and Death*. John Day Company, New York.

Keilin, D., 1959. The Leeuwenhoek Lecture: The Problem of Anabiosis or Latent Life: History and Current Concept. *Proceedings of the Royal Society of London. Series B, Biological Sciences*, 150 (939), 149–191.

Kimmelman, B., 1983. The American Breeders' Association: Genetics and Eugenics in an Agricultural Context. 1903–13. *Social Studies of Science*, 13 (2), 163–204.

Knorr-Cetina, K., 1999. *Epistemic Cultures: How the Sciences Make Knowledge*. Harvard University Press, Cambridge, MA.

Koselleck, R., 2004 [1979]. *Futures Past: On the Semantics of Historical Time*. K. Tribe, trans. Columbia University Press, New York.

Kowal, E., Radin, J. and Reardon, J., 2013. Indigenous Body Parts, Mutating Temporalities, and the Half-Lives of Postcolonial Technoscience. *Social Studies of Science*, 43 (4), 465–483.

Kumamoto, A. T., 1998. The Frozen Zoo: Extending the Living Animal Collection. *CRES Report*, Winter 1–2.

Landecker, H., 2005. Living Differently in Biological Time: Plasticity, Temporality, and Cellular Biotechnologies. *Culture Machine*, 7, no. Biopolitics.

Lears, T. J. Jackson, 1989. *No Place of Grace: Antimodernism and the Transformation of American Culture, 1880–1920*. University of Chicago Press, Chicago.

Livingstone, D. N., 2003. *Putting Science in Its Place: Geographies of Scientific Knowledge*. University of Chicago Press, Chicago.

Lutz, C. and Collins, J. L., 1993. *Reading National Geographic*. University of Chicago Press, Chicago.

Luyet, B. J. and Gehenio, P. M., 1940. *Life and Death at Low Temperatures*. Biodynamica, Normandy, MO.

M'Charek, A., 2005. *The Human Genome Diversity Project: An Ethnography of Scientific Practice*. Cambridge University Press, Cambridge UK/New York.

O'Brien, S. J., 2003. *Tears of the Cheetah: And Other Tales from the Genetic Frontier*. Thomas Dunne Books/St. Martin's Press, New York.

O'Brien, S. J. and Johnson, W. E., 2005. Big Cat Genomics. *Annual Review of Genomics and Human Genetics*, 6, 407–429.

Onion, R., 2008. Reclaiming the Machine: An Introductory Look at Steampunk in Everyday Practice. *Neo-Victorian Studies*, 1 (1), 138–163.

Outram, D., 1996. New Spaces in Natural History. In: Jardine, N., Secord, J. A. and Spary, E. C., eds., *Cultures of Natural History*. Cambridge University Press, Cambridge, 249–266.

Parry, B., 2004. Technologies of Immortality: The Brain on Ice. *Studies in History and Philosophy of Biological and Biomedical Sciences*, 35 (2), 391–413.

Polge, C. and Lovelock, J., 1952. The Preservation of Bull Semen at -79c. *Vet Rec*, 64, 396.

Polge, C., Smith, A. and Parkes, A., 1949. Revival of Spermatazoa after Vitrification and Dehydration at Low Temperatures. *Nature*, 164, 666.

Qureshi, S., 2012. *Peoples on Parade: Exhibitions, Empire, and Anthropology in Nineteenth Century Britain*. University of Chicago Press, Chicago.

Rader, K. A. and Cain, V., 2008. From Natural History to Science: Display and the Transformation of American Museums of Science and Nature. *Museum and Society*, 6 (2), 152–171.

Radin, J., 2013. Latent Life: Concepts and Practices of Human Tissue Preservation in the International Biological Program. *Social Studies of Science*, 43 (4), 483–508.

Radin, J., 2014. Unfolding Epidemiological Stories: How the WHO made Frozen Blood into a Flexible Resource for the Future. *Studies in History and Philosophy of Biological and Biomedical Sciences*, 47, 62–73.

Reardon, J., 2005. *Race to the Finish: Identity and Governance in an Age of Genomics*. In-Formation Series. Princeton University Press, Princeton, NJ.

Reardon, J. and TallBear, K., 2012. Your DNA Is Our History. *Current Anthropology*, 53 (S5), 233–245.

Ritvo, H., 1997. *The Platypus and the Mermaid, and Other Figments of the Classifying Imagination*. Harvard University Press, Cambridge, MA.

Roosth, S., 2010. *Crafting Life: A Sensory Ethnography of Fabricated Biologies*. MIT Press, Cambridge, MA.

Rothfels, N., 2002. *Savages and Beasts: The Birth of the Modern Zoo*. Johns Hopkins University Press, Baltimore.

Ryder, O. and Fleischer, R.C., 1996. Genetic Research and Its Application in Zoos. In: Kleiman, D., Allen, M.E., Thompson, K.V. and Lumpkin, S., eds., *Wild Mammals in Captivity: Principles and Techniques*. University of Chicago Press, Chicago.

Ryder, O., McLaren, A., Brenner, S., Zhang, Y.-P. and Benirschke, K., 2000. DNA Banks for Endangered Animal Species. *Science*, 288 (5464), 275–277.

Sale, K., 1995. *Rebels against the Future: The Luddites and Their War on the Industrial Revolution: Lessons for the Computer Age*. Addison-Wesley Pub. Co., Reading, MA.

Schirber, M., 2009. Reanimating Extinct Genes. *Astrobiology Magazine*, 27 April.

Schrader, A., 2010. Responding to Pfiesteria Piscicida (the Fish Killer): Phantomic Ontologies, Indeterminacy, and Responsibility in Toxic Microbiology. *Social Studies of Science*, 40 (2), 275–306.

Schwarz, C. et al., 2009. New Insights from Old Bones: DNA Preservation and Degredation in Permafrost Preserved Mammoth Remains. *Nucleic Acids*, Advanced Access, 24 March.

Severin, T., 1973. *Vanishing Primitive Man*. American Heritage Publishing Co, New York.

Smith, A., 1950. Prevention of Hemolysis During Freezing and Thawing of Red Blood Cells. *Lancet*, 2, 910–911.

Smith, A.U., 1961. *Biological Effects of Freezing and Supercooling*. Williams & Wilkins, Baltimore.

Star, S.L. and Griesemer, J.R., 1989. Institutional Ecology, 'Translations,' and Coherence: Amateurs and Professionals in Berkeley's Museum of Vertebrate Zoology, 1907–1939. *Social Studies of Science*, 19, 387–420.

Stocking, G.W., 1987. *Victorian Anthropology*. Free Press Collier Macmillan, New York.

Strasser, B., 2010. Laboratories, Museums, and the Comparative Perspective: Alan A. Boyden's Quest for Objectivity in Serological Taxonomy, 1924–1962. *Historical Studies in the Natural Sciences*, 40 (2), 149–182.

Sunder Rajan, K., 2012. The Capitalization of Life and the Liveliness of Capital. In: Sunder Rajan, K., ed., *Lively Capital: Biotechnologies, Ethics, and Governance in Global Markets*. Duke University Press, Durham.

Thompson Cussins, C., 1999. Confessions of a Bioterrorist: Subject Position and Reproductive Technologies. In: Kaplan, E.A. and Squier, S., eds., *Playing Dolly: Technocultural Fictions, Fantasies, and Fictions of Assisted Reproduction*. Rutgers, New Brunswick, 189–219.

Tresch, J., 2012. *The Romantic Machine: Utopian Science and Technology After Napoleon*. University of Chicago Press, Chicago.

Volobouev, V., 2009. The Role of Cryopreserved Cell and Tissue Collections for the Study of Biodiversity and Its Conservation. In: Amato, G., DeSalle, R., Ryder, O. and Rosenbaum, H.C., eds., *Conservation Genetics in the Age of Genomics*. Columbia University Press, New York.

Watson, P.F. and Holt, W.V., eds., 2001. *Cryobanking the Genetic Resource: Wildlife Conservation for the Future?* Taylor & Francis, London.

Wojcik, D., 1997. *The End of the World as We Know It: Faith, Fatalism, and Apocalypse in America*. New York University Press, New York.

Wolf, E., 1982. *Europe and the People without History*. University of California Press, Berkeley.

Coda

Who is the "we" endangered by climate change?[1]

Julia Adeney Thomas

Under the threat of climate change, culture and nature seem to converge. Ethicist Clive Hamilton (2010, 9) argues that "humans have become a 'natural' planetary force." Historian Dipesh Chakrabarty (2009, 201) insists that anthropogenic climate change "spells the collapse of the age-old humanist distinction between natural history and human history." As the divide between the humanities and the sciences melts in the heat of global warming, humanists and scientists might be expected to envision the endangered human figure in similar terms. When asked who is threatened, all disciplines might now reasonably be expected to answer in chorus, producing an understanding of the embodied mind, a naturalized history, a cultured nature, and a figure of the human recognizable in all corners of the university. The unprecedented and enormous threat of anthropogenic climate change would seem to demand the dissolution of artificial barriers between forms of knowledge revealing deep complementarity. With the challenges of intense heat, wilder weather, acidic oceans, increasingly virulent diseases, chemical pollution, decreased biodiversity, failed crops, rising political tensions, revolutions and wars, greater inequality and injustice, massive migration, and strains on practices dedicated to knowledge and beauty, a conjoint understanding of the human-at-risk might seem inevitable and desirable. Indeed, attempts at such harmony have been made by both historians and biologists, two disciplines central to this chapter. Historians such as Ian Morris (2011) and biologists such as E. O. Wilson (1998) have tried to reconcile disciplinary differences and create consilience across the divides. But, as I will argue, we are nowhere near to achieving a rapprochement and its benefits themselves are doubtful.

As this chapter will demonstrate through a brief tour of paleobiology, microbiology, and biochemistry, humanists trying to come to grips with climate change and to think somatically cannot rely on our scientific colleagues to define "the endangered human" for us. Instead, biology's many branches have produced throngs of radically different figures of the "human," not all of them endangered by anthropogenic environmental change according to the criteria crucial to humanists. Rather than simplifying the picture and defining the endangered "we," engaging with biologists complicates the view of who is threatened. Although this chapter celebrates the increasingly sophisticated conceptualization of human reality and endorses efforts by humanists and biologists to pool their resources in the face of global danger, the conclusion I reach is that it is impossible to treat "endangerment" as a simple scientific fact. Instead, endangerment is a question of value and a question of perspective. What we value, what we are in danger of losing under the pall of global climate change, is most fully articulable not through science but in the humanities.

Paleobiology: masters of the planet

From certain vantages in the age of the Anthropocene, the human figure appears gigantic, best understood not as an individual or even as a group but as the whole of the species over millennia. Paleoecologist Curt Stager (2012) describes our species' greenhouse gas emissions as transforming the planet not just for the next several centuries, but even deeper into the future. Stager argues that "we" as "species" have decisively prevented the next ice age, previously "scheduled" for 50,000 years from now. "Thanks to the longevity of our greenhouse gas pollution," Stager argues, "the next major freeze-up won't arrive until our lingering carbon vapors thin out enough, perhaps 130,000 years from now, and possibly much later" (2012, 11). The human species operates on a truly geological scale. Likewise, Chakrabarty (2009) argues that *the* outstanding challenge to his discipline, history, is to understand this new figure of the human: this immense, baleful, aggregate entity now undermining the Earth's life-support systems through a whole array of activities from agriculture to industry, from transportation to communication.[2] Chakrabarty draws on the work of climate scientists, particularly Paul Crutzen and Eugene Stoermer, who in 2000 declared "mankind" a "major geological force," to point to our collective "agency in determining the climate of the planet as a whole, a privilege reserved in the past only for very large-scale geophysical forces."[3] This version of the species pumped up for action on a global scale is, as Chakrabarty urges us to see, what climatologists and paleobiologists are positing in contrast to humanistic understandings of the human figure.

Up to this point, Stager and Chakrabarty describe the planetary situation in homologous terms and name the human species as the culprit of climate change. Looking only this far, it is possible to think that we have arrived at a figure of the human – the species – shared by historians and biologists alike. But here the similarities end. For Stager, thinking in terms of the species is easy – he is, after all, a trained biologist – and his general argument is that most species, including ours, will survive pretty well, especially if we allow for migration. Looking back on the Eocene era 55 million years ago, which produced temperatures 18 to 22°F higher than today's, Stager (2012, 84–85) maintains that the Paleocene-Eocene thermal maximum (PETM) was not so very terrible: "On a relatively bright note, we also know that many plants and animals, including our own primate ancestors, made it through PETM just fine." This depends, of course, on how you define "fine." Looking forward into the deep future, Stager (2012, 34, 41) explores two climate change models: the "moderate" one projecting a rise in atmospheric carbon concentrations to 550–600 ppm (well above the 350 ppm that climatologist James

Hansen and environmental activist Bill McKibben see as safe) with globally aver-
aged temperature increases of 3 to 7°F (2 to 4°C), and the "extreme" one of 2,000
ppm with temperature rises of "at least 9 to 16°F (5 to 9°C)."[4] Either way, Stager
argues (2012, 11), the human species is here to stay, however uncomfortably, and
he even hints that a new "ethics of carbon pollution" may credit us with having
rescued our distant descendants from the freezing, glacial devastation formerly
due in 50,000 years. For Stager, then, the human species is not endangered and
global warming may even have some benefits.

On the other hand, when Chakrabarty (2009) weighs the viability of the concept of "the species" for historians, he finds it wanting. This is not because it
falsely attributes responsibility to humanity or because climate change is not a
threat. Instead, the species is not, argues Chakrabarty, something humanist historians can *understand* through self-reflection in Dilthey's sense – where historical
consciousness is "a mode of self-knowledge" – or in Collingwood's sense – where
historical comprehension rests fundamentally not on reconstructing the past but
on reenacting "in our own minds the experience of the past" (2009, 9). While
"species" may work for paleobiologists collating, say, the fossil records of Eemian
biota from 130,000 years ago against modern organisms, theirs is a labor of reconstruction as opposed to one of self-reflection or mental re-enactment. The scale
both temporally and spatially on which these types of biologists work disallows
the tools of intellectual and emotional imagination honed by historians attempting to penetrate evidence produced by particular minds in the rich context of
particular cultures. For most historians, it is only on this smaller and more particular scale that judgments regarding ethical actions can be made. For one thing,
sheer biological survival is not the ultimate value or highest ethical good of every
individual or every culture. The concept of species remains for humanist historians such as Chakrabarty a galvanizing flash, important in illuminating the new
landscape but unable to provide sustained light. What appears most endangered
in this strobe light flash are our structures of knowledge, the figure of the individual capable of understanding his or her predicament.

Chakrabarty's brilliant double move, both toward the sciences and back again
to theoretical reflection on history, can be performed with other biological sciences. Through this dialectic, Chakrabarty demonstrates the impact of scale in
distinguishing between history as a description of past events (something we share
with many biologists) and history as the formation of self-knowledge. Climatologists and paleobiologists put "species" on the historian's map in unprecedentedly
difficult ways because of the macroscales involved, but other types of biology
compound our difficulties by looking at the human on a microscale. Microbiology and biochemistry contort "the human" into yet other forms, blurring and
even undermining the very notion of "species" and raising different yet equally
perplexing questions about how humanists might think biologically.

As I will discuss in the next section, some microbiologists now describe "the
human" as a coral reef of many "species," mostly microscopic, evolving at different rates and in different ways. From the microbiological perspective, the
very solidarity of the human species is eroded. In the third section, I turn to

yet another science, biochemistry, and its revelations about the industrial toxins that since World War II have flooded our environments and infiltrated our bodies, including our brains. These powerful chemicals, including endocrine disruptors, transform human beings both brutally and subtly, raising questions about the continuity of "the human" in the ways we think and respond to the world. Climatology, microbiology, and toxicology reveal three distinct human figures. As environmental concerns compel humanists to grapple with the biological sciences, what emerges is not a single, shared human figure understood to be endangered in the same way by all disciplines, but new conceptions of the human that challenge the understanding, solidarity, and continuity that serve as the foundation of humanist inquiry.

Microbiology: the body multiple[5]

Biologists work not only with the grand scale of species over hundreds of millennia, but also with minute organisms where mutations and adaptations are measured in minutes. For those of us wishing to understand who is endangered by climate change, this microbiological scale needs to be considered alongside the macrobiological scale of species.

One of the difficulties for humanists and social scientists is keeping up with the rapidly changing field of microbiology. Until very recently, bacteria were considered incidental to ourselves unless they caused disease. Then, in the 1960s, scientists began to understand that our skin was coated with benign bacteria that only occasionally caused problems. This new vision of the human being as accompanied by thousands of tiny life forms sparked the imagination of the poet W.H. Auden. In 1969, responding to a *Scientific American* article by Mary J. Marples (1965), Auden wrote "A New Year Greeting" to the "Bacteria, Viruses, Aerobics and Anaerobics" inhabiting his ectoderm, inviting them to settle themselves on his skin as long as they behaved "as good guests" should, not causing acne, athlete's foot, or boils. This invitation to microbes to live where they chose as long as they minded their manners was entirely congruent with the scientific view forty-odd years ago that microbes were mere "passive riders" on our bodies, as microbiologist Bonnie Bassler recently put it (Kolata 2012). Accepting this view, historians ignored the well-behaved "guests," focusing instead on the badly behaved "parasites," as in William McNeill's path-breaking *Plagues and Peoples*, which traced the effects of illnesses on political, economic, social, colonial, and intellectual developments.[6] Historians of medicine continue to do incisive work on the rampaging ingrates infecting our bodies with new diseases like AIDS (Engel 2006; Hamlin 2009; Johnston 1995).

But the science has now changed dramatically. The relationship between "ourselves" and microbes is not best described as one between genial host and guests, well-behaved or otherwise. According to very recent studies, we are *mostly* bacteria when measured purely by number of cells. With the completion of the Human Microbiome Project (HMP) in the summer of 2012, the estimated number of bacteria in healthy human adults was put at 100 trillion.[7] "Going strictly by the numbers," notes science writer Valerie Brown (2010), "the vast majority – estimated by many scientists at 90 percent – of the cells *in what you think of as your body* are actually bacteria, not human cells." Put a different way by the National Institutes of Health (2012), "The human body contains trillions of microorganisms – outnumbering human cells by 10 to 1." But this figure, jaw dropping though it may be, hardly conveys the drama of the new findings. After all, microbial cells are so very tiny compared with human cells that merely counting them reveals less than may appear. Bacteria make up only 1 to 3% of normal adult body weight, usually a mere two to six pounds (Brown 2010). More significant than their sheer quantity, the HMP reveals that these microbes are neither our "passive passengers" nor our incidental allies aiding digestion and the like. Instead, they are actually, inseparably, "us" by almost any calculation. The HMP concludes that on a genetic level, "this plethora of microbes contribute *more genes* responsible for human survival than humans contribute. Where the human genome carries some 22,000 protein-coding genes, researchers estimate that the human microbiome contributes some 8 million unique protein-coding genes or 360 times more bacterial genes than human genes" (National Institutes of Health 2012). Not only do bacteria participate in our physical processes but also in our mental ones (assuming this distinction still holds), producing "some of the same types of neurotransmitters that regulate the function of the brain" (Brown 2010). For all practical purposes, then, the distinction between "us" and "them" within microbiology has eroded away.

The current biological understanding of "the human" on this scale, as Stanford microbiologist David Relman describes it, is that we are "like coral, 'an assemblage of life-forms living together'" (Kolata 2012).[8] Harvard systems biologist Peter Turnbaugh and his colleagues speak of the "human 'supra-organism,'" "a composite of microbial and human cells, the human genetic landscape as an aggregate of the genes in the human genome and the microbiome, and the human metabolic features as a blend of human and microbial traits" (Turnbaugh et al. 2007, 804). A person is not an individual but a congregation. Analogously, the research done to establish these and other findings was also communal, coordinated among 200 scientists and 80 institutions, the data generated so vast that a single mammoth computer would still not suffice (Kolata 2012). The host and his guests played by Auden and his microbes in the poem are, from the microbiologist's point of view today, indistinguishable: everyone prepares the meal, pours the wine, joins in the laughter, and scrubs the dishes.

Furthermore, our microbes, like our friends, are not neatly categorizable as "good, healthy, and helpful" or "bad, disease-causing, and unwelcome." Just as an assembled company may be surprised when the loutish drunk digs everyone's cars out of the snow, scientists have been surprised to discover the "genetic signatures of disease-causing bacteria in everyone's microbiome. But instead of making people ill, or even infectious, these disease-causing microbes live peacefully among their neighbors" (Kolata 2012). We are good and bad, diseased and healthy, not in essence but in relation to particular situations, not genetically but epigenetically (Carey 2012).

Microbiology's description of "us" as the "body multiple," as a "coral reef" or "supra-organism," begs two major questions for humanists in the age of climate change: how might this inflect our understanding of human solidarity, and what light does this perspective shed on who is endangered? As Chakrabarty has shown, analysis on the macrobiological level presents the human species as a distinguishable, discrete, and immense entity: "mankind" in the words of Crutzen and Stoermer. On this macroscale, there are humans and non-humans, with the human species now emerging as a global geophysical agent, the master of the planet. But through the lens of microbiology, "the human species" is dramatically less coherent. Humanists attempting to grapple with the microbiological view of the human must try to comprehend not just the diseases that cause parts of ourselves to rebel against the symbiotic harmonies of our communal life, but also the idea that each "individual" can be understood as a collectivity of "species" and that "humanity" can be seen as an archipelago of multiple life forms. Although the findings of the Microbiome Project are far too recent to have been absorbed into the literature and thinking of scientists, let alone historians, this figure of the human poses grave challenges to most humanists' ideas. "We" in this microbiological assemblage differ from "one" another more than imagined. While we share about 99.9% of our (human) DNA, our microbial cells may have as little as 50% of their genetic profile in common. From the perspective of human solidarity, this finding is disturbing. If 90% of my cells are bacterial and half of those have a different DNA sequence than yours, then on a cellular level, it is not as clear that we are "the same species" as other branches of biology or the humanities have hitherto defined "species."[9] Self-reflection and mental reenactment as empathetic historical practices assume a cohesion to "the human" not apparent on this cellular level. For the humanities, it would seem just as difficult to think in terms of microbiology's view of the human as of paleobiology's, yet the difficulties are different.

The second question is about how this coral reef entity might be endangered by climate change. Here, too, microbiology offers a picture at odds with paleobiologists and climate scientists. Not only is "the species" *not* the coherent, planet-altering agent it is to those working on the larger scale, it is also *not* the victim of these changes in the same way. Back in 1969, when Auden, with piquant playfulness, questions his own microbiome regarding their perspective on the hurricanes, dousings, and heatwaves of life on the Audenesque ectoderm, he can separate himself and his activities of dressing and bathing from his bacteria. He politely hopes that "he" makes "a not impossible world" for these tiny guests, but muses that his "purposive acts, may turn to catastrophes there." Auden's description of the havoc wrought by his daily activities conveys something of the magnitude of microbial death rates as it was understood 40 years ago, but today's understanding of the horrific slaughter of "ourselves" in our digestive tracts (where more than half the weight of our feces are not undigested food but extruded microbes) and in other arenas of "our" bodies might give rise not to light verse but to epic dirge.[10] But the more important point about these deaths is that they allow for more births and rapid evolution. While Auden muses on the deaths of bacteria, he does not mention their fecundity: the high death rate of parts of our supra-organism is matched by an equally high birth rate.[11] The microbial part of us reproduces with such astounding rapidity that the number of bacteria, in the right conditions, can double every 20 minutes.

Not only that, but these high birthrates are accomplished through a different reproductive strategy and a different evolutionary style. These simple prokaryotic cells, without nuclei, mitochondria, or smaller organelles, can conjugate, techni-cally, with any other bacteria, thereby creating an interactive web of informa-tion evolving in many directions at once. Compared with the laborious process of sexual reproduction embraced by eukaryotes, which results in (fairly) linear evolution – a process often represented by the branching tree of life – prokaryotic cells are like sports cars with the capacity to turn on a dime. Their apparent abil-ity to conjugate with "anyone" means that the concept of distinct species among bacteria is extremely flexible. As new studies show, "bad" human-disease-causing bacteria exist with "good" bacteria throughout a healthy body such that differ-entiating them is, as I have suggested, less a matter of ontology than of situation.

The consequence of this rapid reproduction and evolutionary strategy is that the microbial part of us could evolve at faster rates and in a different way from the non-microbial part of us and could therefore respond more quickly to envi-ronmental changes. That a part of us might be capable of coping with more acidic water, drier climates, and higher temperatures than other parts of us produces a strikingly different version of what might be endangered. Understood in this way, the body multiple is not an *entity* to be protected, but a *system*, an interac-tive process of life and death combined. As such, this "supraorganism," this coral reef, is not threatened by climate change in the same way as other branches of knowledge have imagined "the human" to be threatened when they talk of rising oceans advancing on coastal cities, wars for natural resources, and environmental injustice perpetrated on the poor.[12]

My point here in underscoring the wildly variant visions of "the human" in paleobiology and microbiology is to show that defining what is endangered by climate change cannot be left to biology. There is more than one biology and these biologies produce human figures of varying vulnerability. The question is how to think with these new biological figures about "our" past and "our" pos-sibilities under the threat of climate change, and the answers are far from clear. Who is this "we" that is endangered?

Biochemistry: toxic beings[13]

The wizened, claw-like hand of the Minamata-disease victim in Kuwabara Shi-sei's photograph curls in an impossible shape, more reminiscent of Karl Bloss-feldt's furled ferns than human digits (Blossfeldt n.d.; Kuwabara 2004, 118) (see Figure 10.1). This nearly abstract representation of a portion of a body exposed to methylmercury from the Chisso chemical plant in Minamata, Japan, is politi-cally potent precisely because it divides the normal from the diseased, the healthy from the ill. Kuwabara's photograph documents the effects of a corporation's

Figure 10.1 "Minamata, 1970," reproduced by kind permission of Kuwabara Shisei.

criminally inhumane actions and demands redress. If ever there were an instance of "no caption needed," the provocative title of a book on photography by Robert Hariman and John Louis Lucaites (2007), this image is it, proclaiming at a glance that *this should not be* even before the context is made clear. In *Toxic Archipelago*, historian Brett Walker (2010) describes to devastating effect how Minamatabyo⁻, as it is known in Japanese, this "industrial disease," affected the body and mind of one of its victims, a fisherwoman who had lost everything, including her unborn child, to its predations: "In only four years, methylmercury had destroyed enough cells in Sakagami's brain to deprive her of control of herself almost entirely: mercury devours the brains of adults and stops the development of fetal ones" (140). Walker then details a horrific scene in which Sakagami, in her confusion, imagines that the oily fish on her hospital dinner plate is her by-then-aborted fetus. She tries to eat it in order to save it from the excruciating pain caused by methylmercury, only to have it flop from her chopsticks to the floor where she chases it, stuffing it into her mouth with her spasmodic hands. Methylmercury affects not only human beings. One of the earliest signs of the community's poisoning was that Minamata's cats "danced" crazily just before they died, leading to an explosion of mice that damaged the fishing nets (George 2001, 3). Unquestionably, introducing methylmercury into bodies harms and sometimes destroys life, mental and physical, economic and social, in the womb and out of it, human and otherwise. What Kuwabara's image, Walker's prose, and the death of a hundred convulsing cats clearly show is a situation that "should not be."

However, the divide between the body and its non-organic chemical infiltrators is not as clear as the black-and-white photograph or the heart-wrenching stories of Minamata suggest.[14] As scientists, historians of medicine, and others have come to realize, we must cast aside what Steve Kroll-Smith and Worth Lancaster (2002, 204) have dubbed "the Enlightenment-inspired idea that bodies and environments are genuinely discrete realities." In many cases, there is not even a threshold between "us" and "outside of us," let alone a stalwart barricade preventing penetration by dangerous substances. The new chemical compounds being pumped out in the millions of tons annually, particularly the endocrine disrupters, enter our bodies through multiple and little-understood pathways, as the work of ecologists and historians such as Nancy Langston (2010), Sandra Steingraber (2001), and Florence Williams (2012) shows.[15] As Langston (2010, 17) explains, "Since World War II the production of synthetic chemicals has increased more than thirtyfold. The modern chemical industry, now a global enterprise of $2 trillion annually, is central to the world economy, generating millions of jobs and consuming vast quantities of energy and raw materials. Each year more than seventy thousand different industrial chemicals annually make their way into our bodies and ecosystems. Americans are saturated with industrial chemicals."[16] In the same vein, historian of science Michelle Murphy (2008) speaks of our "chemical embodiment," plainly and powerfully stating that "in the twenty-first century, humans are chemically transformed beings." Of the more than 70 to 80,000 chemicals in commercial use in the United States, notes an editorial in *Scientific American* (2010), "the EPA has been able to force testing for only around 200."[17] Even those of us who have escaped the horrific deformities visible in Kuwabara's photographs appear biochemically altered when examined by other means of imaging and analysis.

As before, scale is crucial here: exponentially more chemicals have been introduced throughout the planet more quickly than ever before in greater amounts, vaster coverage, and shorter time. This point about scale is important because even before the Industrial Revolution, some people lived in chemically altered local environments where such things as lead, mercury, coal, ergot poisoning, and even wood smoke harmed human health and altered human brains and bodies. However, after World War II, the new industrial substances infiltrating our bodies became more plentiful, more potent, and more complex by many orders of magnitude. By 1986, the substances that became part and parcel of every single American included measurable amounts of styrene and ethyl phenol in 100% of the population, toluene in 91%, and polycholorinated byphenols in 83%. "Virtually every person who has lived in the United States since 1951 has been exposed to radiological fallout," as the Environmental Protection Agency admits, and "'all organs and tissues of the body have received some radiation exposure'" (Kroll-Smith and Lancaster 2002, 205). The toxic load carried by almost every human being includes phthalates (a toxin derived from plastics) and mercury, the substance responsible for Minamata disease (Williams 2012).[18] Of particular concern are the endocrine disruptors such as the synthetic estrogen used in cattle feed, putting "masculinity at risk" and raising the rates of intersex conditions and reproductive cancers in human beings and other creatures. For instance, surveys of many British streams reveal that "more than 30 percent of the fish . . . are now intersex" (Langston 2010, 143). With these endocrine disruptors, scale is an issue in two ways. Not only is there concern about the massive amounts of these hormones in the environment at large, but contrary to original assumptions, tiny amounts can be *even more dangerous* than large doses when it comes to altering body chemistry. Today, not only the planet but also we ourselves have been fundamentally transformed by the energy- and resource-intensive activities of agriculture and other industries altering the Earth.

What this "chemical embodiment" means, in short, is that there is not one group of healthy human beings living without toxic – or potentially toxic, yet untested – chemicals and another group of unhealthy (and unlucky) human beings living with them. Our chemical environment *is* us, not just in those extreme cases such as Minamata, but everywhere and with everyone (Langston 2010, 143).[19] The old idea that a line between "the body" and "the environment" could be carefully policed by governments reining in corporations or by individuals making healthy choices no longer pertains as we have come to understand the interpenetrability of bodies and environments. Since the environment is now radically altered through the various industrial processes polluting and warming the planet, the body too is radically altered. As Langston (2010, 136) argues, "Bodies are, in effect complex ecosystems made up of a dynamic interweaving of material and cultural feedbacks that are themselves subjects and sources of environmental degradation. Whatever humans do to the natural world finds its way back inside our bodies, with complex and poorly understood consequences. And in turn, what happens inside our bodies makes its way back into the broader world, often with surprising effects." This conception of our bodies' embeddedness in the surrounding environment resembles the nineteenth-century idea that individual health could only be attained in a healthy climate. Illness was not understood as a pathology existing exclusively within the compromised individual, but instead as a disorder arising between individuals and their surroundings

(Nash 2006). A century ago, it might have been possible for people to move to truly healthy climates; now the possible range of habitats extends only from the not-immediately harmful to the deadly.

Much of the insightful and impassioned research tracing the processes responsible for our toxic bodies and our toxic landscapes has been done by historians, so it would be perverse, ungenerous, and simply wrong to suggest that science alone has contributed to the recognition of this chemically altered human figure. Nevertheless, at the theoretical level, those in history and the humanities have not yet grasped the challenge posed by humanity's unprecedentedly rapid biochemical transformation. History relies, as Reinhardt, Koselleck, Chakrabarty, and many others argue, on the assumption of a certain continuity of experience that permits us to understand not just what happened, but also how and why. This continuity is at root physiological. The figure of "the human" in biochemical terms remains, it has always been assumed, recognizable. Daniel Smail (2008, 114) expresses it as follows: "The existence of brain structures and body chemicals means that predispositions and behavioral patterns have a universal biological substrate that simply cannot be ignored. . . . Basic social emotions are almost certainly universal. Nonetheless – the point is almost too obvious to bear repeating – they do different things in different historical cultures." But the rapid introduction of hitherto unknown industrial chemicals affecting our bodies, including our brains – as illustrated in the extreme example of Sakagami's hallucinations of her dinner being her suffering child – threatens this continuity. Humanists (and biologists) are now confronted with the problem of how the postwar biochemical revolution affects the continuity of our "universal biological substrate."

At the usually less-than-fatal levels at which every person is now imbued with industrial toxins, it is hard to imagine that there are not subtle – and perhaps not so subtle – changes in our thought processes. If history involves self-reflection in order to understand the existential human condition, and yet that self and that condition have been chemically altered, how do we proceed? How could we even measure these effects, given the wide range of human abilities and talents? Can we articulate the way contemporary brains differ in perception and function from those of Auden's generation? In asking these questions, we emphasize what we may be losing in terms of human continuity. On the other hand, if we *are* our chemically altered environment, then who is the "we" endangered by the industrial processes producing climate change? From this perspective, there may be no endangerment.

If we really define the human organism as part and parcel of its environment, we would view adaptation to new chemicals as yet another life process, neither good nor bad. Indeed, in some corporate circles, the malleability of human physiology is presented as a reason to dismiss climate concerns. In 2009, the U.S. Chamber of Commerce advised the Environmental Protection Agency that should the predictions of global transformation be correct, "populations can acclimatize to warmer climates via a range of behavioral, *physiological* and technological adaptations" (McKibben 2012, 8). McKibben wryly observes, "As radical goes, demanding that we change our physiology seems right up there."

With both paleobiology and microbiology, we remain essentially organic – "natural" in the sense of being constituted of organic cells created with simple non-organic molecules, including water (65%), carbon (19%), hydrogen (9.7%), nitrogen (3.2%), and calcium (1.8%). Even though the microbial part of ourselves reproduces differently and evolves by different pathways, it has done so in relation to the human cells, a coevolutionary pattern with its origins at the beginning of life. We and they – or the "us" that is the human body from this perspective – have shared for millennia the same molecular compounds and a joint history that we can reconstruct, even if we cannot reenact it in our minds in Collingwood's fashion. However, the swift introduction of hitherto unknown industrial chemicals and hormone disruptors into our bodies is unlike the earlier coevolutionary processes absorbed for historians through the work of Edmund Russell (2003; 2011). Like "the species" of climate change and the "coral reef" of microbiology, the "toxic self" challenges the historians' tools and craft, but even more so because of the radical transformation, faster than any hitherto discovered event in human evolution. From the perspective of biochemistry, we look back at an historical "us" abruptly discontinuous with the toxic beings we have become, and it is very hard to know how we might begin to grapple with the transformation. Prior to World War II, there is a "we" already lost, not endangered but extinct, if "we" assumes a continuity of organic biology.

Conclusion: a critical friendship

Geneticists Craig Venter and Daniel Cohen (2004, 73) declare that "the 21st century is the century of biology." What is needed for other branches of knowledge in relation to these life sciences is what sociologist Nikolas Rose (2013, 3–4) calls a "critical friendship" centering around "the vitality of the living body." In this concluding section, there are two points I want to make about this "critical friendship," the first having to do with reality and the second with values.

First, reality may be described truthfully and cogently in many ways. Manifold biologies produce manifold descriptions of the human. Each of the biological sciences in my limited sample – paleobiology, microbiology, and biochemistry – has defined "us" in a different way and each one poses a different challenge to historians. The human defined purely as a species defies historical understanding; the human as multi-species coral reefs undermines solidarity; a toxified humanity undermines temporal continuity. Despite these tensions, for those concerned with climate change, these various biological understandings enrich and broaden our conception of what is at stake. They confirm our embeddedness in the global environment on different scales: as an increasingly domineering species operating over vast eons, as a coral reef of many species spreading out in awkward archipelagos of co-dependent beings, and as a semi-industrialized product of the last, brief half century. In doing so, they usefully defamiliarize "the human" as portrayed by most humanists.

In the sense of defamiliarizing the present, biology contributes to the humanist's project of creating a critical distance from the status quo and to our fund of information about climatic conditions, disease patterns, and coevolution. Thinking with biologists reminds us of the biological foundation of all that we are and do. With biologists, we go deeper, beyond the old materialism of the economic "base" to a new, and far richer, biological materialism (Thomas 2008). With them, we trace the limits of our age of abundance and grasp the scale of our exorbitant use of fossil fuels and its implications for life.[20] With them, we learn to think somatically through the body, like Auden thinking through his bacteria-laden epidermis or Kuwabara thinking with images of deformed hands. Biologists work on many scales, and in engaging with them, we fruitfully learn to see the human on different scales as well: as species over eons, as amalgamated with microbes, and permeated by our industrial products. Reality may be accurately described in many ways.

But these engagements also remind us of the limits of biological description. When it comes to climate change, according to most of the work in paleobiology, microbiology, and biochemistry, human beings in one form or another will continue. If our ancestral species survived horrifically hot temperatures in the Eocene era, we are likely, Stager says, to do so again. The vicious struggles for natural resources and high death rates evoked by environmental journalist Mark Lynas in *Six Degrees: Our Future on a Hotter Planet* (2008) will not necessarily wipe us out entirely. Our microbiome's capacity for rapid evolution also suggests that some of us will resist new diseases and be able to adapt to extreme environmental conditions, avoiding the extinction that worries *Scientific American* editor Fred Guterl in *The Fate of the Species: Why the Human Race May Cause Its Own Extinction and How We Can Stop It* (2012). Our biochemistry's mirroring of environmental toxins will produce deformities and cancers, but still allow for adequate reproduction rates so that, as a whole, humans are unlikely to disappear.

And yet this is hardly what most people mean in expressing concern about environmental dangers. It is not mere survival that the humanities teach us to value, nor description that humanistic disciplines teach us to practice. When ethicist Clive Hamilton (2010) mourns our inevitable losses in *Requiem for a Species*, it is not our sheer physicality that he grieves over. Ideas about value are another type of knowledge, rooted in cultural genealogies, conversations, and controversies, and true to the extent that they are persuasive rather than provable.

Scientists do not address the questions of value that are central to the humanities. As the late biologist Stephen Jay Gould argued, "the factual state of the universe, whatever it may be, cannot teach us *how we should live* or *what our lives should mean* – for these ethical questions of value and meaning belong to such different realms of human life as religion, philosophy, and humanistic study. Nature's facts can help us to realize a goal once we have made our ethical decisions on other grounds" (Gould 2002, xxxvi). When humanists turn to science for answers to questions of value and meaning, they often stumble. In response to philosopher Thomas Nagel's (2012) wistful insistence on a natural teleology culminating in human consciousness, evolutionary geneticist Allen Orr (2013) points to the greater success of fungi, observing that "if nature has goals, it certainly seems to have many and consciousness would appear to be fairly far down on the list." Biology, as Orr suggests, has no special fondness for philosophers, poets, photographers, or historians. Nor does it have a penchant for justice, peace, or decency. Nor any particular desideratum. If nature "is trying to get somewhere," Orr asks, "why does it keep changing its mind about the destination?" What conversations with biologists demonstrate first and foremost for historians and others in the humanities is that biology is not going to ease our burden of responsibility for crafting an understanding of the human figure currently threatened by climate change. Instead, the humanities must bear the responsibility of defining and defending threatened commitments to justice, decency, and beauty and to imagining a human figure capable of embracing and sustaining them. Biologists can help us understand our predicament, but they cannot provide the cultural, social, and political imagination to resolve it.

Political theorist Wendy Brown (2011) is particularly eloquent in celebrating humanistic approaches. She sees the rush to embrace scientific models of knowledge as part of the reduction of human experience to "the one dimensional rationality of *homo economicus*" celebrated in neoliberal regimes. Arguing against "the convergent challenges of scientization and neoliberalization within and outside the academy," whereby all knowledge becomes "marketable, immediately applicable, or scientific in method," she defends the humanities: "They speak to, cultivate and elevate precisely what a neoliberal rationality would extinguish in us individually and collectively – not only historical, philosophical and literary consciousness and viewpoints, not only notions of the political exceeding interest and featuring shared power and purpose, but the play of ambiguity, vulnerability, awe, ambivalence, psychic depths, boundary, identity, spirit, and other elements foreign to neoliberal rationality" (126–127).[21] These humanistic qualities and modes of understanding are particularly threatened in the coming world of duress and hardship. They define what is most endangered, which is not our fragile bodies but rather the even frailer edifices of decency, justice, playfulness, and beauty. In what Nikolas Rose (2013, 7) describes as the epistemic shift in both the human and the biological sciences whereby "personhood itself is becoming increasingly somatic," biology is crucial, but only the humanities can articulate the value of what is endangered and produce the wisdom, grace, and humor, and the cultural, political, and social resources available in our records to begin to address the problem.

In short, "in the moment of danger that is climate change" as the figures of the species, the microbiome, and the toxic body flash before us, the most important scale for exploring the human figure remains the one that comes most readily to hand for most humanists: the scale in time and space where individuals and communities have political agency; the scale, in other words, that has long framed our studies. But now there is a difference. In the age of biology, this figure's biological being weighs equally with its conscious actions. Indeed, the two are imbricated. We must not only be "historians of mind" in Collingwood's terms, where mind and body can be neatly separated, but also historians of the eating, sleeping, making love, and much else besides that he dismisses (Collingwood 1946,

216). In so doing, we are politicizing passivity, politicizing the received nature of our environment and bodies without letting go of the need for mindful action (Thomas 2011). For historians, mindful action occurs in the archives tracing not only exponential expansion of populations and economies founded on fossil fuel since the eighteenth century but also the byways taken by those not pursuing the illusion of limitless growth. In revealing multiple viable ways of life, we can offer a somatic politics that counters neoliberalism's naturalization of growth. Biology underscores human malleability, but history and other humanistic disciplines provide a forum for deliberating how we might direct this malleability. Engaging with biology reveals a multiplicity of human figures and delimits the possible answers to humanistic questions of value – but cannot decide them. Ultimately, defining what is most endangered by climate change is the role of the humanists.

Notes

1 An alternative version of this chapter appeared as "History and Biology in the Anthropocene: Problems of Scale, Problems of Value," *American Historical Review*, 119 (5), December 2014.

2 Systems analysis, pioneered in 1972, reminds us that it is not only fossil fuel burning that is responsible for climate change but the whole panoply of human activities, including agriculture, demographic rates, transport systems, and many other things. See Meadows et al. (1974) and Meadows, Randers, and Meadows (2004). The attempt to call people's attention to the problem has taken many forms, including a wildly successful one-man play, "Ten Billion," written and performed by Stephen Emmott, head of computational science at Microsoft Research in Cambridge and professor of computational science at Oxford, which essentially consisted of his reading data in sold-out London performances during the summer of 2012. Reviewing the play, Ian Jack writes, "food production already accounts for 30% of greenhouse gases – more than manufacturing or transport; more food needs more land, especially when the food is meat; more fields mean fewer forests, which means even more carbon dioxide in the atmosphere, which means an even less stable climate, which means less reliable agriculture" (Jack 2012). Regrettably, some people continue to believe that laptops, mobile phones, iPads, and other devices enabling electronic communication and supposedly cutting down on paper use (as in books) is ecologically sound. On the contrary, tantalum, also known as "coltan," as well as other rare minerals necessary for these devices, are mined in terrible conditions with great harm to the environment. See Nest (2011).

3 Crutzen and Stoermer date the beginning of the Anthropocene to the "latter part of the eighteenth century, although we are aware that alternative proposals can be made (some may even want to include the entire Holocene)" (2000, 71). The ramifications of this periodization have been discussed from a number of angles by historians. See, for instance, Steffen, Crutzen, and McNeill (2007) and Chakrabarty (2009). Ruddiman (2005) argues that the Anthropocene began with agriculture and should be dated to 8,000 years ago. Responses to Chakrabarty include Baucom (2012) and During (2012).

4 In relying on a projected high of 600 ppm, Stager is following climatologist David Archer's prediction. Bill McKibben's (2012) standard for survival is 350 ppm, pushing us back down from our current level of almost 400 ppm. According to the Potsdam Institute (2012) in their report for the World Bank, "By the time the concentration reaches 550 ppm (corresponding to a warming of about 2.4°C in the 2060s), it is likely that coral reefs in many areas would start to dissolve. The combination of thermally

induced bleaching events, ocean acidification, and sea-level rise threatens large fractions of coral reefs even at 1.5°C global warming. The regional extinction of entire coral reef ecosystems, which could occur well before 4°C is reached, would have profound consequences for their dependent species and for the people who depend on them for food, income, tourism, and shoreline protection." See also Lynas (2008) and Archer (2010).

5 The phrase "the body multiple" is used as the title of Mol (2002).

6 See, for instance, McNeill (1977), Crosby (1973; 2004), and in my own field of Japanese history, Farris (1985).

7 In June 2009, the estimate of unique bacterial genes in each human gut was about 9 million (see Brown 2010). The NIH newsletter reports that: "In a series of coordinated scientific reports published on June 14, 2012, in *Nature* and several journals in the Public Library of Science (PLoS), some 200 members of the Human Microbiome Project (HMP) Consortium from nearly 80 universities and scientific institutions report on five years of research. HMP has received $153 million since its launch in fiscal year 2007 from the NIH Common Fund, which invests in high-impact, innovative, trans-NIH research. Individual NIH institutes and centers have provided an additional $20 million in co-funding for HMP consortium research."

8 The coral reef metaphor works well. Another used by biologist David George Haskell (2012, 4) musing on the microbes in our guts and the mitochondria in our cells is that of Russian dolls: "We are like Russian dolls, our lives made possible by the other lives within us."

9 This is a problem that troubles microbiologists. Peter Turnbaugh et al. (2007, 804) point to some of the challenges in asking: "How similar are the microbiomes between members of a family or members of a community, or across communities in different environments?"

10 Kolata (2012) quotes Dr. Lita Proctor, program director for the Human Microbiome Project, explaining, "half of your stool is not leftover food. It is microbial biomass."

11 The reproductive powers of microbes are such that, in the right conditions, their numbers can double every 20 minutes, which allows them to survive virus attacks. In the oceans, daily, viruses kill half of all the bacteria only to have them regenerate. And they invade a microbe host 10 trillion times a second around the world (Zimmer 2012).

12 The term "environmental justice" seems to have particular resonance in writing about the United States, where the concept is enshrined in the directives of the Environmental Protection Agency with the intention of protecting poor communities from becoming sites for contaminated waste and other problems. See, for instance, Schlosberg (2007). From a global perspective, the concept applies to the way the wealthy countries of the northern hemisphere (roughly speaking) use a disproportionate amount of the Earth's natural resources and produce a disproportionate amount of the pollution. For this perspective, see, for example, Shiva (2005; 2008).

13 On remaking human bodies, see Roberts and Langston (2008) and the related articles, Vogel (2009), and Langston (2010).

14 Such industrial diseases, as Brett Walker (2010, 139) carefully reminds us, are the "result of hybrid causation, because of complex and largely unanticipated interrelationships among advanced technologies, idiosyncratic social practices, and naturally occurring agencies."

15 See also Colborn, Dumanoski, and Myers (1997).

16 See also Roberts and Langston (2008) and the related articles in that issue, and Vogel (2009).

17 Quoted in Magdoff and Foster, they write, "The United States continues to have one of the worst records among industrial countries concerning protection of its citizens from toxic chemicals found in products in everyday use – from cosmetics to food containers to denture cream" (2011, 24).

18 In her review of Williams (2012), Elizabeth Kolbert notes: "One in 17 women had enough mercury in her blood to risk causing learning disabilities in her children. The Environmental Protection Agency expressed concern that even low-level exposure to perfluorooctanoic acid, used in the manufacture of Teflon, could potentially lead to developmental problems. Flame retardants known as PBDEs, which were known to cause brain damage in rats, were increasingly showing up in human breast milk" (Kolbert 2012, 54).

19 Human bodies are not the only ones affected by the soup of synthetic chemicals. Reproduction in wildlife worldwide is affected. Just to take a few of the most startling examples provided by Nancy Langston (2010, 4), "Male alligators exposed to DDT in Florida's Lake Apopka developed penises that were one-half to one-third the typical size, too small to function. Two-thirds of male Florida panthers had cryptorchidism, a hormonally related condition in which the testes do not descend. Prothonotary warblers in Alabama, sea turtles in Georgia, and mink and otters around the Great Lakes all showed reproductive changes. Male porpoises did not have enough testosterone to reproduce, while polar bears on the Arctic island of Svarlbard developed intersex characteristics. In one particularly disturbing example, Gerald A. LeBlanc of North Carolina State University in Raleigh found that more than a hundred species of marine snails were experiencing a condition known as imposex, a pollution-induced masculinization. Affected females could develop a malformed penis that blocked their release of eggs. Engorged by eggs that could not get out, many snails died."

20 There is a dispute between deep historians who have argued that the intensive use of fossil fuel since the eighteenth century is the unremarkable continuation of millennia-old patterns of resource exploitation and those who see not only an abrupt quantitative change but a qualitative change as well. For the former position, see Shyrock and Smail (2012), and Smail (2008). For the latter argument, see for instance Burke (2009, 49), who refers to modernity as "deeply aberrant." For an excellent exposition of the stakes of this debate, see Jonsson (2012).

21 For an expanded discussion on the ideas framing knowledge in the last few decades, see Rodgers (2011).

Bibliography

Archer, D., 2010. *The Long Thaw: How Humans Are Changing the Next 100,000 Years of Earth's Climate*. Princeton University Press, Princeton.

Baucom, I., 2012. The Human Shore: Postcolonial Studies in an Age of Natural Science. *History of the Present*, 2 (1) Spring, 1–23.

Blossfeldt, K., n. d. *Karl Blossfeldt Photography*. Wilde, A. and Wilde, J., eds. Hatje Cantz Verlag, Ostfildern, Germany.

Brown, V., 2010. Bacteria 'R' Us, Smart Journalism. Real Solutions. *Miller-McCune*, 2 December (http://www.miller-mccune.com/science-environment/bacteria-r-us-23628/#). Accessed 2 December 2014.

Brown, W., 2011. Interventions: Neoliberalized Knowledge. *History of the Present: A Journal of Critical History*, 1 (1) Summer, 113–129.

Burke, E. III, 2009. The Big Story: Human History, Energy Regimes, and the Environment. In: Pomeranz, K. and Burke, E. III, eds., *The Environment and World History*. University of California Press, Berkeley and Los Angeles, 33–53.

Carey, N., 2012. *The Epigenetics Revolution: How Modern Biology is Rewriting Our Understanding of Genetics, Disease, and Inheritance*. Columbia University Press, New York.

Chakrabarty, D., 2009. The Climate of History: Four Theses. *Critical Inquiry*, 35 (2) Winter, 197–222.

Colborn, T., Dumanoski, D. and Myers, J. P., 1997. *Our Stolen Future: Are We Threatening Our Fertility, Intelligence, and Survival? A Scientific Detective Story*. Penguin Books, New York.

Collingwood, R. G., 1946. *The Idea of History*. Oxford University Press, Oxford.

Crosby, A. W. Jr., 1973. *The Columbian Exchange: Biological and Cultural Consequences of 1492*. Greenwood Press, Westport, CT.

Crosby, A. W. Jr., 2004. *Ecological Imperialism: The Biological Expansion of Europe, 900–1900*, 2nd ed. Cambridge University Press, Cambridge.

Crutzen, P. J. and Stoermer, E. F., 2000. Anthropocene: An Epoch of Our Making. *Global Change: International Geosphere-Biosphere Programme Newsletter*, 12–15. Republished in: McKibbon, B., ed., 2011. *The Global Warming Reader: A Century of Writing about Climate Change*. Penguin Group, New York, 69–74.

During, S., 2012. Empire's Present. *New Literary History*, 43 (2), 331–340.

Engel, J., 2006. *The Epidemic: A Global History of AIDS*. Smithsonian, Washington, DC.

Farris, W. W., 1985. *Population, Disease, and Land in Early Japan, 645–900*. Harvard University Press, Cambridge, MA.

George, T. S., 2001. *Minamata: Pollution and the Struggle for Democracy in Postwar Japan*. Harvard University Asia Center, Cambridge, MA.

Gould, S. J., 2002. Introduction. In: Zimmer, C., *Evolution: The Triumph of an Idea*. Harper Collins, New York.

Guterl, F., 2012. *The Fate of the Species: Why the Human Race may Cause its Own Extinction and How We Can Stop It*. Bloomsbury, New York.

Hamilton, C., 2010. *Requiem for a Species: Why We Resist the Truth about Climate Change*. Earthscan, New York and Abingdon, UK.

Hamlin, C., 2009. *Cholera: The Biography*. Oxford University Press, Oxford.

Hariman, R. and Lucaites, J. L., 2007. *No Caption Needed: Iconic Photographs, Public Culture, and Liberal Democracy*. University of Chicago Press, Chicago.

Haskell, D. G., 2012. *The Forest Unseen: A Year's Watch in Nature*. Viking, New York.

Jack, I., 2012. The implications of overpopulation are terrifying. But will we listen to them? *The Guardian*, 3 August 2012 (http://www.theguardian.com/commentisfree/2012/aug/03/ian-jack-overpopulation-ten-billion). Accessed 23 August 2012.

Johnston, W., 1995. *The Modern Epidemic: A History of Tuberculosis in Japan*. Harvard University Asia Center, Cambridge, MA.

Jonsson, F. A., 2012. Review: The Industrial Revolution in the Anthropocene. *The Journal of Modern History*, 84 (3) September, 679–696.

Kolata, G., 2012. In Good Health? Thank Your 100 Trillion Bacteria. *New York Times*, 13 July 2012 (http://www.nytimes.com/2012/06/14/health/human-microbiome-project-decodes-our-100-trillion-good-bacteria.html?pagewanted=all&module=Search&mabReward=relbias%3Ar%2C%7B%222%22%3A%22RI%3A17%22%7D&_r=0). Accessed 5 December 2014.

Kolbert, E., 2012. The Nature of Breasts. Florence Williams. *Breasts: A Natural and Unnatural History* [review] *OnEarth*, 18 June (http://www.onearth.org/article/anatomy-lessons). Accessed 5 December 2014.

Kroll-Smith, S. and Lancaster, W., 2002. Bodies, Environments, and a New Style of Reasoning [review article]. *Annals of the American Academy of Political and Social Science*, 584, 203–212.

Kuwabara, S., 2004. *Kuwabara shisei shashin zenshu*, Vol. 1 *Minamata*. Kusanone Publishing Company, Tokyo.

Langston, N., 2010. *Toxic Bodies: Hormone Disruptors and the Legacy of DES*. Yale University Press, New Haven.

Lynas, M., 2008. *Six Degrees: Our Future on a Hotter Planet*. National Geographic Society, Washington, DC.

Magdoff, F. and Foster, J. B., 2011. *What Every Environmentalist Needs to Know about Capitalism*. Monthly Review Press, New York.

Margulis, L., Sagan, D. and Thomas, L., 1997. *Microcosmos: Four Billion Years of Microbial Evolution*. University of California Press, Berkeley and Los Angeles.

Marples, M. J., 1965. *The Ecology of the Human Skin*. Charles C. Thomas, Springfield, IL.

McKibben, B., 2012. Global Warming's Terrifying New Math: Three simple numbers that add up to global catastrophe – and that make clear who the real enemy is. *Rolling Stone*, 2 August 2012 (http://www.rollingstone.com/politics/news/global-warmings-terrifying-new-math-20120719?page=5). Accessed 4 August 2014.

McNeill, W., 1977. *Plagues and Peoples*, Anchor Books, Random House, New York.

Meadows, D. H. et al., 1974. *The Limits to Growth: A Report for the Club of Rome's Project of Mankind*. Universe Books, New York.

Meadows, D. H., Randers, J. and Meadows, D. L., 2004. *The Limits to Growth: The 30-Year Update*. Chelsea Green Publishing Company, White River Junction, VT.

Mol, A., 2002. *The Body Multiple: Ontology in Medical Practice*. Duke University Press, Durham and London.

Morris, I., 2011. *Why the West Rules – for Now: The Patterns of History and What They Reveal about the Future*, Reprint ed. Picador, New York.

Murphy, M., 2008. Chemical Regimes of Living. *Environmental History*, 13 (4) October, 695–703.

Nagel, T., 2012. *Mind and Cosmos: Why the Materialist Neo-Darwinian Conception of Nature is Almost Certainly False*. Oxford University Press, Oxford.

Nash, L., 2006. *Inescapable Ecologies: A History of Environment, Disease, and Knowledge*. University of California Press, Berkeley and Los Angeles.

National Institutes of Health, U.S. Department of Health & Human Services, 2012. *NIH Human Microbiome Project defines normal bacterial makeup of the body: Genome sequencing creates first reference data for microbes living with healthy adults* [press release], 13 June. Contact: Raymond MacDougall NHGRI (http://www.nih.gov/news/health/jun2012/nhgri-13.htm). Accessed 19 June 2012.

Nest, M., 2011. *Coltan*. Polity Press, Cambridge, UK.

Orr, H. A., 2013. Awaiting a New Darwin. *New York Review of Books*, 7 February, 28.

Potsdam Institute for Climate Impact Research and Climate Analytics, 2012. 4°: Turn Down the Heat: Why a 4°C Warmer World Must be Avoided. Executive Summary. *International Bank for Reconstruction and Development / The World Bank* (http://www.climateanalytics.org/sites/default/files/attachments/publications/1305917_4degrees_ExecutiveSummary.pdf). Accessed 4 August 2012.

Roberts, J. A. and Langston, N., 2008. Toxic Bodies/Toxic Environments: An Interdisciplinary Forum. *Environmental History*, 13 (4), 629–703.

Rodgers, D. T., 2011. *Age of Fracture*. Harvard University Press, Cambridge, MA.

Rose, N., 2013. The Human Sciences in a Biological Age. *Theory, Culture, & Society*, 30 (1) January, 3–34.

Ruddiman, W. F., 2005. *Plows, Plagues, and Petroleum: How Humans Took Control of Climate*. Princeton University Press, Princeton.

Russell, E., 2003. Evolutionary History: Prospectus for a New Field. *Environmental History*, 8 (2), 204–228.

Russell, E., 2011. *Evolutionary History: United History and Biology to Understand Life on Earth*. Cambridge University Press, Cambridge.

Schlosberg, D., 2007. *Defining Environmental Justice: Theories, Movements, and Nature*. Oxford University Press, New York.

Scientific American, 2010. Chemical Controls. *Scientific American*, April (http://www.scientificamerican.com/article/chemical-controls/). Accessed 5 August 2012.

Shiva, V., 2005. *Earth Democracy: Justice, Sustainability, and Peace*. South End Press, New York.

Shiva, V., 2008. *Soil Not Oil: Environmental Justice in the Age of Climate Change*. South End Press, New York.

Shyrock, A. and Smail, D., 2012. *Deep History: The Architecture of Past and Present*. University of California Press, Berkeley and Los Angeles.

Smail, D. Lord, 2008. *On Deep History and the Brain*. University of California Press, Berkeley and Los Angeles.

Stager, C., 2012. *Deep Future: The Next 100,000 Years of Life on Earth*. St. Martin's Press, New York.

Steffen, W., Crutzen, P. J. and McNeill, J. R., 2007. The Anthropocene: Are Humans now Overwhelming the Great Forces of Nature? *Ambio: A Journal of the Human Environment*, 36 (8), 614–621.

Steingraber, S., 2001. *Having Faith: An Ecologist's Journey to Motherhood*. Perseus Press, Cambridge, MA.

Thomas, J. A., 2008. Atarashii Busshitsu Shugi [The New Materialism]. Preface to *Kindai no saikochiku: Nihon seiji ideorogii ni okeru shizen no gainen* [Reconfiguring Modernity: Concepts of Nature in Japanese Political Ideology]. Hosei University Press, Tokyo.

Thomas, J. A., 2011. From Modernity with Freedom to Sustainability with Decency: Politicizing Passivity. In: Coulter, K. and Mauch, C., eds., *The Future of Environmental History: Needs and Opportunities*, March. Rachel Carson Center Perspectives, University of Munich, München, Germany, 53–56.

Turnbaugh, P. et al., 2007. The Human Microbiome Project. *Nature*, 449 October, 804–810.

Venter, C. and Cohen, D., 2004. The Century of Biology. *New Perspectives Quarterly*, 21 (4) Fall, 73–77.

Vogel, S. A., 2009. The Politics of Plastics: The Making and Unmaking of Bisphenol A 'Safety'. *American Journal of Public Health*, 99, 559–566.

Walker, B. L., 2010. *Toxic Archipelago: A History of Industrial Disease in Japan*. University of Washington Press, Seattle.

Williams, F., 2012. *Breasts: A Natural and Unnatural History*. Norton, New York.

Wilson, E. O., 1998. *Consilience: The Unity of Knowledge*. Vintage, New York.

Zimmer, C., 2012. *A Planet of Viruses*. University of Chicago Press, Chicago.

Index